IntechOpen

Supporting open minds since 2005

Microalgae - From Physiology to Application
http://dx.doi.org/10.5772/intechopen.83737
Edited by Milada Vítová

Contributors
Facundo J. Marquez-Rocha, Pedro Garcia-Alamilla, Sandra Santiago-Morales, Jenny Lopez-Hernandez, Abelardo Flores-Vela, Diana Palma-Ramírez, Jithu Jacob, Mária Čížková, Milada Vítová, Vilém Zachleder, Juan C. Castro, Marianela Cobos, Hicler N. Rodríguez, Segundo L. Estela, Juan Saldaña, J. Dylan Maddox, Jae D. Paredes, Sheyla Pérez, Alvaro Tresierra, Jorge L. Marapara, Pedro M. Adrianzén, Rosana Gonzales, Manoj Kamalanathan, Antonietta Quigg, Natalia Yu. Grigoryeva, Nilesh Joshi, Ilza Mor, Edmundo Lozoya-Gloria, Xochitl Morales -De La Cruz, Takehiro Alfredo Ozawa-Uyeda, Lee-Feng Chien, Wen-Hao Lin, Ernesto Jose Gonzalez, Gabriel Roldán, Fábio Neves, Giustino Tribuzi, Mariana Demarco, Maiara Souza, Andrea Sanchez-Barrios, Tiele Medianeira Rizzetti, Lisianne Brittes Benitez, Michele Hoeltz, Rosana De Cassia De Souza Schneider

© The Editor(s) and the Author(s) 2020
The rights of the editor(s) and the author(s) have been asserted in accordance with the Copyright, Designs and Patents Act 1988. All rights to the book as a whole are reserved by INTECHOPEN LIMITED. The book as a whole (compilation) cannot be reproduced, distributed or used for commercial or non-commercial purposes without INTECHOPEN LIMITED's written permission. Enquiries concerning the use of the book should be directed to INTECHOPEN LIMITED rights and permissions department (permissions@intechopen.com).
Violations are liable to prosecution under the governing Copyright Law.

(cc) BY

Individual chapters of this publication are distributed under the terms of the Creative Commons Attribution 3.0 Unported License which permits commercial use, distribution and reproduction of the individual chapters, provided the original author(s) and source publication are appropriately acknowledged. If so indicated, certain images may not be included under the Creative Commons license. In such cases users will need to obtain permission from the license holder to reproduce the material. More details and guidelines concerning content reuse and adaptation can be found at http://www.intechopen.com/copyright-policy.html.

Notice
Statements and opinions expressed in the chapters are these of the individual contributors and not necessarily those of the editors or publisher. No responsibility is accepted for the accuracy of information contained in the published chapters. The publisher assumes no responsibility for any damage or injury to persons or property arising out of the use of any materials, instructions, methods or ideas contained in the book.

First published in London, United Kingdom, 2020 by IntechOpen
IntechOpen is the global imprint of INTECHOPEN LIMITED, registered in England and Wales, registration number: 11086078, 7th floor, 10 Lower Thames Street, London, EC3R 6AF, United Kingdom
Printed in Croatia

British Library Cataloguing-in-Publication Data
A catalogue record for this book is available from the British Library

Additional hard and PDF copies can be obtained from orders@intechopen.com

Microalgae - From Physiology to Application
Edited by Milada Vítová
p. cm.
Print ISBN 978-1-83880-035-2
Online ISBN 978-1-83880-036-9
eBook (PDF) ISBN 978-1-83880-271-4

For EU product safety concerns:
IN TECH d.o.o., Prolaz Marije Krucifikse Kozulić 3, 51000 Rijeka, Croatia, info@intechopen.com or visit our website at intechopen.com.

We are IntechOpen, the world's leading publisher of Open Access books

Built by scientists, for scientists

4,500+
Open access books available

119,000+
International authors and editors

135M+
Downloads

151
Countries delivered to

Our authors are among the
Top 1%
most cited scientists

12.2%
Contributors from top 500 universities

CLARIVATE ANALYTICS BOOK CITATION INDEX INDEXED

WEB OF SCIENCE™

Selection of our books indexed in the Book Citation Index
in Web of Science™ Core Collection (BKCI)

Interested in publishing with us?
Contact book.department@intechopen.com

Numbers displayed above are based on latest data collected.
For more information visit www.intechopen.com

Meet the editor

Prof. (FH) RNDr. Milada Vítová, PhD, works at the Laboratory of Cell Cycles of Algae, Centre Algatech, Institute of Microbiology of the Czech Academy of Sciences. She studied at Charles' University and obtained her PhD in molecular and cell biology. She started her research career in SAC, UK and continued at the University of Vienna, Austria. She has 20 years of experience in the research of algal physiology and biotechnology. She leads the group involved in the research of metal metabolism in algae and its environmental aspects. Her scientific fields concern namely: a) metal transport and metabolism in algae, b) cell cycle regulation of algae by external factors, and c) bioremediation, biomining, and recycling. She is a lecturer at the University of Applied Sciences, Krems, Austria.

Contents

Preface

I am pleased to introduce the book "Microalgae - From Physiology to Application" published by IntechOpen publisher. The book summarizes current reviews and original articles on microalgae. The term microalgae is commonly used to describe microscopic algae and cyanobacteria. It is thus a very diverse group of photosynthetic microorganisms living in a variety of environments including extreme habitats. Not only are these organisms fascinating, they can also be exploited by humans for biotechnologies. Except for the variety of applications, this book aims to address at least some aspects of microalgae physiology, as applications in biotechnology cannot be successful without a solid knowledge of physiology.

The book is organized into four sections and twelve chapters. The sections *Phytoplankton, Microalgae as a nutrition, Microalgae applications,* and *Microalgae secondary metabolites* are not strictly defined and the topics partly overlap, reflecting the complexity of the topic.

The first section *Phytoplankton* presents Chapter 1, which describes the potential of a novel fluorescent spectroscopic technique for a microalgae species discrimination. The method is based on special data processing of a set of fluorescent spectra, obtained from a single photosynthetic cell of cyanobacteria. The universality of the considered technique makes it possible to use for investigation of any phytoplankton species irrespective of their habitat or cultivation. The automatization of the cyanobacterial species differentiation is a key problem in both industrial biomass production and environmental monitoring.

Chapter 2 discusses the eutrophication of inland waters in many countries from North to South America due to the increase of the internal load of nitrogen and phosphorus. As a consequence, the phytoplankton community in freshwater lakes and reservoirs is mainly dominated by Cyanobacteria. The taxonomic groups of phytoplankton can be used to indicate the trophic level and water quality status of freshwater systems.

Chapter 3 is a very nice tutorial summarizing various techniques and laboratory protocols of the isolation of native microalgae and their characterization standardized by the authors. The biotechnological potential (biodiesel and nutraceuticals production) of the native microalgae isolated from Peruvian Amazon is evaluated. The isolation and culture techniques of native microalgae, culture media, growth evaluation process and techniques for molecular characterization are listed and described in detail.

The second section *Microalgae as a nutrition* includes Chapter 4, which reviews the effect of traditional and innovative drying methods on quality and stability of microalgae biomass powders with potential use in human alimentation. Microalgae are dried to allow easy storage and transportation as well as to facilitate their use in biorefinery and the food and feed industry. The understanding of the degradative phenomena that occur during storage, namely high sensitivity to light, heat and oxygen, of dried microalgal biomass is important.

Chapter 5 deals with the use of microalgae as food supplements. The diversity of microalgae, their components, applications, and potential health benefits are discussed. The major economically important species like *Spirulina, Chlorella, Dunaliella, Haematococcus,* and *Aphanizomenon* are described in detail in single subchapters.

The third section *Microalgae applications* forms the biggest part of the book and deals with the exploitation of microalgae in biotechnology.

Chapter 6 introduces the unicellular red alga *Galdieria* of the class Cyanidiophyceae. The authors review the origin and taxonomy of *Galdieria* as well as its morphology, genome, and unique extremophilic features. The algae can grow in extreme conditions like high acidity, very low pH, or extremely high metal concentrations. It also exhibits a broad metabolic flexibility to grow photoautotrophically, heterotrophically, or mixotrophically, and to utilize a whole range of unusual carbon sources. This chapter also focuses on applications in biotechnology. Wastewater treatment, selective metal precipitation, or scarce metals recovery from secondary sources is discussed.

In Chapter 7, the authors evaluate the potential of microalga *Botryococcus braunii* (Trebouxiophyceae) to be used as a biorefinery organism. The authors clearly describe the physiology and biochemistry of different *Botryococcus braunii* races and summarize all value-added compounds produced by the alga. The profitability of the production of hydrocarbons, lipids, pigments, polysaccharides, and other polymers is mentioned too.

Chapter 8 deals namely with poly unsaturated fatty acids (PUFAs) production by marine algae and the use of lipases as a tool for their enrichment. Marine algae rich in n-3 PUFA are a natural and readily available resource of PUFA and alternative to fish oil derived n-3 PUFA. The enrichment of microalgae using biolipase from the yeast *Candida cylinderacea* source is of particular interest. The analysis of fatty acids in marine algae and the PUFA application effects are explained.

In Chapter 9, the authors explain the composition of carbohydrates in microalgae biomass, focusing on separation methods, chemistry and molecular characterization as well as their application in several areas. Steps for extraction and purification are discussed as well as the relationship between the type of microalgae and its composition. The current and prospective trends and methodologies for the use of microalgae carbohydrates are summarized.

Chapter 10 is an original research article dealing with adaptation of two *Chlorella* strains to chilling temperature and irradiance. *Chlorella* is used for large-scale outdoor production in Taiwan so the understanding of how green algae overcome chilling temperatures during winter is useful from a biotechnological point of view. The authors studied changes in photochemical efficiency and differential induction of superoxide dismutase as a response to the combined stresses of chilling temperature and high irradiation.

The last section, *Microalgae secondary metabolites*, is devoted to the cultivation of microalgae for the production of a variety of secondary metabolites for possible commercialization.

Chapter 11 describes the cultivation of microalgae for secondary metabolite production. The authors describe in detail the metabolism of microalgae and the kinetics of mixotrophic growth. The comparison of open and closed photobioreactors is introduced and the problematics discussed. The topic of scale up production of microalgae is mentioned. The secondary metabolites like fine chemicals, pigments, antioxidants, lipids, and proteins are listed and the microalgae and the conditions suitable for their production are listed.

The closing chapter, Chapter 12, is an excellent review of various applications of microalgae. The authors focus on the physiological limitations behind the synthesis of microalgal products, highlight the crucial unknowns behind the role and synthesis of these products and hint at strategies to overcome the limitations to realize the commercial dream of microalgal products.

As the editor, I am happy to present to the readers such a broad collection of topics by a large number of experts from the field of microalgae research. Both authors and readers can enjoy the open access to all chapters.

I thank all the contributors for their effort and Ms. Lada Bozic, Author Service Manager, for the smooth cooperation and help.

Welcome to the amazing world of microalgae!

Prof. (FH) RNDr. Milada Vítová, PhD.
Laboratory of Cell Cycles of Algae,
Centre Algatech,
Institute of Microbiology of the Czech Academy of Sciences,
Czech Republic

Section 1

Phytoplankton

Chapter 1

Self-Fluorescence of Photosynthetic System: A Powerful Tool for Investigation of Microalgal Biological Diversity

Natalia Grigoryeva

Abstract

It is well-known that photosynthetic cells of various microalgae species display distinct fluorescent properties. The efficiency of self-fluorescence excitation and emission at different wavelengths depends on the structure of photosynthetic system and particularly on the structure of antenna complex of specific strains. The peculiar structure of blue-green algae light-harvesting complex allows to discriminate and classify known and new cells up to species/strain level by means of microscopic spectroscopy. In this chapter, a novel fluorescent spectroscopic technique for microalgae species discrimination will be presented. This method is based on a special data processing of a set of fluorescent spectra, obtained from a single photosynthetic cell of microalgae, particularly from cyanobacterial cells. According to the presented technique, single-cell self-fluorescence spectra are recorded by means of confocal laser scanning microscopy (CLSM), and data processing is conducted via linear discriminant analysis (LDA) and artificial neural networks (ANN).

Keywords: microalgae, cyanobacteria, photosynthetic system, biological diversity, fluorescence, microscopic spectroscopy, artificial neural networks

1. Introduction

Cyanobacteria are the most ancient photosynthetic microorganisms on Earth. Nowadays, cyanobacteria are one of the most widespread organisms in nature, and the ecological aspect in their investigation is quite valuable. On the other hand, thousands of strains belonging to different species are cultivated in biolaboratories all over the world for different cyanobacterial biotechnological applications such as biofuel cells, food production, pharmaceuticals, fertilizers, etc. [1–3]. Thus the noninvasive spectroscopic methods are quite requisite for monitoring of physiological state of cyanobacterial cultures and natural communities.

It is well-known that the analysis of self-fluorescence of photosynthetic system is a powerful noninvasive tool for investigation of microalgae in vivo. It reports on the energy transfer and trapping and, thus, reflects the metabolic mechanisms in photosynthetic cells and their photosynthetic efficiency. The detected self-fluorescence finally reflects the diversity in morphological and physiological states of photosynthetic cells [4–6].

Self-fluorescence originates from excited states that were lost before photochemistry took place and usually represents a small fraction of the excited state decay in a functional photosynthetic complex. Nevertheless, this small fraction can be easily detected by confocal laser scanning microscopy (CLSM). With the confocal fluorescence microscopy, a very small excitation and detection areas can be investigated, so that single cells under non-damage conditions can be studied in vivo. Single-cell detection can provide the information on small peculiarities that is regularly buried in normal ensemble average experiments. This is thus a good way to study the time evolution process and spectroscopic properties of individual cells. Both steady-state and time-resolved fluorescence measurements can be used for probing the organization and functioning of photosynthetic systems by means of CLSM.

Till now the best taxonomic differentiation is still obtained using classical inverted microscopy. Unfortunately, this method is time-consuming, human based, and requires appropriate technical skills; this eliminates the possibility of its application for continuous online monitoring. Nearly single-cell flow cytometric analysis, based on light scattering by the cells and fluorescence of the chlorophylls and the phycobilins, can be easily automated, but it is appropriate only for unicellular species and is useless for numerous industrially cultured filamentous strains [7, 8]. The main problem of all chemical methods (e.g., high-performance liquid chromatography (HPLC) [9, 10]) is that during the chemical sample preparation, the most of the information about the peculiarities of individual species is lost and the residual part of the information is not enough for species/strain discrimination inside cyanobacterial genera and is suitable only for the rude differentiation of big classes of phytoplankton. Thus, the analysis of the in vivo fluorescence spectra is the only one noninvasive technique for obtaining qualitative information about the phytoplankton abundance and composition, which is continuously demonstrated by various publications [10–17]. The relative phytoplankton abundance can be calculated once initial assumptions about the phytoplankton classes are presented and their pigment compositions have been made [7, 12, 13].

Maybe the first attempt to use phycoerythrins as chemotaxonomic markers was done by Glazer et al. [18] for red algae in 1982, but until now fluorescence spectra of phycobilins do not appear to be useful at familial, ordinal, and class levels in taxonomic studies. Although the investigation in [18] concerns only purified high-molecular-weight phycoerythrin from red algae, this work clearly demonstrates the possibility of the correct taxonomic analysis on the basis of phycobiliproteins structural differences, which can serve as intrinsical fingerprints for taxons and genera in phytoplankton diversity. Later, the correlation between the distribution of the biliproteins and the genera of Cryptophyceae was discussed in [19]. In 1985, Yentsch and Phinney [20] proposed an ataxonomic technique that utilized the spectral fluorescence signatures of major ocean phytoplankton. Seppälä [16] used spectral fluorescence signals to detect changes in the phytoplankton community. In 2002, Beutler et al. reported a reduced model of the fluorescence from the cyanobacterial photosynthetic apparatus designed for the in situ detection of cyanobacteria and presented a commercially available diveable instrument for online monitoring of phytoplankton structure [11].

However, the correct classification of cyanobacterial species on the basis of their bulk fluorescence signature is hampered by alterations in pigment composition within one strain, which depends on the physiological state of the culture (community) and environmental conditions [21]. On the other hand, several researchers show that the nutrient and light limitations do not significantly change the initial fluorescence spectra and cannot impede the species discrimination [17, 22].

Recent rapid development of confocal microscope functionality initiates new directions in subcellular biology research [23, 24]. Confocal laser scanning microscopes are distinguished by their high spatial and temporal resolution. Modern laser scanning microscopes are unique tools for visualizing cellular structures and analyzing dynamic processes inside single cells. One of the specific fields of CLSM application is the investigation of self-fluorescence of living cells. CLSM single-cell microscopic spectroscopy is undoubtedly the most powerful tool for in vivo investigation of physiological processes in photosynthetic organisms (cyanobacteria, algae, and higher plants). The investigation of self-fluorescence of single living cells reveals the relation between the physiological state and the operational activity of photosynthetic system. A lot of interesting static and dynamic effects can be studied by means of CLSM. The investigation of self-fluorescence gives the information about single-cell processes as well as about the collaboration in cell communities. Changes in spectral characteristics of living photosynthetic cells indicate changes in their physiological state and can be applied for the studies of the results of stress states and external actions [4–6]. Moreover, the diversity in single-cell self-fluorescence for different species and strains can serve the basis for ataxonomic discrimination of cyanobacterial genera.

In this chapter, a novel ataxonomic approach to differentiation of cyanobacterial cells based on the numerical analysis of in vivo single-cell fluorescence spectra, recorded by means of CLSM, is presented. The differentiation is conducted according to the structure and operation of their photosynthetic apparatus. An optimal set of the parameters is selected, which is sufficient for determination of the taxonomic position of cyanobacteria by means of mathematical statistics. On the basis of the linear discriminant analysis, the obtained spectroscopic data for 23 cyanobacterial strains from CALU collection were analyzed. It was shown that the presented technique allows an accurate differentiation of cyanobacteria up to the species/strain level and enables to distinguish automatically potentially harmful strains. All presented results were obtained using cyanobacterial strains from CALU collection of the Core Facility Center “Centre for Culture Collection of Microorganisms” of the Research Park of St. Petersburg State University as model objects for CLSM studies.

2. Materials and methods

2.1 Cyanobacterial strains and cultivation conditions

All work on preparing cyanobacteria cultures for this research was carried out at the Core Facility Center “Centre for Culture Collection of Microorganisms” of the Research Park of St. Petersburg State University. In the CALU collection [25] at the core facility center, cyanobacterial strains were maintained in semiliquid agar (0.8%) medium no. 6 after Gromov [26] in test tubes of volume 5–6 mL under cotton plugs. The strains were stored at 14°C under a constant illumination of 2000 lux and were recultivated with a periodicity of 2–3 months.

Cyanobacteria used in this investigation were grown on liquid medium no. 6. A stock culture was preliminarily prepared, for which it was cultivated in 30 mL of medium and incubated for 2 weeks at room temperature under continuous illumination from fluorescent lamps. To maintain a constant volume, 5 mL of medium were added to the stock culture every 2 weeks. All experiments in this study were conducted with cultures presumably in the logarithmic phase of their growth.

In this work, 23 cyanobacterial strains from CALU collection were used:

1. *Anabaena variabilis* Kutz. CALU 824, ponds near Old Petergof, Saint Petersburg, Russia.

2. *Arthrospira* sp. CALU 1712, Gulf of Finland, Saint Petersburg, Russia.

3. *Geitlerinema* sp. CALU 1315, Lake Kyzyl-Tash, Ozersk, Chelyabinsk, Russia.

4. *Geitlerinema* sp. CALU 1718 Lake Kamenetz, Pskov, Russia.

5. *Leptolyngbya* sp. CALU 1713, river Tikhaya, Saint Petersburg, Russia.

6. *Leptolyngbya* sp. CALU 1715, Gulf of Finland, Saint Petersburg, Russia.

7. *Leptolyngbya* CALU 1750 sp., Lake Tarasovskoe, Saint Petersburg, Russia.

8. *Lyngbya* sp. CALU 1804, Lake Valdai, Novgorod, Russia.

9. Merismopedia sp. CALU 666 punctata Meyen f., Pinar del Rio, Rio de Soroa, Cuba.

10. Microcystis firma sp. CALU 398 (Breb. et Lenorm) Schmidle, Turkmenbashi Canal, Turkmenistan.

11. *Myxosarcina chroococcoides* Geitl. sp. CALU 601, Russia.

12. *Nostoc* sp. CALU 1763, Lake Ladoga, Saint Petersburg, Russia.

13. *Nostoc* sp. CALU 1817, springs on the Island Big Solovetsky, White Sea, Russia.

14. *Oscillatoria* sp. CALU 1415, ponds in Vorkuta region, Russia.

15. *Oscillatoria* sp. CALU 1416, ponds in Vorkuta region, Russia

16. *Phormidium favosum* CALU 624, brook Ammersbek, Hamburg, Germany.

17. *Plectonema* sp. CALU 457, pond in Strelna, Saint Petersburg, Russia.

18. *Pleurocapsa* sp. CALU 1126, Lake Ladoga, Saint Petersburg, Russia.

19. *Spirulina platensis* sp. CALU 550 (Nordst.), Czech Republic.

20. *Synechococcus* sp. CALU 535, ponds near Old Petergof, Saint Petersburg, Russia.

21. *Synechococcus* sp. CALU 756, Czech Republic.

22. *Synechococcus* sp. CALU 1409, ponds in Vorkuta region, Russia.

23. *Synechocystis* sp. CALU 1336 *aquatilis*, Gulf of Finland, Saint Petersburg, Russia.

Fluorescent and corresponding transmission photomicrographs, obtained via CLSM, for several strains from CALU collection are presented in **Figure 1**. In further illustrations only the CALU numbers for corresponding strains will be used for the clarity of the narration.

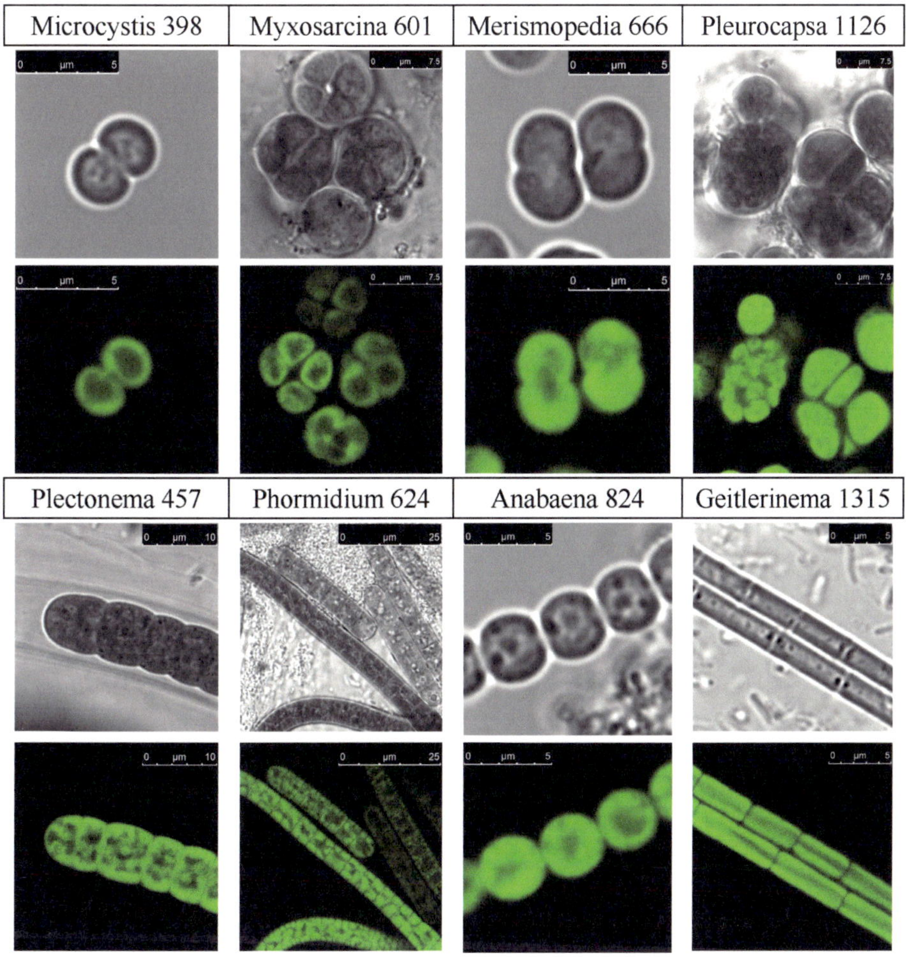

Figure 1.
CLSM fluorescent and transmission photomicrographs for eight cyanobacterial strains from CALU collection. The white bar indicates the object scale.

2.2 Confocal laser scanning microscopy

Confocal laser scanning microscopes are distinguished by their high spatial and temporal resolution [23, 24]. Modern laser scanning microscopes are unique tools for visualizing cellular structures and analyzing dynamic processes inside single cells. They exceed classical light microscopes especially in their axial resolution, which enables to acquire optical sections (slices) of a specimen. Apart from simple imaging, confocal laser scanning microscopes are designed for the quantification and analysis of image-coded information. Among other things, they allow easy determination of fluorescence intensities, distances, areas, and their changes over time. New acquisition CLSM tools include the detection of quantitative properties of the emitted light such as spectral signatures and fluorescence lifetimes. The most impressive feature of modern CLSMs is their capability for single-cell microscopic spectroscopy, which allows to obtain spectroscopic information inside single cells and small regions.

In the present investigation, Leica TCS-SP5 was used for the investigation of living cyanobacterial cells. Fluorescence emission spectra of the intact cells were measured at eight excitation wavelengths corresponding to all available laser lines. The excitation wavelengths are 458, 476, 488, 496, and 514 nm are the lines of Ar laser,

405 nm is the line of diode UV laser, and 543 and 633 nm are the lines of HeNe laser. In all presented experiments, laser power settings are as follows: 29% of Ar laser power was reflected onto sample with acousto-optical tunable filter (AOTF), and further power percentage for its laser lines was 30% of 458 nm laser line and 10% for all other lines. 405 nm line of diode UV laser was reflected onto sample with 3%; HeNe laser lines 543 and 633 nm were reflected with 10 and 2%, respectively. An acousto-optical beam splitter (AOBS) was used to transmit sample fluorescence to the detector. Emission spectra between 520 and 785 nm were recorded using the lambda scan function of the "Leica Confocal Software" by sequentially acquiring a series ('stack') of 38–45 images, each with a 6 nm fluorescence detection bandwidth and with 6 nm wavelength step. For obtaining fluorescence-intensity information, images of 512 × 512 pixels were collected with a 63× Glycerol immersion lens (Glycerol 80% H_2O) with a numeric aperture of 1.3 (objective HCX PL APO 63.0 × 1.30 GLYC 37°C UV) and with additional digital zoom factor 5–9 (depending on a cyanobacterial strain). One pixel corresponds to 53.5 × 53.5 nm. The photomultiplier (PMT) voltages were used in range from 900 to 1100 V. The fluorescence emission images were accompanied with the transmission images (in the parallel channel), collected by a transmission detector with the photomultiplier voltages ranged from 300 to 500 V. For better signal yield, lambda scans were performed with "low speed" setting (400 Hz) in bidirectional scan mode and with a pinhole setting of 1 Airy unit (the inner light circle of the diffraction pattern of a point light source corresponds to a diameter of 102.9 μm with the lens used (see [23]). Regions of interest (ROIs) representing single cells or subcellular regions were used to calculate fluorescence spectra.

For 2D imaging, to raise the sensitivity and contrast, images were recorded at 405 nm excitation wavelength (diod UV laser) and by Leica HyD hybrid detector, which strongly improves contrast in comparison to PMTs. HyD gain was taken as 100 V. The images of 1024 × 1024 and 2048 × 2048 pixels were collected with a 63× Glycerol immersion lens (Glycerol 80% H_2O) with a numeric aperture of 1.3 (objective HCX PL APO 63.0 × 1.30 GLYC 37°C UV) and with additional digital zoom factor 10–35. The fluorescence emission images were accompanied with the transmission images (in the parallel channel). The images were recorded with a pinhole setting of 1 Airy unit.

In CLSM applications, the laser light density in the focus point is high. But, generally, it is deposited in short "dwell times" during the laser scanning process. Dwell time and the intervals between the illuminations may influence photodamage and saturation of photosynthetic apparatus of living cells. Thus, since most chromophores bleach under the high laser excitation energies, a bleach test should be performed [27]. It was shown experimentally that especially phycoerythrin (PE) and phycocyanin (PC), as accessory pigments, were very sensitive to photobleaching, while the fluorescence of chlorophyll a (Chl a) and allophycocyanin (APC) remained stable in the intact living cells [27]. During the detection the fluorescence of the main accessory pigments for each cyanobacterial strain should be controlled and the changes in their fluorescence should not exceed 10–20%. In this investigation the power of individual laser lines was chosen according to the photodamage they cause. The repeated spectra were obtained under selected excitation power at a fixed point in a cell to check whether the excitation would affect the cells. It was shown that at the above chosen excitation energies (laser line percentage) the fluorescence spectra did not vary within the experimental error during 10–15 records. When excitation energy was increased, both the height and the center of the bands varied enormously with time because of photodamage or structure breakdown in photosynthetic systems. In the experiments, where several laser lines were involved for the investigation, the first spectrum was recorded again at the end of each series to control the initial state of the cell. It should be pointed out that the whole procedure of fluorescence spectra recording, used in this study, was designed to minimize preparatory manipulation, so as to conduct a noninvasive investigation of small amounts of experimental material and to prevent any damage of living cells.

To exclude unpredictable variations in physiological state of investigated cultures, the fluorescence spectra were taken from the cells of one strain several times, at different days and for various developmental stages of the culture. And it was established that the variations in spectrum shape and intensity among cells of one strain are not considerable.

2.3 Data processing

2.3.1 Data preprocessing

The main difficulty of the considered discrimination problem resulted from the high nonuniformity of the initial data and different numbers of observation for different strains. A small size of initial dataset as well as the sophisticated nature of the experimental data required a complex preprocessing procedure. The original experimental data represents 307 sets of self-fluorescence spectra obtained from cyanobacterial cells, belonging to 23 different strains. Each observation from a data set is described by a series of seven spectra taken from a single cell by means of CLSM. Each initial spectrum is an array of 38–45 numbers, which correspond to the fluorescence intensities on specific emission frequencies of visible light in the range from 520 to 785 nm. In contrast to the previous investigations, which utilized for classification a full spectrum of the samples [12–14], we used a set of integral and statistical characteristics, describing the shape of each spectrum. To extract a set of classification parameters from initial data, a computer program has been developed in a mathematics system MATLAB. By means of this program, normalization, interpolation, extrapolation, and smoothing of the raw spectra were carried out, to eliminate the random noise and metering fluctuations. All spectra were reduced to the same scale and size of data array, the first derivative was taken over initial spectra, and the fast Fourier transform (FFT) was performed, to exclude random noise, owing to the low intensity of the exciting and emitting light. The specific values characterizing the shape of obtained curves and the spectral composition of their derivatives were calculated. All selected classification parameters can be divided into three groups: asymmetry and excess, fluorescence emission percentage for individual pigments in four main spectral regions (phycoerythrin, 573–586 nm; phycocyanin and allophycocyanin, 649–661 nm; chlorophyll a PSII, 674–689 nm; chlorophyll a PSII, 711–727 nm), and the frequency characteristics of the corresponding first-derivative Fourier transforms for each plot (mean values in three specified frequency domains: 43–58 μm-1, 95–110 μm-1, 123–135 μm-1). The detailed description of the extraction of classification parameters is given in Zhangirov et al. [28].

2.3.2 Linear discriminant analysis

Linear discriminant analysis (LDA) is well-known and often applied in biology for various classification problems [15, 17, 29, 30]. Linear discriminant analysis (LDA) is a statistical technique for classifying samples into two or more groups (classes) [31, 32]. It utilizes linear combinations of independent variables to form a basis for a classification scheme. In our case, the independent variables are 63 classification parameters extracted from each set of single-cell self-fluorescent spectra.

LDA builds n linear discriminant functions, where n is a number of classes and a row vector with a number of parameters describes each observation. The decision of the sample belonging to the class is based on the selecting of the maximal discriminant function for the sample row vector. Discriminant analysis has two very useful applications. First, it identifies a set of classification parameters that are needed to discriminate between known groups, that is, sets of classification parameters can be identified that are necessary to discriminate between known cyanobacteria strains. Second, the analysis can be used to classify an unknown sample (within a certain probability) into a known group of species or strains. The high classification accuracy of LDA is due to the fact that it works with distribution functions for classification parameters and their statistical characteristics, which allows to build better classification model. However, LDA has strong restrictions on the presence of correlations between classification parameters.

In addition, LDA allows to reduce dimension of the feature space. This so-called linear Fisher discriminant analysis (LFDA) is a data classification method, which classifies the samples by dividing them into groups. The boundaries of these groups are determined by threshold coordinate values. The goal of this method is to find the informative projections by maximizing the function constructed of the projective matrix, the between-class scatter matrix, and the within-class scatter matrix. In this procedure, the first largest component (canonical discriminant function) is the maximal, and the classifications are performed using the three-dimensional space defined by the three largest components. The selection of the best classification parameters is based on the criterion that the dissimilarity between classification parameters of different species/strains should be greater than between those of the same group. Actually, LFDA bases on a solution of eigenvalue problem. The eigenvectors with the first highest eigenvalues are used to construct a lower dimensional space, while the other dimensions are neglected.

Also a stepwise discriminant analysis (SDA) was used in this investigation at the stage of selection of the most valuable classification parameters to determine which parameters discriminate better between the specified groups of observations. Standardized coefficients for each variable in each discriminated function represent the contribution of the respective parameter to the discrimination between groups.

The calculations were performed in MATLAB software using custom-built programs [33].

2.3.3 Artificial neural network

Artificial neural networks (ANNs) are currently being used in a variety of applications with great success [8, 34–36]. In contrast with conventional programs for data analysis, neural networks follow an adaptive approach. They are flexible and eminently suited for application to complex data structures that are not apt for other data analysis methods like cluster analysis or principal component analysis. Their first main advantage is that they do not require a user-specified problem solving algorithm (as is the case with classic programming), but instead they "learn" from examples, much like human beings. Their second main advantage is that they possess an inherent generalization ability. This means that they can identify and respond to patterns that are similar but not identical to the ones on which they have been trained.

ANN can be described as a mathematical model of a specific structure, consisting of a number of the single processing elements (called artificial neurons), arranged in interconnected layers. An active neuron multiplies each input vector by its weight, sums the products, and passes the sum through a transfer function to produce the output [37]. The ANN is made up of a group of interconnected artificial neurons, belonging to different layers, while inside one layer neurons are independent. ANN consists of an input, hidden, and output layers. Each neuron transforms input and sends outputs to other neurons to which it is connected.

There are many different types and architectures of neural networks varying fundamentally. In this paper a feed-forward ANN (FFANN) is used for solving considered classification problem [34, 37]. **Figure 2** illustrates the model of the ANN used in this work. Due to the simplicity of the classification problem to be solved, a multilayer feed-forward neural network (NN) with one hidden layer was considered. As an activation function, a hyperbolic tangent was used at the hidden layer and Softmax function at the output layer, which allows interpreting the output layer as the distribution of probabilities of belonging to each of the classes. On both layers a bias neuron with a signal equal to unity is added. The size of the input layer (N_{in}) depends on the number of classification parameters. The number of neurons on the output layer was fixed and equal to the number of classes (N_{out} = 16). The number of neurons on hidden layer was estimated by the following equation $N_h \sim \sqrt{N_{in} N_{out}}$ [37].

Learning in ANNs is accomplished through special training algorithms developed based on learning rules presumed to mimic the learning mechanisms of biological systems. According to supervised learning, the network is trained with a dataset of observations and optimized basing on its ability to predict a set of known outcomes. The deviation of the network solution from the target (true) value is computed, and the calculation of the error is propagated backward from the output layer to adjust the connection weights. Since in our case the activation function at the output layer was determined as Softmax, the loss function was calculated via cross-entropy method. A lot of special training algorithms were developed according to learning rules. In this investigation the method of adaptive moment estimation (Adam) was chosen for further calculations [38].

In the training phase, a sample set of classification parameters and the known solution (the strain number of the corresponding cell) are forced iteratively upon the network. The neuron's weights (ANN parameters) are adjusted in small steps until the network has learned the training examples. In the experiments described

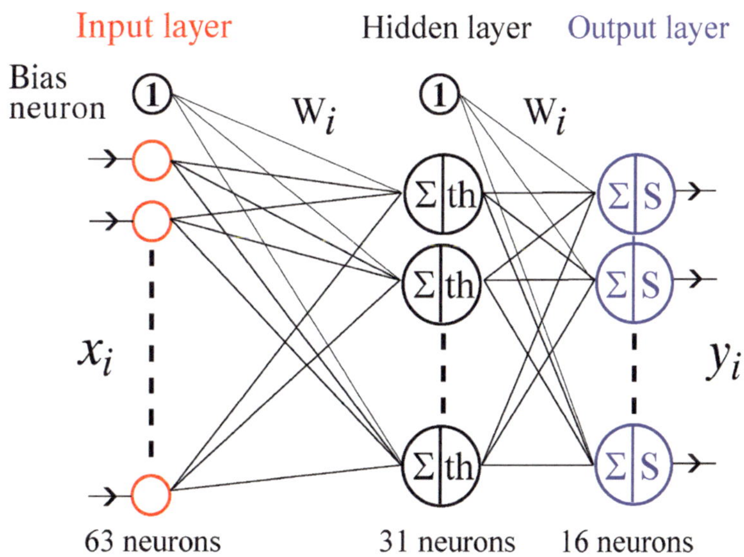

Figure 2.
The multilayer feed-forward artificial neural network. The input classification parameters are fed to the input layer of ANN, and signals are propagated through the network via internal neurons to the output layer. In this way input signal pattern and output signal pattern are associated with each other.

in this study, the training procedure has 500 iterations (epochs). After training, the network is tested. In this test phase, the characteristics of a number of cyanobacterial cells with known identities are fed to the network, and the solutions are compared with these known identities. In this study, after training the network was capable of recognizing about 96% of cyanobacterial cells in the test set. The analysis of generalization quality of ANN is identical to the test procedure; only the identities of the cells are not known beforehand.

The ratio of training sample to the test sampling in this investigation was taken 70:30%. Other parameters of the selected training algorithm were as follows: acceptable error threshold is 0.01, the bandwidth parameter (size of error control window) is 20, the moment parameter is 0.1, and the regularization parameter is 0.001. The selected learning rate was chosen 0.01, and the number of training epochs lays in the range from 300 to 800.

The main criteria for assessing the quality of ANN operation is the value of classification accuracy. There are several approaches to evaluate the accuracy of classification. In the considered case, the classification accuracy is calculated for each class separately, as the ratio of the number of correctly classified class observations to the total number of observations in a given class. Then the average classification accuracy for all classes was obtained. In such case it is possible to build a matrix of errors with size N × N (N—number of classes) and present the results in a bar chart, on which a classification accuracy for each class can be visualized (see **Figure 8** in "Ataxonomic differentiation of cyanobacterial strains on the base of single-cell fluorescence spectra").

On the base of the classification accuracy analysis, it is possible to evaluate the quality of ANN training as well as the quality of internal and external generalization. In our case, the evaluation of the quality of external generalization was obtained on the base a priori knowledge about new species, which was taken from an expert. To validate the correctness of the neural network operation, the results of the NN classification were compared with the results of the LDA.

The ANN architecture presented in this paper, as well as the learning algorithm and its parameters were determined during the study of various configurations. The selected model after training consistently gives a classification accuracy of at least 95%. In this study, ANN was simulated using MATLAB software [33].

3. Light-harvesting system of cyanobacteria

In cyanobacteria, the antenna complexes for photosystem II (PS II) and to some extent for photosystem I (PS I) are extrinsic and formed as large multiprotein organelles, which are located on the stromal side of the thylakoid membranes. These supramolecular pigment-protein complexes, so-called phycobilisomes (PBSs), first described by Gantt [39], are the main light-harvesting antennae in cyanobacteria.

Phycobilisomes are primarily composed of phycobiliproteins, a colored family of water-soluble proteins. Their chemical and spectroscopic properties are determined by their structure and function that they perform in the photosynthetic process. The three classes of phycobiliproteins are allophycocyanin (APC), phycocyanin (PC), and phycoerythrin (PE). However, in some cyanobacteria phycoerythrin can be replaced by phycoerythrocyanin (PEC), or both pigments can be lacking; phycocyanin and allophycocyanin are constitutively present in all cyanobacteria. Actually, there are very slight species differences between detached phycobiliproteins, even between prokaryotic cyanobacteria and eukaryotic red algae [40].

Usually PBSs are assembled from 12 to 18 different types of polypeptides, which may be grouped into three classes: (1) phycobiliproteins, (2) linker polypeptides, and (3) PBS-associated proteins. The amino acid sequences of all components constituting the phycobilisomes of some cyanobacterial strains have been determined, and analysis of these data has revealed phylogenetic relationships [41].

The polypeptide composition of PBS varies widely among strains of cyanobacteria. It should be noted that the degree of PBS compositional variability, which reflects the ability of an organism to adapt to environmental changes, varies from strain to strain. Moreover, for a single strain it sometimes depends upon the environmental conditions such as nutrient availability, temperature, light quality, and light intensity.

It is well-known that total biliprotein content of cyanobacterial cells is inversely related to the quality and quantity of irradiance. A comprehensive review given in Refs. [41–46] details the various degrees of such chromatic adaptation. However, for cyanobacteria cultured under white light of reasonable intensity and in the medium with habitual nutrient composition, no chromatic adaptation can occur, and PBS structure remains invariable within each strain. Thus the unique spectroscopic properties of different cyanobacterial strains, while analyzing in vivo, may become promising fingerprints for practical and laboratory applications.

Phycobilisomes are constructed from two main structural elements: a core substructure and peripheral rods that are arranged in a hemidiscoidal fashion around that core (**Figure 3**). Each core cylinder is made up of four disc-shaped phycobiliprotein trimers, allophycocyanin (APC), allophycocyanin B (APC-B), and APC core-membrane linker complex (APC-L_{CM}). By the core-membrane linkers, PBSs are attached on thylakoids and structurally coupled with PSII. The peripheral cylindrical rods (six or eight) radiate from the lateral surfaces of the core substructure and are usually not in contact with the thylakoid membrane. The rods are made up of hexamers, disc-shaped phycobiliproteins (PE, PEC, and PC), and corresponding rod linker polypeptides [41–44]. Most linker polypeptides are colorless proteins, but some also contain phycobilin chromophores, endowing them with the ability to harvest light as well as aid in the assembly of the phycobilisomes [46]. For more details about phycobilisome structure, see [18, 39, 47].

The phycobilisome is attached to the membrane by multiple weak charge-charge interactions, either with proteins or with lipid head-groups. Binding is rather unstable. The core-membrane linker polypeptide provides a flexible surface, allowing interaction with a range of structurally distinct membrane complexes, including photosystem II (PSII) and photosystem I (PSI) (see **Figure 4**). The stability of each interaction may be modulated by covalent modification and/or the presence of accessory subunits.

Recently, it was established that phycobilisomes diffuse rapidly on the surface of the thylakoid membrane, while PS II reaction centers are normally almost immobile. Fluorescence recovery after photobleaching (FRAP) has been used to measure

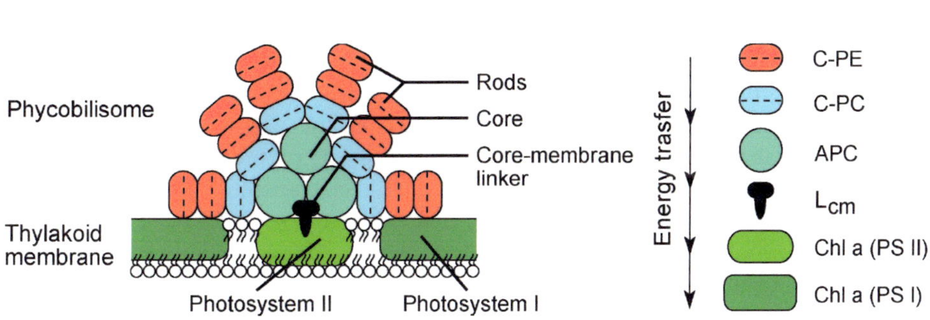

Figure 3.
Schematic drawing of phycobilisome and photosynthetic energy transfer to reaction center.

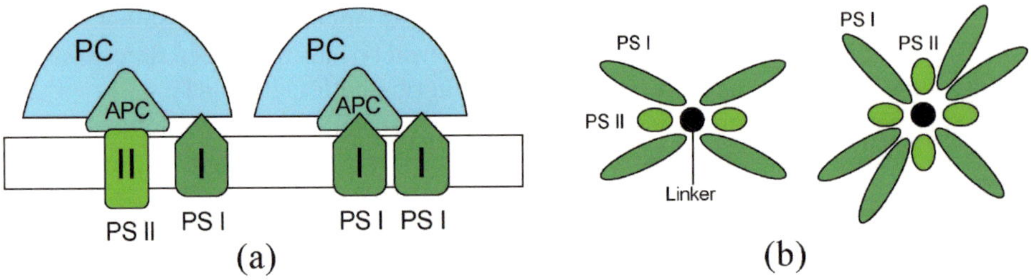

Figure 4.
Model for the interaction between phycobilisomes and membrane components. (a) Phycobilisomes are shown as semicircles, with the core as a darker rectangle [48]. (b) A parallel section through the membrane with the terminal phycobilisome pigment in relation to photosystems I and II [51].

the mobility of phycobilisomes in the intact cyanobacterial cells [48, 49], and it was clearly demonstrated that a significant proportion of phycobilisome-absorbed energy is delivered to PS I as well as to PS II [45, 49, 50].

The high mobility of phycobilisomes along the thylakoid membrane gives the opportunity of the occasional direct interaction of phycobilisome rods or core with PS I (**Figure 4**). Two ways that energy could be transferred from phycobilisomes to photosystem I are shown in **Figure 4**; "Spillover" from photosystem II with an attached phycobilisome (supposed by Su et al. [51]) (**Figure 4a**, left photosynthetic complex) and direct association of the phycobilisome core with photosystem I (**Figure 4a**, right photosynthetic complex).

Another possible variant of the interaction between phycobilisome and reaction centers of two photosystems was proposed by Gantt in the Chapter 6.3 of the book [40]. The author assumed that the special close arrangement of both photosystems around the base of the phycobilisome provides the partial transfer of the absorbed energy to PSII and PSI simultaneously (**Figure 4b**).

4. Fluorescence spectra of intact cyanobacterial cells

The intrinsic fluorescence of photosynthetic organisms originates from excited states that were trapped by light-harvesting system and lost before photochemistry took place. Photoexcitation energy absorbed at the outer surface of phycobilisomes is transported sequentially through several rod chromoproteins to an inner core and then to core-membrane linker (the terminal pigment) that acts as the final energy transmitters from the phycobilisome to Chl a heterodimers of two photosystems (PSII and PSI), incorporated in the thylakoid membrane. This excitation transfer is recognized as due to the Förster dipole-dipole interaction with an extremely high efficiency, near unity.

The more distal parts of the antenna system, a peripheral antenna complex (phycobilisome), maximally absorb photons at shorter wavelengths (higher energies) than do the pigments in the antenna complexes that are proximal to the reaction center. Subsequent energy transfer processes are from these high-energy pigments physically distant from the reaction center to low-energy pigments that are physically closer to the reaction center (**Figure 5**). With each transfer, a small amount of energy is lost as heat, and the excitation is moved closer to the reaction center, where the energy is stored by photochemistry. Note that the probability of excitation energy escape from the trap in the form of fluorescence at all transfer steps is non-zero and depends on the intensity and wavelength of the excitation light.

During the energy transfer process, the occasional quenching of the absorbed light by fluorescence can occur, and this becomes the essential property for

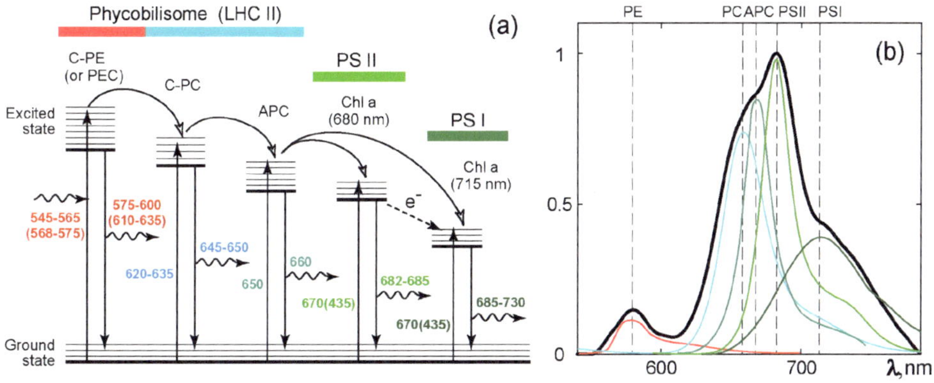

Figure 5.
Schematic illustration of the energy transfer in light-harvesting system of cyanobacteria (a). Panel (b) represents the example of normalized in vivo single-cell fluorescence emission spectra at excitation wavelength 488 nm. Dashed lines and letters over them indicate emission wavelengths of PE, PC, APC, and Chl a of PSII and PSI, correspondingly. Fine lines represent fluorescent spectra of corresponding detached pigment-protein complexes.

fluorescent spectroscopy. It usually represents a small fraction of the excited states and diminishes in a functioning photosynthetic complex. Nevertheless, the fluorescence is an extremely informative quantity, because it reports on the energy transfer and trapping. Both steady-state and time-resolved fluorescence measurements are widely used methods for probing the organization and functional state of photosynthetic systems.

The fluorescence of intact living cyanobacterial cells is originated from the inefficiency of the energy transfer between all components of the energy transfer chain including the final step, the delivery to PSII or PSI (**Figure 5a**). Due to the occasional quenching by fluorescence each transfer step result in peak or shoulder on the corresponding spectrum (**Figure 5b**). This is due to the fact that when phycobilisomes are bound to the thylakoid membrane, most of the energy from phycobilisome is channeled to chlorophylls in the thylakoid membrane and thus did not shade the fluorescence of the previous steps in energy transfer chain. In the course of the energy transfer from the initially photoexcited phycobiliprotein to the reaction center of photosystems PSI and PSII, fluorescence is emitted from almost every type of pigment and can be used as a probe to examine the mechanism of energy transfer within the light-harvesting system [43, 44, 52].

A convenient way to monitor this energy transfer process is to irradiate a sample with light that is selectively absorbed by one set of pigments and then monitor fluorescence that originates from a different set of pigments. Obviously, if the energy transfer is taken place between pigments, the light absorbed by one set of pigments is emitted by another set differently, depending on the excitation wavelength. This type of fluorescence excitation experiment can be used to measure quantitatively the efficiency of energy transfer from one set of pigments to another [43]. Moreover, different species of cyanobacteria contain different accessory pigment proteins and specific linker proteins between them; therefore a set of fluorescence emission spectra excited by different wavelengths have its own unique shape for the cells of one strain and are quite distinguishable from other species and strains. Such sets of fluorescence emission spectra can be used for automatic differentiation of cyanobacterial species.

Figure 6 shows several characteristic sets of single-cell fluorescence spectra corresponding to Microcystis CALU 398, Merismopedia CALU 666, Leptolyngbya CALU 1715, and Phormidium CALU 624, obtained by confocal laser scanning

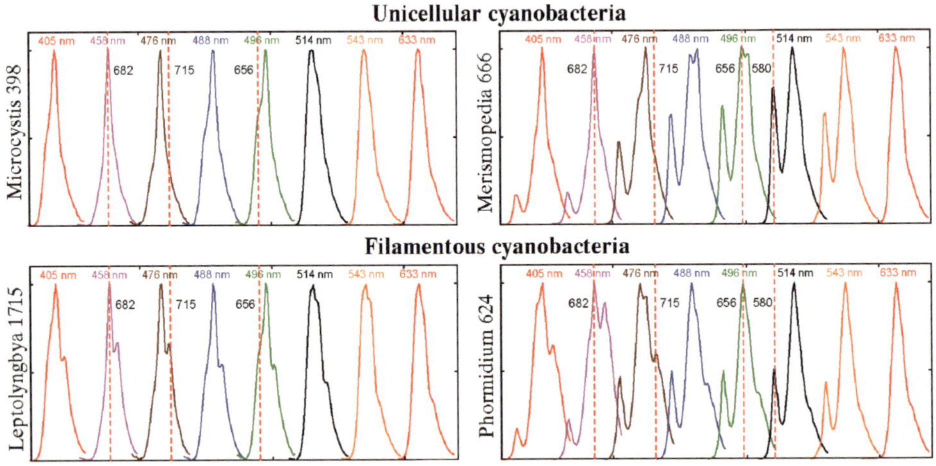

Figure 6.
Four characteristic sets of single-cell fluorescence spectra. The excitation wavelengths (405, 458, 476, 488, 496, 514, 543, and 633 nm) are given over the curves. All spectra are normalized to the maximum intensity and shifted along x-axis for convenience of observation. The dashed lines indicate fluorescence maxima of the individual pigments (PE, 580 nm; PC, 656 nm; Chl a, 682 and 715 nm).

microscope (CLSM) Leica TCS-SP5, which are placed near each set. Each spectrum in the set was obtained using different laser lines for excitation: 405, 458, 476, 488, 496, 514, 543, and 633 nm. Corresponding excitation wavelengths are given over each spectrum. All spectra are normalized to the maximum intensity and shifted along x-axis for convenience of observation. It can be easily noticed that laser line 458 nm excites mostly in vivo fluorescence of Chl a in both photosystems PSII and PSI around 682 and 715 nm, correspondingly, and the emission spectrum by cyanobacterial cells shows no appreciable emission of PC or APC. In cyanobacteria, the 458 nm excitation is preferentially absorbed by PSI that contains more Chl a than by PSII and is stoichiometrically more abundant than PSII. However, because reaction center of PSI turns over faster than the PSII, it has lower fluorescence intensity than the PSII antenna. This is indicated by PSI emission band at 715 nm which is much weaker than the PSII emission band at 682 nm. The excitation by intermediate (blue and green) wavelengths (405, 488, and 496 nm) reveals fluorescent maxima of all photosynthetic pigments, as the light in this range is absorbed by all pigment-protein complexes almost in equal portions and fluorescence emits by all steps of energy transfer chain (**Figure 5**). The direct excitation of cells in the PE absorption region at 514 and 543 nm results in emission spectrum with two main peaks at 580 and 656 nm, which are due to PE, PC, and APC emission, and for species that lack PE, the emission accumulates mostly near 656 nm. Two chlorophyll fluorescence components can be resolved for some species in a number of spectra. The spectra of the 633 nm excitation directly give a prominent emission band at 656 nm that originates from C-PC, omitting band at 580 nm, which cannot be excited by 633 nm, even for species that have PE (see **Figure 6**). Other small emission bands, corresponding to fine pigment structure of antenna complex, are not resolved at the room temperature.

These in vivo fluorescence emission spectra reflect the structure of light-harvesting complex of corresponding species and correct or incorrect functioning of its energy transfer chain. Four characteristic wavelengths, corresponding to the fluorescence maximum or shoulder, can be easily distinguished: (1) peak near 580 nm corresponds to the fluorescence of phycoerythrin, (2) peak near 656–560 nm corresponds to the fluorescence of phycocyanin and allophycocyanin in common (they are undistinguishable at room temperature), (3) peak near 682 nm corresponds to the fluorescence of chlorophyll a in photosystem II, and (4) peak or shoulder near 720 nm represents the fluorescence from photosystem I [10, 53].

Comparative analysis of the series of fluorescence spectra for different cyanobacterial species and strains reveals visible variations in their shape. If the fluorescence spectra were taken from live cells in normal physiological state, which are cultured in the same growth environmental conditions, then the interspecies variations in pigment/Chl a ratios are more pronounced than variations within the individual species. And species/strain differentiation could be carried out on the basis of conventional multivariate analysis.

5. Ataxonomic differentiation of cyanobacterial strains on the base of single-cell fluorescence spectra

Fluorescence spectra have been used to classify phytoplankton populations since approximately the early 1970s [54, 55]. However, because of the generally low device precision and poor availabilities, the rate of species discrimination was relatively low. Recently new attempts to conduct the discrimination among microalga on the base of absorption or fluorescence spectra were reported [7, 10, 13, 15]. But again in published experiments only big algal groups with a considerable differences in pigment composition can be successfully separated (e.g., cryptophytes, chlorophytes, cyanobacteria, etc.). Moreover, all the authors pointed out that the discrimination among cyanobacterial species is quite complex and ambiguous. Actually, the correct discrimination of cyanobacterial species on the base of fluorescence signature is usually hampered by alterations in the pigment composition within one strain, which depends on the environmental conditions and physiological state of the culture. These difficulties can be overcome by using single-cell fluorescence spectra instead of bulk ones and by recording 7–8 spectra with different excitation wavelengths for each cell instead of one or two as usually is done.

In the presented investigation, 307 sets of 8 single-cell fluorescent spectra for 23 cyanobacterial strains, belonging to 15 genera, were analyzed. An optimal set of classification parameters was considered that is sufficient for determining the generic membership of cyanobacterial cells by means of mathematical statistics. The results of this study show that LDA and ANN are able to recognize cyanobacteria up to species/strains according to the data recorded by means of CLSM. This implies that the classifier (LDA or ANN) is capable of defining a unique niche in a multiparameter space for each of 23 cyanobacterial strains, used in this investigation.

The results of LDA, evaluated over 63 parameters extracted from 307 single-cell fluorescence spectra, are presented in **Figure 7** as 3D-plots in the space of canonical discriminating functions. It is clear that the discrimination between species is sufficiently good. Moreover, the closely related species (e.g., *Spirulina* and *Oscillatoria*, *Synechococcus* and *Chlorogloea*, *Microcystis*, *Synechocystis* and *Myxosarcina*) appear close to each other. Such species as *Leptolyngbia*, *Geitleninema*, and *Oscillatoria*, which includes several strains, form big groups. However, inside these groups single strains also can be discriminated, which is demonstrated on the right panel, where the corresponding scaled region 1 is presented. This is confirmed by a classification diagram plotted in **Figure 7C**. The classification accuracy in the presented example was near 97.4%. The high classification accuracy is due to the fact that LDA works with distribution functions for classification parameters and their statistical characteristics, which allows to build a good classification model.

In the legend all used cyanobacterial strains are named and enumerated according to CALU collection. Solid curves bounded the regions, occupied by seven strains

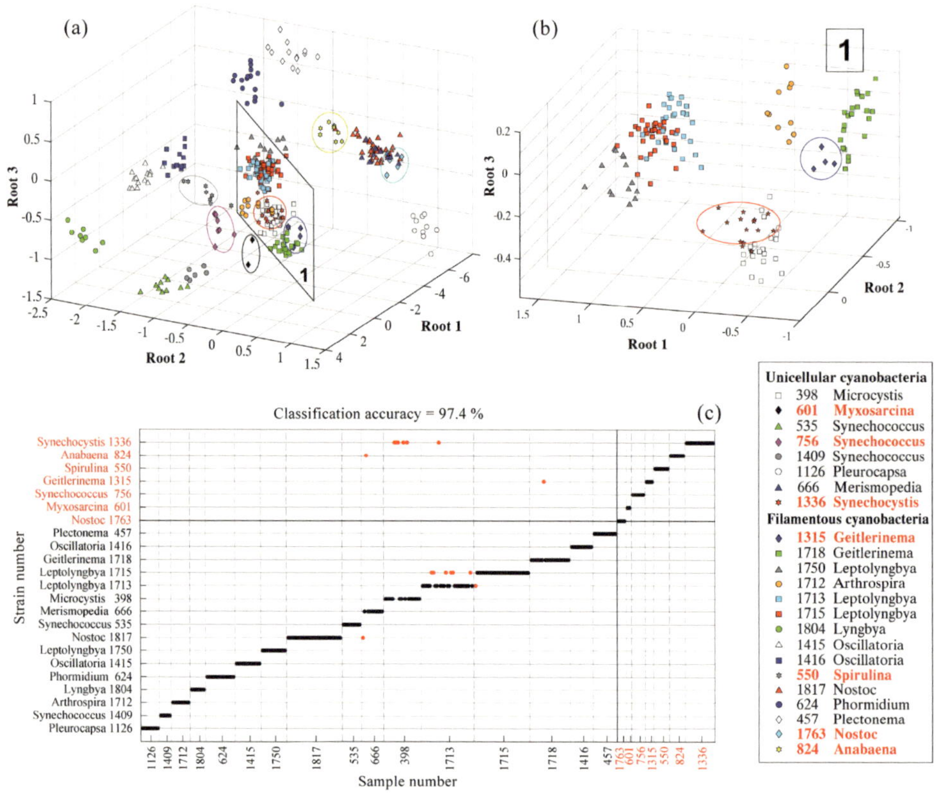

Figure 7.
The results of linear discriminant analysis. (a) and (b) Observations in 3D space of first three maximal canonical discriminant functions (root 1,2,3). Solid curves bounded the regions, occupied by seven new species. In the legend new species are indicated with red. (c) Classification diagram for 23 cyanobacterial strains from CALU collection. Red dots indicate false results.

(*Anabaena variabilis* Kutz. sp. CALU 824, *Geitlerinema* sp. CALU 1315, *Myxosarcina chroococcoides* sp. CALU 601, *Nostoc* sp. CALU 1763, *Spirulina platensis* (Nordst.) sp. CALU 550, *Synechococcus* CALU 756, and *Synechocystis aquatilis* sp. CALU 1336) used for testing ANN classificator (in the legend they are indicated by red color).

In the considered classification problem, the quality of the ANN operation should be determined not only by the absolute value of the classification accuracy but also by the ability of the designed ANN to recognize and properly classify unknown species that did not participate in the training process. Thus, the performance of ANN was tested first with the aim only to discriminate between 16 known cyanobacterial species (**Figure 8a**). Another seven strains were identified as test ones, to verify the correctness of ANN in recognizing new strains (so-called generalization quality). Analysis of a test set with data from the same monocultures confirmed that the parameters extracted from the fluorescent spectrum sets contained enough information to correctly identify cyanobacterial cells at the species/strain level. The trained neural network presented here showed not the highest rate of correct classification—only about 95.7%—but it shows the best recognition quality for new strains. The results of the ANN recognition are presented in **Figure 8b**.

Bar charts in **Figure 8a** represent the results of the classification of 268 experimental measurements by 16 classes. Each bar represents the classification results as the probability distributions. Each color in the bar corresponds to 1 of 16 target classes (known cyanobacterial strains). The percentage rate of colors in the bar

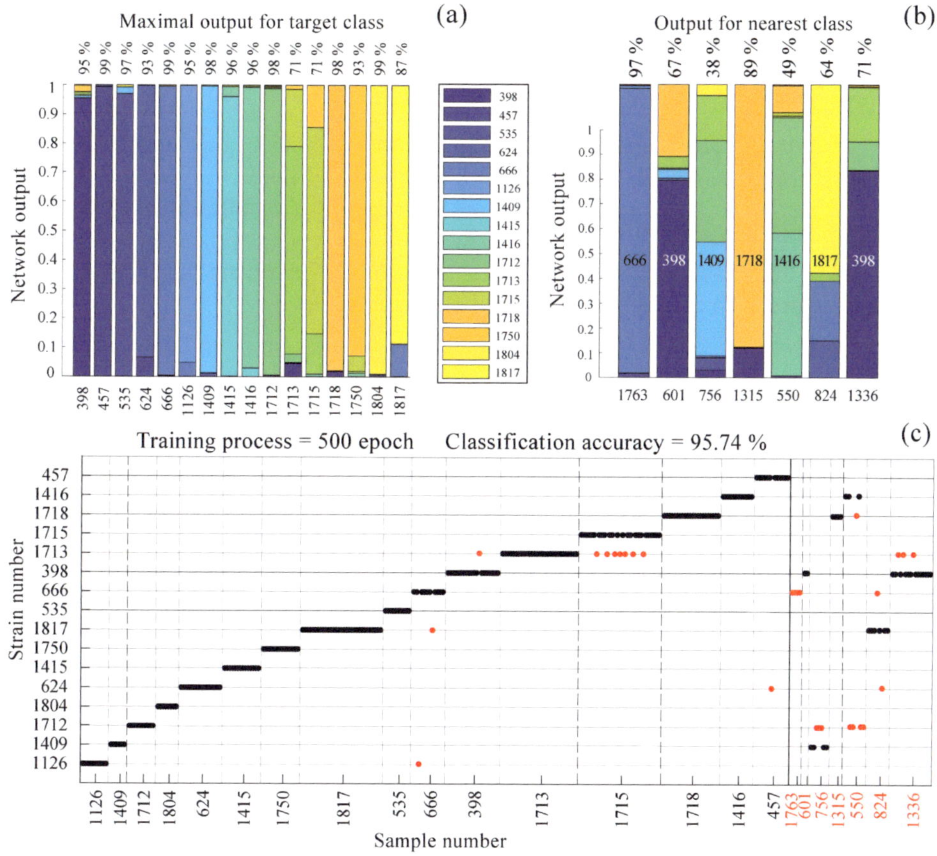

Figure 8.
The results of ANN classification. (a) The results of recognition of 16 known cyanobacterial strains. (b) The results of recognition of seven unknown strains. Numbers over each bar indicate maximal class probability for each strain. (c) General classification results. Red dots indicate false results. Strains are numbered according to CALU collection names.

shows the probability distribution of belonging to the target classes. Maximal eigen-class probability is indicated above each bar.

In contrast to standard classifiers, a classifier built on the base of ANN has a so-called generalization ability. It means that ANN is able to recognize new cyanobacterial strains that were previously unknown for it and suggest possible variants of their generic affiliation to known classes. In **Figure 8c**, the ANN classification results for 16 target classes and 7 strains that were not presented in the training set are shown. The aim of ANN classifier was to determine which of the 16 known classes and 7 unknown strains could be attributed. The results of ANN classification correlate well with the results predicted by LDA (**Figure 7**). The closely related strains in this case were 1763–666, 601–398, 756–1409, 1315–1718, 550–1416, 824–1817, and 1336–398 (in the pairs, the first strain is unknown for ANN, and the second is the one of the nearest target classes). The strains of 1336 *Synechocystis*, 601 *Myxosarcina*, and 1315 *Geitlerinema* ANN classifiers relate to the close genera *Microcystis* and *Geitlerinema*, correspondingly. And for the remaining strains, it proposed possible classification options. Minor errors in classification of strains 756 *Synechococcus*, 824 *Anabaena*, and 550 *Spirulina*, in which the classifier relates to genera *Synechococcus*, *Nostoc*, and *Oscillatoria*, correspondingly, can be explained by the fact that in the space of classification parameters they lie in the wide free regions between the groups of the known strains, approximately, at equal distances from 2 or 3 nearest ones (see **Figure 7**). Therefore the ANN cannot make a correct decision. And the false result of ANN classificator in classification of 1763 *Nostoc* may due to the incorrect initial dataset or false a priori information about 1763 strain affiliation.

To validate the correctness of the neural network operation, the results of the ANN classification were compared with the results of the LDA. The neural network-based classification agrees well with the expected results and with the results of LDA. The identification performance of the network for cyanobacterial strains from the same species is slightly less than for the cells from different species, but anyway they can also be distinguished perfectly well.

6. Conclusion

The automatization of the cyanobacterial species differentiation is a key problem in both industrial biomass production and environmental monitoring. Unfortunately, all presently utilized methods cannot be implemented in online monitoring procedures due to various reasons. In this work, an example of the use of LDA and ANN technologies for online differentiation of cyanobacterial strains according to their in vivo single-cell fluorescence spectra is presented. The novel discrimination technique demonstrated here includes a strict procedure for recording and processing single-cell fluorescence emission spectra, which eliminates most of usual data processing difficulties and, as a result, has a quite high classification accuracy. And the initial information is obtained via fluorescent spectroscopy; the experimental data can be processed automatically. Moreover, due to the use of CLSM microscopic spectroscopy instead of conventional fluorimetry, the initial data have less variations and can be accurately sorted. Any objectionable and unpredictable impact is eliminated at the first step of obtaining fluorescence spectra. Since noninvasive and nondistructive method is used, the information about vital cell operation (e.g., light harvesting) can be additionally taken into account, to obtain the desirable precision of discrimination.

The universality of the considered technique makes it possible to use it for investigation of any phytoplankton species irrespective of their habitat or cultivation. Utilizing data from several fluorescence spectra, instead of one, results in more fingerprint information which leads to the taxonomic differentiation on a finer scale. Differentiation procedure, presented here, was carried out by means of statistical analysis on the base of mathematical characteristics of intrinsic fluorescence spectra of living single cells; therefore it is free from usual subjectivity, which can occur while using methods of direct optical microscopy. Moreover, formalization of data processing gives a wide opportunity for automating of the classification procedure of cyanobacterial strains in field samples, while online monitoring of water bodies is conducted.

Undoubtedly, the data set should be expanded to include more species and phytoplankton classes/divisions, grown under different nutrient and light conditions. However, this work already demonstrates the potential of the discrimination of phytoplankton classes by means of fluorescence microscopic spectroscopy. Combining the knowledge of phytoplankton structure along with taxon-specific measurements of photosynthetic activity and biochemical cell composition can lead to new models which increase the reliability of online monitoring.

Author details

Natalia Grigoryeva
St. Petersburg Scientific Research Centre for Ecological Safety of RAS,
Saint-Petersburg, Russia

*Address all correspondence to: renes3@mail.ru

IntechOpen

© 2019 The Author(s). Licensee IntechOpen. This chapter is distributed under the terms of the Creative Commons Attribution License (http://creativecommons.org/licenses/by/3.0), which permits unrestricted use, distribution, and reproduction in any medium, provided the original work is properly cited.

References

[1] Abed RM, Dobretsov S, Sudesh K. Applications of cyanobacteria in biotechnology. Journal of Applied Microbiology. 2009;**106**(1):1-12. DOI: 10.1111/j.1365-2672.2008.03918.x

[2] Vijayakumar S, Menakha M. Pharmaceutical applications of cyanobacteria—A review. Journal of Acute Medicine. 2015;**5**(1):15-23. DOI: 10.1016/j.jacme.2015.02.004

[3] Grewe CB, Pulz O. The biotechnology of cyanobacteria. In: Ecology of Cyanobacteria II. Netherlands: Springer; 2012. pp. 707-739. DOI: 10.1007/978-94-007-3855-3_26

[4] Grigoryeva N, Chistyakova L. Fluorescence microscopic spectroscopy for investigation and monitoring of biological diversity and physiological state of cyanobacterial cultures. In: Tiwari A, editor. Cyanobacteria. London: IntechOpen; 2018. pp. 11-44. DOI: 10.5772/intechopen.78044

[5] Rumyantsev VA, Grigor'eva NY, Chistyakova LV. Study of changes in the physiological state of cyanobacteria caused by weak ultrasonic treatment. Doklady Earth Sciences. 2017;**475**(2):939-941. DOI: 10.1134/S1028334X17080190

[6] Grigoryeva NY, Chistyakova LV, Liss AA. Spectroscopic techniques for estimation of physiological state of blue-green algae after weak external action. Oceanology. 2018;**58**(6):896-904. DOI: 10.1134/S0001437018060061

[7] Becker A, Meister A, Wilhelm C. Flow cytometric discrimination of various phycobilin-containing phytoplankton groups in a hypertrophic reservoir. Cytometry Part A. 2002;**48**(1):45-57. DOI: 10.1002/cyto.10104

[8] Balfoort HW, Snoek J, Smiths JR, Breedveld LW, Hofstraat JW, Ringelberg J. Automatic identification of algae: Neural network analysis of flow cytometric data. Journal of Plankton Research. 1992;**14**(4):575-589. DOI: 10.1093/plankt/14.4.575

[9] Olson RJ, Sosik HM. A submersible imaging-in-flow instrument to analyze nano- and microplankton: Imaging FlowCytobot. Limnology and Oceanography: Methods. 2007;**5**(6): 195-203. DOI: 10.4319/lom.2007.5.195

[10] Johnsen G, Samset O, Granskog L, Sakshaug E. In vivo absorption characteristics in 10 classes of bloom-forming phytoplankton: Taxonomic characteristics and responses to photoadaptation by means of discriminant and HPLC analysis. Marine Ecology Progress Series. 1994;**105**:149-157. DOI: 10.3354/meps105149

[11] Beutler M, Wiltshire KH, Meyer B, Moldaenke C, Lüring C, Meyerhöfer M, et al. A fluorometric method for the differentiation of algal populations in vivo and in situ. Photosynthesis Research. 2002;**72**(1):39-53. DOI: 10.1023/A:101602660

[12] Escoffier N, Bernard C, Hamlaoui S, Groleau A, Catherine A. Quantifying phytoplankton communities using spectral fluorescence: The effects of species composition and physiological state. Journal of Plankton Research. 2014;**37**(1):233-247. DOI: 10.1093/plankt/fbu085

[13] Millie DF, Schofield OM, Kirkpatrick GJ, Johnsen G, Evens TJ. Using absorbance and fluorescence spectra to discriminate microalgae. European Journal of Phycology. 2002;**37**(3): 313-322. DOI: 10.1111/j.1529-8817.2001.jpy37303-89.x

[14] MacIntyre HL, Lawrenz E, Richardson TL. Taxonomic

discrimination of phytoplankton by spectral fluorescence. In: Chlorophyll a Fluorescence in Aquatic Sciences: Methods and Applications. Dordrecht: Springer; 2010. pp. 129-169

[15] Bruckman LS, Richardson TL, Swanstrom JA, Donaldson KA, Allora M Jr, Shaw TJ, et al. Linear discriminant analysis of single-cell fluorescence excitation spectra of five phytoplankton species. Applied Spectroscopy. 2012;**66**(1):60-65. DOI: 10.1366/11-06294

[16] Seppälä J, Olli K. Multivariate analysis of phytoplankton spectral in vivo fluorescence: Estimation of phytoplankton biomass during a mesocosm study in the Baltic Sea. Marine Ecology Progress Series. 2008;**370**:69-85. DOI: 10.3354/meps07647

[17] Zhang QQ, Lei SH, Wang XL, Wang L, Zhu CJ. Discrimination of phytoplankton classes using characteristic spectra of 3D fluorescence spectra. Spectrochimica Acta Part A: Molecular and Biomolecular Spectroscopy. 2006;**63**(2):361-369. DOI: 10.1016/j.saa.2005.05.041

[18] Glazer AN, West JA, Chan C. Phycoerythrins as chemotaxonomic markers in red algae: A survey. Biochemical Systematics and Ecology. 1982;**10**(3):203-215. DOI: 10.1016/0305-1978(82)90031-X

[19] Hill DR, Rowan KS. The biliproteins of the Cryptophyceae. Phycologia. 1989;**28**(4):455-463. DOI: 10.2216/i0031-8884-28-4-455

[20] Yentsch CS, Phinney DA. Spectral fluorescence: An ataxonomic tool for studying the structure of phytoplankton populations. Journal of Plankton Research. 1985;7(5):617-632. DOI: 10.1093/plankt/7.5.617

[21] Wilhelm C, Becker A, Toepel J, Vieler A, Rautenberger R. Photophysiology and primary production of phytoplankton in freshwater. Physiologia Plantarum. 2004;**120**(3):347-357. DOI: 10.1111/j.0031-9317.2004.00267.x

[22] Bodemer U. Variability of phycobiliproteins in cyanobacteria detected by delayed fluorescence excitation spectroscopy and its relevance for determination of phytoplankton composition of natural water samples. Journal of Plankton Research. 2004;**26**(10):1147-1162. DOI: 10.1093/plankt/fbh105

[23] Pawley JB, editor. Handbook of Biological Confocal Microscopy. Boston, MA: Springer; 1995. 346 p. DOI: 10.1007/978-1-4757-5348-6

[24] Wilson T. Confocal microscopy. London: Academic Press; 1990. 426 p. DOI: 10.1007/978-1-4615-7133-9

[25] Pinevich AV, Mamkaeva KA, Titova NN, Gavrilova OV, Ermilova EV, Kvitko KV, et al. Petersburg culture collection (CALU): Four decades of storage and research with microscopic algae, cyanobacteria and other microorganisms. Nova Hedwigia. 2004;**79**(1-2):115-126. DOI: 10.1127/0029-5035/2004/0079-0115

[26] Gromov BV, Titova NN. Collection of algae cultures of the Laboratory of Microbiology of Biological Institute of Leningrad State University. In: Culturing of Collection Algae Strains. Leningrad: Leningrad State Univ.; 1983. pp. 3-27

[27] Wolf E, Schussler A. Phycobiliprotein fluorescence of Nostoc punctiforme changes during the life cycle and chromatic adaptation: Characterization by spectral confocal laser scanning microscopy and spectral unmixing. Plant, Cell and Environment. 2005;**28**(4):480-491. DOI: 10.1111/j.1365-3040.2005.01290.x

[28] Zhangirov TR, Perkov AS, Ivanova SA, Liss AA, Grigoryeva NY,

Chistyakova LV. Comparison of the efficiency of solution of classification task by methods of linear discriminant analysis and artificial neural networks. Izvestiya SPbGETU “LETI” (In Russ.). 2019;**5**:64-73

[29] Roelke DL, Kennedy CD, Weidemann AD. Use of discriminant and fourth-derivative analyses with high-resolution absorption spectra for phytoplankton research: Limitations at varied signal-to-noise ratio and spectral resolution. Gulf of Mexico Science. 1999;**17**(2):2. DOI: 10.18785/goms.1702.02

[30] Zhang F, He J, Su R, Wang X. Assessing phytoplankton using a two-rank database based on excitation-emission fluorescence spectra. Applied Spectroscopy. 2011;**65**(1):1-9. DOI: 10.1366/10-05927

[31] Fisher RA. The use of multiple measurements in taxonomic problems. Annals of Eugenics. 1936;**7**(2):179-188. DOI: 10.1111/j.1469-1809.1936.tb02137.x

[32] Huberty CJ. Issues in the use and interpretation of discriminant analysis. Psychological Bulletin. 1984;**95**(1):156. DOI: 10.1037/0033-2909.95.1.156

[33] MATLAB MV. 8.0. 0.783 (R2012b). Natick. Massachusetts: The MathWorks Inc.; 2012. p. 82

[34] Samanta BI, Al-Balushi KR, Al-Araimi SA. Artificial neural networks and support vector machines with genetic algorithm for bearing fault detection. Engineering Applications of Artificial Intelligence. 2003;**16**(7-8):657-665. DOI: 10.1016/j.engappai.2003.09.006

[35] da Silva CE, Filardi VL, Pepe IM, Chaves MA, Santos CM. Classification of food vegetable oils by fluorimetry and artificial neural networks. Food Control. 2015;**47**:86-91. DOI: 10.1016/j.foodcont.2014.06.030

[36] Manel S, Dias JM, Ormerod SJ. Comparing discriminant analysis, neural networks and logistic regression for predicting species distributions: A case study with a Himalayan river bird. Ecological Modelling. 1999;**120**(2-3):337-347. DOI: 10.1016/S0304-3800(99)00113-1

[37] Haykin SS. Neural Networks: A Comprehensive Foundation. 2nd ed. Pearson Education: Delhi, India; 2005. 823 p. ISBN: 0131471392

[38] Kingma DP, Ba J. Adam: A method for stochastic optimization. arXiv preprint arXiv:1412.6980. 2014. Available at: https://arxiv.org/abs/1412.6980 [Accessed: Jul 19, 2019]

[39] Gantt E. Phycobilisomes. Annual Review of Plant Physiology. 1981;**32**(1):327-347. DOI: 10.1146/annurev.pp.32.060181.001551

[40] Staehelin LA, Arntzen CJ, editors. Photosynthesis III: Photosynthetic Membranes and Light Harvesting Systems, Encyclopedia of Plant Physiology. Vol. 19. Berlin: Springer; 1986. p. 802. DOI: 10.1007/978-3-642-70936-4

[41] Bryant DA, editor. The Molecular Biology of Cyanobacteria, Advances in Photosynthesis. Vol. 1. Dordrecht: Kluwer Academic Press; 1994. 881 p. DOI: 10.1007/978-94-011-0227-8

[42] Carr NG, Whitton BA, editors. The Biology of Cyanobacteria. Berkeley: University of California; 1982. 688 p. ISBN: 0-520-04717-6

[43] Blankenship RE. Molecular Mechanisms of Photosynthesis. 2nd ed. Chichester: John Wiley & Sons; 2014. 314 p. DOI: 10.1002/9780470758472

[44] Papageorgiou GC. Fluorescence emission from the photosynthetic apparatus. In: Eaton-Rye J, Tripathy B, Sharkey T, editors. Photosynthesis.

Advances in Photosynthesis and Respiration. Vol. 34. Dordrecht: Springer; 2012. pp. 415-443. DOI: 10.1007/978-94-007-1579-0_18

[45] Gault PM, Marler HJ, editors. Handbook on Cyanobacteria: Biochemistry, Biotechnology and Applications, Bacteriology Research Developments Series. New York: Nova Science Publishers; 2009. 538 p. ISBN: 978-1-60741-092-8

[46] Ke B. In: Govindjee U, editor. Photosynthesis: Photobiochemistry and Photobiophysics, Advances in Photosynthesis. Vol. 10. Dordrecht: Kluwer Academic Publishers; 2001. 763 p. DOI: 10.1007/0-306-48136-7

[47] MacColl R. Cyanobacterial phycobilisomes. Journal of Structural Biology. 1998;**124**(2-3):311-334. DOI: 10.1006/jsbi.1998.4062

[48] Mullineaux CW. Phycobilisome-reaction centre interaction in cyanobacteria. Photosynthesis Research. 2008;**95**:175-182. DOI: 10.1007/s11120-007-9249-y

[49] Yang S, Su Z, Li H, Feng J, Xie J, Xia A, et al. Demonstration of phycobilisome mobility by the time-and space-correlated fluorescence imaging of a cyanobacterial cell. Biochimica et Biophysica Acta (BBA)-Bioenergetics. 2007;**1767**(1):15-21. DOI: 10.1016/j.bbabio.2006.11.012

[50] Liu H, Zhang H, Niedzwiedzki DM, Prado M, He G, Gross ML, et al. Phycobilisomes supply excitations to both photosystems in a megacomplex in cyanobacteria. Science. 2013;**342**(6162):1104-1107. DOI: 10.1126/science.1242321

[51] Su X, Fraenkel PG, Bogorad L. Excitation energy transfer from phycocyanin to chlorophyll in an apcA-defective mutant of Synechocystis sp. PCC 6803. Journal of Biological Chemistry. 1992;**267**(32):22944-22950. DOI: 10.0000/jbc.1992.267-32-244

[52] Green BR, Parson WW, editors. Light-Harvesting Antennas in Photosynthesis, Advances in Photosynthesis and Respiration. Vol. 13. Dordrecht: Springer Netherlands Kluwer Academic Publishers; 2003. p. 514. DOI: 10.1007/978-94-017-2087-8

[53] Bidigare RR, Ondrusek ME. Spatial and temporal variability of phytoplankton pigment distributions in the central equatorial Pacific Ocean. Deep Sea Research Part II: Topical Studies in Oceanography. 1996;**43**(4-6):809-833. DOI: 10.1016/0967-0645(96)00019-7

[54] Yentsch CS, Menzel DW. A method for the determination of phytoplankton chlorophyll and phaeophytin by fluorescence. Deep Sea Research. 1963;**10**:221-231. DOI: 10.1016/0011-7471(63)90358-9

[55] Yentsch CS, Yentsch CM. Fluorescence spectral signatures: The characterization of phytoplankton populations by the use of excitation and emission spectra. Journal of Marine Research. 1979;**37**:471-483

Chapter 2

Eutrophication and Phytoplankton: Some Generalities from Lakes and Reservoirs of the Americas

Ernesto J. González and Gabriel Roldán

Abstract

Eutrophication is one of the most widespread problems of inland waters in the world. In many countries from North to South America, eutrophication is due to several impacts resulting from the inefficient or nonexistent wastewater treatment; the agricultural expansion with inadequate soil uses and application of chemical fertilizers; the urbanization of watersheds, the increase of intensive husbandry of cattle, pigs, and chicken; the increase of aquaculture; the construction of reservoirs; and the destruction of natural ecosystems. Basically, the increase in the internal load of nitrogen and phosphorus in lakes and reservoirs produces an increase in the biological productivity of the water bodies. As consequence, phytoplankton community in freshwater systems is mainly dominated by cyanobacteria. Despite differences in continental climate regimes, this seems to be a regular pattern along the Americas, where there are various practices related to the use of lower and upper taxonomic groups of phytoplankton for the indication of the trophic level and water quality status of freshwater systems. As was reported in numerous studies in the Americas, increase in nutrient concentrations mainly due to rich in phosphorus cause larger phytoplankton biomass and predominance of cyanobacteria such as genera *Microcystis*, *Anabaena*, *Planktothrix*, *Oscillatoria*, and *Cylindrospermopsis* than ever before.

Keywords: America, eutrophication, phytoplankton, cyanobacterial dominance

1. Introduction

Eutrophication, defined as the nutrient enrichment process (mainly nitrogen and phosphorus) of any water body which results in an excessive growth of phytoplankton and macrophytes [1–3], has become a major cause of concern in developing as well as developed countries [1]. Also, it was recognized as a pollution problem in the European and North American lakes and reservoirs in the mid-twentieth century. Since then, it has become more widespread in the whole world.

Eutrophication is due to several impacts resulting from the inefficient or nonexistent wastewater treatment; the agricultural expansion with inadequate soil uses and application of chemical fertilizers; the urbanization of watersheds; the increase of intensive husbandry of cattle, pigs, and chicken; the increase of aquaculture; the construction of reservoirs; and the destruction of natural ecosystems [4]. The eutrophication event describes formation of a set of symptoms in a lake system exposed to excessive nutrient increase [2]. Common symptoms due to eutrophication include excessive algal blooms, tremendous organic and inorganic material accumulation, and lower biodiversity, high turbidity, excessive sedimentation, and high anoxia conditions, particularly in the deeper parts of lakes. The increase in anoxia condition can cause fish deaths in midsummer. One of the first and worst symptoms of eutrophication has been formation of planktonic algal blooms. In freshwaters, former of these algal blooms are mostly nitrogen (N)-fixing cyanobacteria [5].

Eutrophic water bodies are richly supplied with plant nutrients (N, P, as well other nutrients of less acute demand), and consequences include the increase of biological productivity and turbidity of water because of dense growths of phytoplankton [6, 7]. Thus, phytoplankton community structures and their relevant participants could be used as a biological indicator of negative environmental impacts formed in lakes and reservoirs, as was eutrophication event [8].

In the Americas, as was reported in previous studies, the increase in nutrient concentrations leads to greater biomass of phytoplankton in freshwater systems. In this new region of the world, there are numerous experiences relating the effect of eutrophication on the phytoplankton community.

In this book chapter, we aimed to provide a short overview based on the sparse and scattered literature sources and fixed practices in the American continent related to the proliferation of certain groups of phytoplankton in lakes and reservoirs in terms of eutrophication. We try to depict some generalizations that have arisen from this review, in relation to dominant phytoplankton in the eutrophic lakes and reservoirs in the Americas.

2. Eutrophication and phytoplankton

Excessive nutrient accumulation in aquatic ecosystems by carrying of anthropogenic sources, mainly rich in phosphorus (P) and nitrogen (N), creates a series of changes in their structure and function in the direction of deterioration of water quality, known as eutrophication [9]. Among the structural changes caused by the eutrophication, there is the dominance of the “r” selective species in the community structure of phytoplankton known as tiny primary pelagic producers, particularly in the predominance of cyanobacteria in the freshwater ecosystems such as lakes and reservoirs.

As were reported by Bellinger and Sigee [10], the detection of excessive harmful blooms of some algae that are biological indicators of environmental pollution, particularly of nutrient pollution, reveals anthropogenic activities in freshwater systems and a rapid change in their trophic status. It is known that in the mid-twentieth century, some researchers such as Thunmark [11], Nygaard [12], and Stockner [13] developed trophic status indexes by using typical algal groups of oligotrophic (particularly desmids, a group of green algae) or typical algal groups of eutrophic conditions (chlorococcal, cyanobacterial, and euglenoid species). Although these indices provide useful information about trophic status of the lakes, generally they are not enough to indicate a lot of environmental problems since a lot of algal species have been living in both eutrophic and oligotrophic freshwater systems. In the other word, there are a lot of similarities in view of species homogeneity and seasonal succession of species for both systems. The rehabilitation of previous methods on sampling and taxonomic analysis, and development of new methods in this framework have provided the development in indices based on more specific indicator algal species from different taxonomic groups. Thus, Bellinger and Sigee

[10] revealed in their books what the indicator species of the trophic status would be in mid-summer in temperate lakes (**Table 1**).

In the Americas, there are some examples about how phytoplankton groups and species have been used for the determination of the trophic status and quality of surface waters. Some of these experiences will be presented in the following sections.

2.1 North America

2.1.1 Canada

The dominant indicator species list for trophic status of various lake types in the western region of Canada is revealed in **Table 2**. This list bases on 25 years of observations by Rawson [14].

In the Experimental Lake Area (ELA), located in Ontario, Schindler [15] and Schindler et al. [5] showed that water fertilization (N and P) causes quantitative increase of all phytoplankton groups, especially the cyanobacteria species *Aphanizomenon schindleri* Kling, Findlay and Komárek 1994, and *Limnothrix redekei* (Goor) Meffert 1988.

2.1.2 The United States of America

In 1978, the Environmental Protection Agency (EPA) published a study about eutrophication relating aquatic plant response to nutrient loading to lakes and reservoirs [16]. There was good correlation between phosphorus loading and the average chlorophyll *a* and water transparency. In general, the correlations between phosphorus-loading concentrations and eutrophication response data are better than those observed between nitrogen-loading concentrations and the same eutrophication parameters, supporting the phosphorus limitation of most of the United States water bodies. Summarizing, the characteristic algal species in relation to the phytoplankton in eutrophic lakes are represented by *Anabaena* spp., *Aphanizomenon* spp., *Microcystis* spp., and *Oscillatoria rubescens* De Candolle ex Gomont 1892.

Lake types	Algal indicators
Oligotrophic	Diatoms: *Cyclotella comensis* Grunow in Van Heurck 1882, *Rhizosolenia* spp. Green algae: *Staurodesmus* spp.
Mesotrophic	Diatoms: *Tabellaria flocculosa* (Roth) Kützing 1844 Chrysophytes: *Dinobryon divergens* O. E. Imhof 1887, *Mallomonas caudata* Iwanoff (Ivanov) 1899 Green algae: *Sphaerocystis schroeteri* Chodat 1897, *Dictyosphaerium elegans* Bachmann 1913, *Cosmarium* spp., *Staurastrum* spp. Dinoflagellates: *Ceratium hirundinella* (O. F. Müller) Dujardin 1841 Cyanobacteria: *Gomphosphaeria* spp.
Eutrophic	Diatoms: *Aulacoseira* spp., *Stephanodiscus rotula* (Kützing) Hendey 1964 Green algae: *Eudorina* spp., *Pandorina morum* (O. F. Müller) Bory 1897, *Volvox* spp. Cyanobacteria: *Anabaena* spp., *Aphanizomenon flos-aquae* Ralfs ex Bornet and Flahault 1886, *Microcystis aeruginosa* (Kützing) Kützing 1846
Hypereutrophic	Diatoms: *Stephanodiscus hantzschii* Grunow 1880 Green algae: *Scenedesmus* spp., *Ankistrodesmus* spp., *Pediastrum* spp. Cyanobacteria: *Aphanocapsa* spp., *Aphanothece* spp., *Synechococcus* spp.

Table 1.
Phytoplankton indicative species of trophic status in temperate lakes in mid-summer, modified from Bellinger and Sigee [10].

Lake types	Algal indicators
Oligotrophic	Diatoms: *Asterionella formosa* Hassall 1850, *Melosira islandica* O. Müller 1906, *Tabellaria fenestrata* (Lyngbye) Kützing 1844, *Tabellaria flocculosa* (Roth) Kützing 1844, *Fragilaria capucina* Desmazières 1830, *Stephanodiscus niagarae* Ehrenberg 1845, *Staurastrum* spp., *Melosira granulata* (Ehrenberg) Ralfs 1861 Chrysophytes: *Dinobryon divergens* O. E. Imhof 1887
Mesotrophic	Diatoms: *Fragilaria crotonensis* Kitton 1869 Green algae: *Pediastrum boryanum* (Turpin) Meneghini 1840, *Pediastrum duplex* Meyen 1829 Dinoflagellates: *Ceratium hirundinella* (O.F.Müller) Dujardin 1841 Cyanobacteria: *Coelosphaerium naegelianum* Unger 1854, *Anabaena* spp., *Aphanizomenon flos-aquae* Ralfs ex Bornet and Flahault 1886, *Microcystis aeruginosa* (Kützing) Kützing 1846
Eutrophic	Cyanobacteria: *Microcystis flos-aquae* (Wittrock) Kirchner 1898

Table 2.
The list of dominant algal indicator species for trophic status of various lake types in the western region of Canada, modified from Rawson [14].

2.1.3 Mexico

In several lakes suffering the eutrophication process, green algae and diatoms have been replaced by cyanobacteria, particularly *Anabaena* spp., *Microcystis aeruginosa* (Kützing) Kützing 1846, *Oscillatoria* spp., and *Lyngbya* spp. ([17–22], among others).

Cyanobacteria dominance in the eutrophic Lake Chapala is described by de Anda and Shear [23]. The high TN and TP concentrations contained in the large quantities of domestic, agricultural, and industrial sewage that enter to the lake through its main tributary, the Lerma river, increased the phytoplankton biomass and resulted in the dominance of *Anabaena flos-aquae* Brébisson ex Bornet and Flauhault 1886.

Tomasini-Ortiz et al. [24] reported the dominance of *Aphanizomenon gracile* Lemmermann 1907, followed by *M. aeruginosa*, *Microcystis pulverea* (HC Wood) Forti 1907, and *Anabaena affinis* Lemmermann 1898 in the eutrophic Lake Pátzcuaro, Michoacán State. The authors pointed out that many cyanobacterial blooms have been reported in eutrophic lakes along the Mexican states of Jalisco, Michoacán, Veracruz, San Luis Potosí, Querétaro, Guanajuato, Puebla, Oaxaca, and Hidalgo and in Mexico City.

Valle de Bravo reservoir (State of Mexico) provides drinking water to about 2,500,000 inhabitants in Mexico City [25]. This water body also shows frequent cyanobacterial blooms as consequence of the high nutrient load in its waters, posing health risks for human population. The common genera found during blooms are *Microcystis* sp., *Oscillatoria* sp., *Anabaena* sp., *Cylindrospermopsis raciborskii* (Woloszynska) Seenayya and Subba Raju 1972, and *Nostoc* sp.

2.2 Central America

2.2.1 Guatemala

Unregulated land use and lack of wastewater treatment have led to eutrophication in many lakes of Guatemala [26]. Some examples of this situation are the following studies.

Basterrechea [27] found the prevalence of cyanobacteria in the Lake Amatitlán due to eutrophic conditions. Similarly, Rejmánková et al. [28] recorded blooms of the *Lyngbya* species complex (cyanobacteria) as a consequence of the change in land use in the Lake Atitlán basin. Brocard et al. [26] pointed out that eutrophication has had a dramatic impact on the lake Chichój environment; among other effects, fertilization of lake waters produced severe hypoxia, massive development of the water hyacinth *Eichhornia crassipes* (Mart.) Solms 1883, and "blue-green" algae dominance.

2.2.2 Honduras

The Lake Yojoa is the largest natural lake in the country and represents an important natural resource for Hondurans [29]. The lake is used extensively for commercial production of tilapia fish; fishes are raised to full maturity in floating cages in the lake, and subsequently, high nutrient load is directly supplied to the water body. Other sources of nutrients are: (a) significant amount of wastewater from the ineffective product of water treatment plant, (b) wastewater from restaurants around the lake, and (c) agricultural practices in the neighboring lands, where fertilizers are commonly used, thus contributing with nutrients to the system, having an impact on water quality [29].

Because of the high input of nutrients, cyanobacteria are the dominant phytoplankton group that accounted for 59.0% of total phytoplankton in the lake [30]. Dominant species in the lake are *M. aeruginosa*, *Aphanocapsa delicatissima* West and GS West 1912, and *Oscillatoria limosa* C. Agardh ex Gomont 1892, all of them common in eutrophic tropical and temperate lakes. Other species that present high densities in lake are the green algae *Staurastrum leptocladum* Nordstedt 1870 and *Sphaerocystis schroeteri* Chodat 1897 and the diatom *Aulacoseira granulata* (Ehrenberg) Simonsen 1979.

2.2.3 El Salvador

Wastewater effluents and similar runoffs with high nutrient concentrations derived from agricultural fertilizers, which are increased by the susceptibility to erosion, deforestation, and sediment trawling, have induced the eutrophication process and, consequently, produced the proliferation of *M. aeruginosa* (cyanobacteria) in the volcanic Lake Coatepeque [31].

The Cerrón Grande reservoir also suffered the eutrophication process, and its waters are classified as hypereutrophic. The dominant cyanobacteria species is *Microcystis* spp. [32].

2.2.4 Nicaragua

In the eutrophic Lake Xolotlán (Lake Managua), Hooker and Hernández [33] and Erikson [34] found high phosphorus concentrations (≈150 μg/L), turbid waters (0.40 m of transparency), and high algal biomass of mainly "blue-greens" (cyanobacteria). Phytoplankton community was dominated by cyanobacteria throughout the entire year [35], and *Lyngbya contorta* Lemmermann 1898 accounted for more than 35.0% of total phytoplankton in the lake, followed by the diatom *Cyclotella meneghiniana* Kützing 1844.

Vammen et al. [36] pointed out that increased eutrophication in Lake Cocibolca (Lake Nicaragua) had resulted in the increase of phytoplankton density and a marked dominance of two cyanobacterial species (*M. aeruginosa* and *C. raciborskii*). Cyanobacteria accounted for almost 99% of total phytoplankton in the lake.

Hernández González et al. [37] also found that the most representative phytoplankton genera detected during most of the period sampled in the eutrophic lakes Cocibolca, Tiscapa, and Masaya, were cyanobacteria, among which are distinguished *Anabaenopsis*, *Merismopedia*, *Chroococcus*, and *Lyngbya*.

2.2.5 Costa Rica

Umaña et al. [38] stated that there are few long-term works in lakes in Costa Rica, which have shown a wide annual variation of their characteristics. In the Talamanca region of the province of Limón, Jones et al. [39] found a gradient of trophic states, varying from high-altitude lakes with a tendency to be oligotrophic, to lower-altitude lakes with a tendency to be eutrophic, despite being located in woodland regions where it is away from any human disturbances. These researchers express that, because they do not have a high burden of anthropic phosphorus, their planktonic communities do not show the classic dominance of cyanobacteria against other planktonic groups, but rather a higher prevalence of green algae, a few dinoflagellates and a few cryptomonadales, all of which suggest a more balanced availability between nitrogen and phosphorus. On the other hand, the Arenal reservoir, the largest water body in Costa Rica, located between the provinces of Guanacaste and Alajuela, has been classified as mesotrophic by Jones et al. [39], representing varied phytoplanktonic community that is dominated by green algae, some diatoms, and cyanobacteria (*Microcystis* spp.).

2.2.6 Panama

Reservoirs of the Panama channel show a mesotrophic status, with a predominance of diatom populations that are represented by 40.0% of total phytoplankton in Gatún reservoir, 55.1% in Alajuela reservoir, and 58.0% in Miraflores reservoir [40].

2.2.7 Cuba

Gómez Luna et al. [41] identified the phytoplankton communities in three reservoirs which are used for drinking water supply to 80.0% of the inhabitants in the City of Santiago de Cuba. High concentrations of nutrients were detected in Chalóns, Charco Mono, and Paradas reservoirs, where the phytoplankton communities were dominated by the cyanobacterial species *Microcystis* spp., *Aphanothece minutissima* (West) J. Komárková-Legnerová and G. Cronberg 1994, and *Oscillatoria chalybea* Mertens ex Gomont 1892.

2.2.8 Puerto Rico

Pantoja Agreda [42] conducted a limnological characterization of Guajataca reservoir, which was classified as mesotrophic. The dominant phytoplankton group were Euglenophyta (43.8% of total phytoplankton), followed by Pyrrhophyta (34.9%) and Chlorophyta (10.7%); cyanobacteria accounted for less than 5.00% of total algal density. This fact is common in water bodies with high content of organic matter.

2.3 South America

2.3.1 Colombia

According to Roldán [43] and Roldán and Ramírez [44], water bodies with more signs of eutrophication (Porce II, El Peñol, Prado, and Tominé) have a predominance of cyanobacteria, especially of the genera *Anabaena* spp. and *Oscillatoria* spp. The main source of eutrophication is domestic wastewater that reaches the rivers and streams without any treatment. The use of agrochemicals also contributes to eutrophication. The most outstanding case of eutrophication is that of the Porce II reservoir, which receives the waters of the Medellín river, carrying pollutants of a city of about 3,000,000 inhabitants. Currently, there are two wastewater treatment plants, one of which has started process a short time ago. It is expected that this reservoir will begin to recover in the future. Unfortunately, more than 95.0% of the towns and cities in Colombia do not have wastewater treatment plants.

2.3.2 Ecuador

Composition of phytoplankton in the Lake Yahuarcocha is dominated by the following species: *Cylindrospermopsis* sp., *Anabaena* sp., *Microcystis* sp. (cyanobacteria), *Monoraphidium* sp. (chlorophyta), and *Fragilaria* sp. (diatom) [45]. Eutrophication in this lake is due to the entrance of wastewater caused by the tourism industry around the lake.

2.3.3 Peru and Bolivia

In the Lake Titicaca, in the corresponding Bolivian basin, Fonturbel and Castaño-Villa [46] considered nutrient concentrations and phytoplankton groups as a whole to determine that the families Oscillatoriaceae and Nostocaceae (cyanobacteria) respond positively to the increase in pH (alkalinization) and negatively to the increasing nutrients, while the families Naviculaceae (diatoms), Closteriaceae, and partly Zygnemataceae (green algae) showed an inverse tendency with proliferating in acidic water enriched with nutrients. The diatoms seem to respond negatively, both to the acidification of the water and to the excessive nutrient enrichments. Studies revealed that they are the most sensitive groups to the eutrophication of waters.

Between the 1970s and 1990s, both sections of the Lake Titicaca, deep Lago Mayor and large part of shallow Lago Menor, were oligotrophic with high water transparency and strong nitrogen limitation. Chlorophyta and cyanobacteria (particularly *Anabaena* spp.) dominated the phytoplankton with low biomass and primary production, except for diatoms during the dry season [47]. Currently, the deep pelagic areas of the Lake Lago Mayor remain oligotrophic. However, shallow littoral areas of the Lake Lago Mayor and the Lago Menor turn to eutrophic from mesotrophic. In the northern littoral area of the Lago Menor, there are a lot of villages which have domestic pollution sources, while El Alto is responsible for the heavy contamination of the Cohana bay. In 2015, the extended rainy season produced the first major phytoplankton bloom event in dominance of *Carteria* sp., which is a harmless unicellular green algae in the northern part of the Lago Menor in the period of March–April. Phytoplankton blooms in the region have been spotted since the 2000s. Cyanobacteria *Limnoraphis* (syn. *Lyngbya*) predominates in the Puno bay.

2.3.4 Brazil

Numerous studies on eutrophication of freshwater ecosystems have been conducted along the Brazilian territory. Prevalence of cyanobacteria under this eutrophic condition has also been reported.

Tundisi [48] reported blooms of the cyanobacteria *Microcystis* spp. and *Anabaena* spp. in the reservoirs of the State of São Paulo, where they are characterized by a severe eutrophication in their waters due to industrial and agricultural wastewater.

Huszar et al. [49] and Dantas et al. [50] also found the dominance of *Microcystis* in highly eutrophic shallow lakes and small reservoirs.

Lake Vaca Brava, in the State of Goiás (Central Western Brazil), is an urban water body that suffers the eutrophication process as a consequence of the human settlement in its neighboring areas [51]. Increased cyanobacterial density accompanies the eutrophication process, where *Planktolyngbya limnetica* (Lemmermann) Komárková-Legnerová and Cronberg 1992 is the dominant species.

Likewise, Chellappa et al. [52] studied the dynamics of phytoplankton in the Armando Ribeiro Gonçalves reservoir, located in the state of Rio Grande do Norte (Northern Brazil), which is used for drinking water supply; this reservoir was classified as eutrophic due to its high nutrient concentrations, and the toxic cyanobacteria dominated the phytoplankton composition, particularly *Planktothrix agardhii* (Gomont) Anagnostidis and Komárek 1988 in drought period and *M. aeruginosa* in rainy season.

Due to inadequate treatment, sewage with high levels of P and N reaches to Pampulha and Ibirité reservoirs in South-Eastern Brazil [53, 54], causing changes in the composition and diversity of the plankton community. *Microcystis* spp. were the main species registered during the blooms in these reservoirs.

Despite resulting in known impacts, such as loss of aquatic biodiversity, the emergence of potentially toxic cyanobacterial blooms, overgrowth of aquatic macrophytes, anoxia and fish mortality, increased eutrophication of Brazilian reservoirs has also, as an additional consequence, the increase of greenhouse gas emission that aggravates the global warming process [55, 56].

2.3.5 Chile

In this country, it has also been reported that the increase in the concentration of nitrogen and phosphorus leads to an increase in phytoplankton biomass and to the dominance of cyanobacteria [57]. In lakes which have heavy eutrophication process, the dominant cyanobacteria which occur blooms are usually of *Anabaena* spp., *Microcystis* spp., and *Oscillatoria* spp. [58–60], some of them presenting toxic strains [58]. On the other hand, Parra et al. [61] found that green algae (Chlorophyta), particularly Desmidiaceae family members, are the group more sensitive to negative changes in environmental conditions, especially to those associated with pollution and eutrophication.

2.3.6 Argentina

Quirós [62], analyzing the empirical relationships between nutrient concentrations and biological communities in more than 100 Argentine lakes and reservoirs, found that total phosphorus concentration is the main factor in the control of the phytoplankton biomass. Likewise, he also found that the applied empirical order grouped lakes in two groups which are the lakes located in the lower latitudes which are shallow, warm, and eutrophic where phytoplankton are limited by nitrogen, and the lakes located in higher latitude, temperate-cold, and oligotrophic where phytoplankton are limited by phosphorus [63]. Also, Quirós et al. [64] found that nutrient enrichment of the Pampa's surface water and its multiple negative effects on its lagoons have increased the internal phosphorus load allowing the increase in the frequency of cyanobacterial blooms, especially during relatively dry years.

2.3.7 Uruguay

The largest reservoirs (Salto Grande, Bonete, Baygorria, and Palmar) have suffered the process of eutrophication, which has led to an intense growth of phytoplankton [65]. De León and Chalar [66], Chalar et al. [67], and Chalar [68, 69] studied the phytoplankton dynamics of the Salto Grande reservoir and recorded the dominance of typical diatoms in eutrophic environments and high densities of cyanobacteria which are predominant with *M. aeruginosa* during algal blooms.

Vidal et al. [70] had detected cyanobacterial blooms, mainly supported by *C. raciborskii* in many eutrophic water bodies of the country, as were the artificial lakes in Canelones, the lake Laguna Blanca in Maldonado, and small dam lakes in Rocha.

RAP-AL [71] also detected that in the eutrophic system of Laguna del Sauce which has a predominance of cyanobacteria throughout the year, there was a marked increase in the frequency and duration of microalgae blooms, particularly of *M. aeruginosa*.

2.3.8 Paraguay

The rapid growth of anthropogenic or human sourced activities has led to the environmental degradation of the Lake Ypacaraí, the most renowned water body in Paraguay [72]. Increasing nutrient concentrations over the last decades have recently resulted in intense cyanobacterial blooms; dominant species in the lake are *C. raciborskii*, *M. aeruginosa*, and *Anabaena* spp.

2.3.9 Venezuela

González and Quirós [73], when they considered trends in 16 reservoirs that have different trophic status, found both linear relationship, between total phosphorus (TP) and total nitrogen (TN) with the phytoplankton biomass, and empirical relationship, between total phosphorus concentration and the nitrate/ammonia quotient which determine the dominance of cyanobacteria. According to the authors' empirical ordination, Venezuelan reservoirs were separated in three groups: group 1 includes reservoirs with low TP (<20 μg/L), while groups 2 and 3 include those reservoirs with moderate to high TP concentrations (>20 μg/L) (**Figure 1**). In group 1, green algae (chlorophyta) are dominant. Group 2 is composed by those reservoirs where nitrate is the dominant inorganic nitrogen compound over ammonia (high nitrate/ammonia quotient), with short water residence time, and the dominant phytoplankton taxa are different from cyanobacteria, while

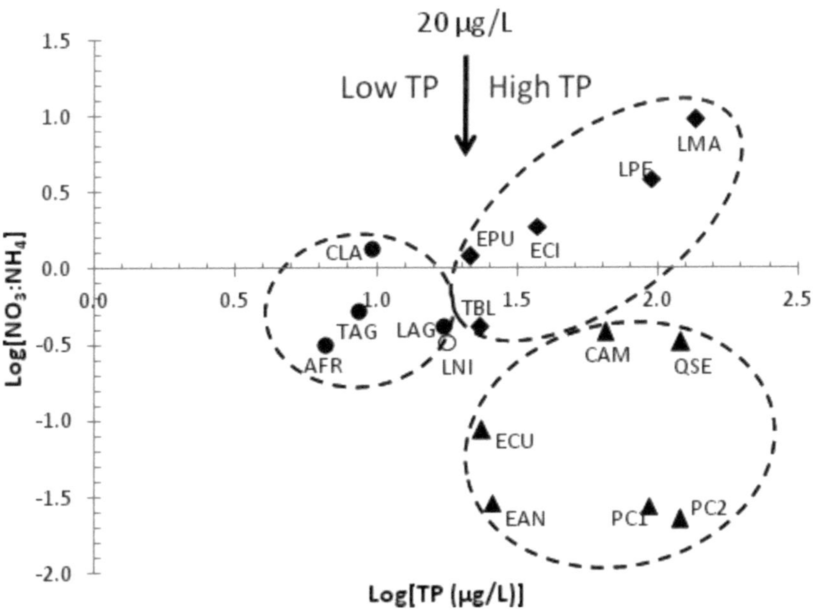

Figure 1.
Relationships between TP and the NO_3:NH_4 ratio in Venezuelan reservoirs: group 1—TP < 20 µg/L—black circles; group 2—TP > 20 µg/L and non-cyanobacteria dominance—black diamonds; and group 3—TP > 20 µg/L and cyanobacteria dominance—black triangles. The Loma de Níquel reservoir is represented by a white circle, because it represented an intermediate situation between groups. Abbreviation of reservoir names: AFR: Agua Fría, TAG: Taguaza, LAG: Lagartijo, CLA: Clavellinos, TBL: Tierra Blanca, LNI: Loma de Níquel, ECI: El Cigarrón, EPU: El Pueblito, ECU: El Cují, EAN: El Andino, LMA: La Mariposa, LPE: La Pereza, CAM: Camatagua, QSE: Quebrada Seca, PC1: Pao-Cachinche—western wing with uptake point and outlet, and PC2: Pao-Cachinche—eastern wing without outlet, modified from González and Quirós [73].

group 3 is composed by those reservoirs with low nitrate/ammonia quotient, high residence time of their waters, and the dominant phytoplankton is cyanobacteria.

3. Discussion

Surveys have shown that more than 40.0% of lakes and reservoirs in the Americas are in eutrophic trophic level [1, 74], and this is a major cause of concern in the developing as well as developed countries.

According to Pratts et al. [75], one of the main problems that affect lakes and reservoirs is the eutrophication. It induces undesirable ecological consequences for the water bodies [1, 3], such as excessive phytoplankton and macrophyte growth. This reduces light penetration and restricts the reoxygenation of water, therefore generating anoxic conditions in the hypolimnetic layers of lakes and reservoirs, as well as high decomposition rates of organic matter that produces a foul smell and makes the water more turbid. Other negative consequences are the proliferation of algal blooms and toxic phytoplankton, fish mortality by suffocation due to drastic oxygen concentration drop during the overturning of waters, proliferation of adequate habitats for vectors of tropical diseases, and loss of biodiversity. These problems are especially important if the water bodies are used for drinking water supply: if these problems are inadequately treated, they may involve serious health risks for human populations.

Regarding the main primary producers in lakes and reservoirs, phytoplankton communities respond quickly to environmental change (as the fertilization of waters with nitrogen and phosphorus) and are indicators of eutrophication [76]. They also show different community dynamics in ecosystems with contrasting trophic states, where high nutrient levels generally favor species belonging to cyanobacteria group.

In most of the studies on lakes and reservoirs in the Americas discussed in the present work, and despite their regional, latitudinal, altitudinal, and climatic differences, eutrophic conditions have led to be dominated by specific species of cyanobacteria. Then, species from the genera *Microcystis*, *Anabaena*, *Planktothrix*, *Oscillatoria*, and *Cylindrospermopsis* seem to be the more widespread dominant organisms under eutrophic conditions in lakes and reservoirs in the American continent. Thus, the cyanobacteria dominance in anthropogenically eutrophic water bodies is an increasing problem that impacts recreation, ecosystem integrity, and human and animal health [77, 78], by deterioration of water quality [79, 80]. The cyanobacterial dominance, on the other side, is under effect of several interacting abiotic (temperature, N/P ratio, other factors) and biotic (intraspecies and interspecific competition in community) factors that usually show different reactions in different environments [81, 82].

According to Reynolds et al. [83] and Bellinger and Sigee [10], cyanobacteria adapts to all types of freshwater environment, including extreme conditions and frequently have the ability to compete with other phytoplankton groups under eutrophic conditions in surface waters. Shapiro [84] and Dokulil and Teubner [81] stated that the ability of "blue-greens" to outcompete other freshwater algae has been attributed to a range of characteristics, including:

- increasing trend in temperature due to climate change;
- optimum growth at high temperatures not preferred by other phytoplankton groups, as in diatoms;
- high survival ability in the water column compared to other species under low light tolerance caused by extensive algal bloom;

- tolerance to low N/P ratios, which is the characteristic for eutrophic lakes—allowing continued growth when N becomes limited;
- depth regulation by buoyancy—avoiding photoinhibition during the early phase of population increase, and allowing algae to obtain inorganic nutrients from the hypolimnion layer when the nutrients decrease in the epilimnion layer from mid- to late-summer;
- resistance to zooplankton grazing by both mechanical and chemical interference;
- tolerance to high pH and low CO_2 concentrations, allowing continued growth of "blue-greens" (but not other algae) at the lake surface during the extensive bloom formation; and
- symbiotic association with aerobic bacteria—bacterial symbionts at the heterocyst surface provide the local reducing atmosphere required for nitrogen fixation that causes inorganic nutrients accumulation in surface waters, poor in nutrients.

The dominance of cyanobacteria in eutrophic water bodies can also be explained according to the main meaningful functional groups proposed by Reynolds [85, 86] and Reynolds et al. [87], based on nutrient availability and stability of the water column. Thus, cyanobacteria can dominate in the phytoplanktonic community in an array of warm, mixed, and fertilized (mainly phosphorus) water bodies, with low transparency values and limitation in C and N [87, 88].

4. Conclusions

In most of the eutrophic lakes and reservoirs in the Americas, as were in various researches on freshwater systems in the temperate regions, there are important increases in level of nutrients, mainly in level of phosphorus, which leads to excessive phytoplankton biomass firstly in a predominance of cyanobacteria. Likewise, in most cases, the dominant cyanobacterial species belong to the genera *Microcystis*, *Anabaena*, *Planktothrix*, *Oscillatoria*, and *Cylindrospermopsis*, which have toxic strains that can cause potential health problems, particularly if the water bodies are used for drinking water supply.

Acknowledgements

Authors would like to thank the Inter American Network of Academic of Sciences (IANAS) for all its help in gathering the information.

We also thank Dr. Diego Rodríguez and an anonymous reviewer for improving the language translation.

Conflict of interest

The authors declare no conflict of interest.

Author details

Ernesto J. González[1]* and Gabriel Roldán[2]

1 Central University of Venezuela, Institute of Experimental Biology, Caracas, Venezuela

2 Colombian Academy of Exact, Physical and Natural Sciences, Medellín, Colombia

*Address all correspondence to: ernesto.gonzalez@ciens.ucv.ve

IntechOpen

© 2019 The Author(s). Licensee IntechOpen. This chapter is distributed under the terms of the Creative Commons Attribution License (http://creativecommons.org/licenses/by/3.0), which permits unrestricted use, distribution, and reproduction in any medium, provided the original work is properly cited.

References

[1] Ansari AA, Gill SS, Khan FA. Eutrophication: Threat to aquatic ecosystems. In: Ansari AA, Gill SS, Lanza GR, Rast W, editors. Eutrophication: Causes, consequences and control. Dordrecht: Springer; 2011. pp. 143-170. DOI: 10.1007/978-90-481-9625-8_7

[2] Hutchinson GE. Eutrophication. American Scientist. 1973;**61**:269-279

[3] Ryding SO, Rast W. The Control of Eutrophication of Lakes and Reservoirs. Paris: Unesco, The Parthenon Publishing Group; 1989. p. 314

[4] González EJ, Challar G, Bicudo DC, Silva e Souza AT, Luzia AP, Sidagis Galli C, et al. Eutrophication in South America: Synthesis. In: Tundisi JG, Matsumura-Tundisi T, Sidagis Galli C, editors. Eutrofização na América do Sul: Causas, Conseqüências e Tecnologias de Gerenciamento e Controle. São Carlos: Instituto Internacional de Ecologia, Instituto Internacional de Ecologia e Gerenciamento Ambiental, Academia Brasileira de Ciências, Conselho Nacional de Desenvolvimento Científico e Tecnológico, InterAcademy Panel on International Issues, InterAmerican Network of Academies of Sciences; 2006. pp. 529-531

[5] Schindler DW, Hecky RE, Findlay DL, Stainton MP, Parker BR, Paterson MJ, et al. Eutrophication of lakes cannot be controlled by reducing nitrogen input: Results of a 37-year whole-ecosystem experiment. Proceedings of the National Academy of Sciences of the United States of America. 2008;**105**:11254-11258. DOI: 10.1073/pnas.0805108105

[6] Salameh E, Harahsheh S. Eutrophication processes in arid climates. In: Ansari AA, Gill SS, Lanza GR, Rast W, editors. Eutrophication: Causes, Consequences and Control. Dordrecht: Springer; 2011. pp. 69-90. DOI: 10.1007/978-90-481-9625-8_3

[7] Wetzel RG. Limnology: Lake and River Ecosystems. 3rd ed. San Diego: Academic Press; 2001. p. 1006

[8] Vázquez Silva G, Castro Mejía G, González Mora I, Pérez Rodríguez R, Castro BT. Bioindicadores como herramientas para determinar la calidad del agua. ContactoS. 2006;**60**:41-48

[9] Quirós R. Cianobacterias en lagos y embalses de Argentina: Década de los 80. Serie de documentos de trabajo del área de Sistemas de Producción Acuática. Documento No 2. Buenos Aires: Departamento de Producción, Facultad de Agronomía, Universidad de Buenos Aires; 2004. p. 23

[10] Bellinger EG, Sigee DC. Freshwater Algae: Identification and Use as Bioindicators. Chichester: Wiley-Blackwell; 2010. p. 271

[11] Thunmark S. Zur Soziologie des susswasserplanktons: Eine methodologisch-okologische Studie. Folia Limnologica Scandinavia. 1945;**3**:1-66

[12] Nygaard G. Hydrobiological studies on some Danish ponds and lakes, II: The quotient hypothesis and some little known or new phytoplankton organisms. Kunglige Danske Vidensk, Selskab. 1949;7:1-242

[13] Stockner J. Paleolimnology as a means of assessing eutrophication. Verhandlungen des Internationalen Verein Limnologie. 1972;**18**:1018-1030

[14] Rawson DS. Algal indicators of trophic lakes types. Limnology and Oceanography. 1956;**1**:18-25

[15] Schindler DW. Eutrophication and recovery in experimental lakes: Implications for lake management. Science. 1974;**184**:897-899

[16] Environmental Protection Agency (EPA). Summary analysis of the North American (U.S. portion) OECD Eutrophication Project: Nutrient loading—Lake response relationships and trophic state indices. In: Ecological Research Series EPA-600/3-78-008. Corvallis: Environmental Research Laboratory, Office of Research and Development, U.S. Environmental Protection Agency; 1978. p. 455

[17] Alcócer DJ, Lugo A. The urban lakes of Mexico City (Lago Viejo de Chapultepec). Lakeline. 1995;**15**:14-31

[18] Bravo-Inclán LA, Saldaña-Fabela MP, Sánchez-Chávez JJ. Long-term eutrophication diagnosis of a high altitude body of water, Zimapan Reservoir, Mexico. Water Science and Technology. 2018;**57**:1843-1849

[19] Díaz-Pardo E, Vázquez G, López-López E. The phytoplankton community as a bioindicator of health conditions of Atezca Lake, México. Aquatic Ecosystem Health and Management. 1998;**1**:257-266

[20] López-López E, Sedeño-Díaz JE, Ortiz-Ordóñez E, Rosas Colmenárez M, Abeja Pineda O. Health condition assessment in Lake Xochimilco (Mexico). Romanian Journal of Biology-Zoology. 2010;**55**:69-80

[21] Oliva Martínez MG, Rodríguez Rocha A, Lugo Vázquez A, Sánchez Rodríguez MR. Composición y dinámica del fitoplancton en un lago urbano hipertrófico. Hidrobiologica. 2008;**18**:1-13

[22] Quiroz Castelán H, Mora Zúñiga LM, Molina Astudillo I, García RJ. Variación de los organismos fitoplanctónicos y la calidad del agua en el Lago de Chapala, Jalisco, Mexico. Acta Universitaria. 2004;**14**:47-58

[23] de Anda J, Shear H. Nutrients and Eutrophication in Lake Chapala. In: Hansen AM, van Afferden M, editors. The Lerma-Chapala Watershed. Boston, MA: Springer; 2001. pp. 183-198

[24] Tomasini-Ortiz AC, Moellez-Chávez G, Sánchez Chávez JJ, Bravo Inclán LA. Cianobacterias y cianotoxinas en el Lago de Pátzcuaro, Michoacán, México. Revista AIDIS de Ingeniería y Ciencias Ambientales: Investigación, Desarrollo y Práctica. 2012;**5**:93-101

[25] Figueroa-Sánchez MA, Sarma N, Sarma SSS. Zooplankton community structure in the presence of low levels of cyanotoxins: A case study in a high altitude tropical reservoir (Valle de Bravo, Mexico). Journal of Limnology. 2014;**73**:157-176. DOI: 10.4081/jlimnol.2014.784

[26] Brocard G, Bettini A, Pfeifer HR, Adatte T, Morán-Ical S, Gonneau C, et al. Chronology of eutrophication and chromium contamination in Lake Chichój, Guatemala. Revista Guatemalteca de Ciencias de la Tierra. 2016;**3**:20-43

[27] Basterrechea M. Enfoque global del lago de Amatitlán y su cuenca. In: Estudios Recientes Sobre la Contaminación del Lago de AMATITLÁN. Ciudad de Guatemala: Centro Agronómico Tropical de Investigación y Enseñanza (CATIE). 1987. pp. 9-31

[28] Rejmánková E, Komárek J, Dix M, Komárková J, Girón N. Cyanobacterial blooms in Lake Atitlán, Guatemala. Limnologica. 2010;**41**:296-302. DOI: 10.1016/j.limno.2010.12.003

[29] Trate TM. Nutrient load analysis of Lake Yojoa, Honduras [thesis].

Massachuset: Massachuset Institute of Technology; 2006

[30] Hernández Oviedo AI, Marín M, Henríquez L, Gray M. Variación espacial y temporal de la diversidad y abundancia del fitoplancton del Lago de Yojoa en un año hidrológico 2014-2015. Revista de Ciencia y Tecnología. 2016;**19**:40-77

[31] Espinoza-Navarrete JJ, Amaya Monterrosa OA, Rivera Torres WE, Ruíz Rodríguez GA, Escobar Muñoz JD. Intensa proliferación de cianobacterias en el Lago de Coatepeque, Santa Ana; ensayos de toxinas paralizantes y organismos causantes. Bioma. 2013;**2**:43-46

[32] Universidad Centroamericana "José Simeón Cañas" (UCA). Monitorización de Variables Físico-Químicas en Humedales RAMSAR de El Salvador. San Salvador: Universidad Centroamericana; 2013

[33] Hooker E, Hernandez S. Phytoplankton biomass in Lake Xolotlán (Managua): Its seasonal and horizontal distribution. Hydrobiological Bulletin. 1991;**25**:125-131

[34] Erikson R. Algal respiration and the regulation of phytoplankton biomass in a polymictic tropical lake (Lake Xolotlán, Nicaragua). Hydrobiologia. 1999;**382**:17-25

[35] Erikson R, Pum M, Vammen K, Cruz A, Ruíz M, Zamora H. Nutrient availability and the stability of phytoplankton biomass and production in Lake Xolotlán (Lake Managua, Nicaragua). Limnologica. 1997;**27**:157-164

[36] Vammen K, Pitty Tercero J, Montenegro Guillén S. Evaluación del proceso de eutroficación del Lago Cocibolca, Nicaragua y sus causas en la Cuenca. In: Tundisi JG, Matsumura-Tundisi T, Sidagis Galli C, editors. Eutrofização na América do Sul: Causas, Conseqüências e Tecnologias de Gerenciamento e Controle. São Carlos: Instituto Internacional de Ecologia, Instituto Internacional de Ecologia e Gerenciamento Ambiental, Academia Brasileira de Ciências, Conselho Nacional de Desenvolvimento Científico e Tecnológico, InterAcademy Panel on International Issues, InterAmerican Network of Academies of Sciences; 2006. pp. 35-58

[37] Hernández González SE, Ahlgreen E, Ahlgreen G. Fluctuaciones temporales en el fitoplancton y contenido de microcistinas intracelulares en cuatro lagos Nicaragüenses. Revista Científica Agua y Conocimiento. 2018;**3**:1-9

[38] Umaña G, Haberyan KA, Horn SP. Limnology in Costa Rica. In: Wetzel RG, Gopal B, editors. Limnology in Developing Countries 2. New Delhi: International Association of Theoretical and Applied Limnology (SIL), International Scientific Publications; 1999. pp. 33-62

[39] Jones JR, Lohman K, Umaña G. Water chemistry and trophic state of eight in Costa Rica. Verhandlungen des Internationalen Verein Limnologie. 1993;**25**:899-905

[40] Autoridad del Canal de Panamá, Universidad de Panamá. Diatomeas del Canal de Panamá: Bioindicadores y otros Estudios Pioneros. Panamá: Facultad de Ciencias Exactas, Matemáticas y Tecnología, Departamento de Botánica y Centro de Ciencias del Mar y Limnología, Departamento de Ambiente, Agua y Energía, División de Agua, Unidad de Calidad de Agua; 2012. p. 253

[41] Gómez Luna CLM, Álamo Díaz B, Rodríguez Tito JC. Riesgo de contaminación con cianobacterias en tres embalses de Santiago de Cuba. Medisan. 2010;**14**:175-183

[42] Pantoja AF. Dinámica fisicoquímica y fitoplanctónica del embalse Guajataca, Puerto Rico [thesis]. Mayagüez: University of Puerto Rico; 2006

[43] Roldán G. Limnología y eutrofización de embalses en Colombia. In: Fernández-Cirelli A, Chalar-Marquisá G, editors. El agua en Iberoamérica. De la Limnología a la Gestión en Sudamérica. Buenos Aires: Programa Iberoamericano de Ciencia y Tecnología para el Desarrollo, CYTED XVII. Aprovechamiento y Gestión de Recursos Hídricos, Centro de Estudios Transdisciplinarios del Agua, Facultad de Ciencias Veterinarias de la Universidad de Buenos Aires; 2002. pp. 107-122

[44] Roldán G, Ramírez JJ. Fundamentos de Limnología Neotropical. 2nd ed. Medellín: Editorial Universidad de Antioquia; 2008. p. 440

[45] Saelens P. Ecological functioning of a eutrophic, high-altitude shallow lake in Ecuador, Laguna Yahuarcocha [thesis]. K.U. Leuven: Kulak; 2015

[46] Fonturbel FE, Castaño-Villa GJ. Relationships between nutrient enrichment and the phytoplankton community at an Andean oligotrophic lake: A multivariate assessment. Ecología Aplicada. 2011;**10**:75-81

[47] Lazzaro X, Rybarczyk H, Meziane T, Hubas C, Lamy D, Point D, et al. Accelerated eutrophication in Lake Titicaca: Historical evolution, mechanisms, monitoring, and observatory approach. In: Proceedings of the Coloquio internacional sobre la contaminación actual e histórica en los ecosistemas acuáticos andinos; 3-5 May 2016. La Paz: Universidad Mayor de San Andrés; 2016. pp. 23-24

[48] Tundisi JG. Management of reservoirs in Brazil. In: Jorgensen SE, Vollenweider RA, editors. Guidelines of Lake Management. Vol. 1. Principles of Lake Management. Shiga: International Lake Environment Committee (ILEC), United Nations Environment Programme (UNEP); 1988. pp. 155-169

[49] Huszar VLM, Silva LHS, Marinho M, Domingos P, Sant'Anna CL. Cyanoprokaryote assemblages in eight productive tropical Brazilian waters. Hydrobiologia. 2000;**424**:67-77

[50] Dantas EW, Moura AN, Bittencourt-Oliveira MC, Arruda Neto JDT, Cavalcanti ADC. Temporal variation of the phytoplankton community at short sampling intervals in the Mundau reservoir, Northeastern Brazil. Acta Botânica Brasílica. 2008;**22**:970-982. DOI: 10.1590/S0102-33062008000400008

[51] Nardini MJ, De Souza Nogueira I. O processo antrópico de um lago artificial da eutrofização e florações e o desenvolvimento de algas azuis em Goiânia. Estudos. 2008;**35**:23-52

[52] Chellappa NT, Camara FRA, Rocha O. Phytoplankton community: Indicator of water quality in the Armando Ribeiro Gonçalves Reservoir and Pataxó Channel, Rio Grande do Norte, Brazil. Brazilian Journal of Biology. 2009;**69**:241-251

[53] Barbosa FAR, García FC, Marques MMGSM, Nascimento FA. Nitrogen and phosphorus balance in a eutrophic reservoir in Minas Gerais: A first approach. Brazilian Journal of Biology. 1998;**58**:233-239

[54] García FC, Barbosa FAR, Braz S, Petrucio MM, Faria B. Water quality of an urban reservoir subjected to periodic applications of copper sulphate: The case of Ibirité reservoir, Southeast Brazil. Acta Limnologica Brasiliensia. 2009;**21**:235-243

[55] Abe DS, Tundisi JG, Matsumura-Tundisi T, Tundisi JEM, Sidagis Galli C, Teixeira-Silva V, et al.

Monitoramento da qualidade ecológica das águas interiores superficiais e do potencial trófico em escala continental no Brasil com o uso de tecnologias inovadoras. In: Tundisi JG, Matsumura Tundisi T, Sidagis Galli C, editors. Eutrofização na América do Sul: Causas, Conseqüências e Tecnologias de Gerenciamento e Controle. São Carlos: Instituto Internacional de Ecologia, Instituto Internacional de Ecologia e Gerenciamento Ambiental, Academia Brasileira de Ciências, Conselho Nacional de Desenvolvimento Científico e Tecnológico, InterAcademy Panel on International Issues, InterAmerican Network of Academies of Sciences; 2006. pp. 225-239

[56] Sidagis Galli C, Abe DS. Disponibilidade, poluição e eutrofização das águas. In: Scheuenstuhl MC, editor. Águas do Brasil: Análises Estratégicas. São Paulo: Instituto de Botânica; 2010. pp. 165-174

[57] Almanza-Marroquín V, Figueroa R, Parra O, Fernández X, Baeza C, Yáñez J, et al. Bases limnológicas para la gestión de los lagos urbanos de Concepción, Chile. Latin American Journal of Aquatic Research. 2016;**44**:313-326

[58] Campos V, Muñoz D, Straube M, Lisperguer S, Weckesser J. Péptidos tóxicos y no tóxicos de cianobacterias en cuerpos de agua dulce de La V Región, Chile. Boletín Micológico. 2007;**22**:95-100

[59] Mühlhauser HA, Vila I. Eutrofización, impacto en un ecosistema acuático montañoso. Archivos de Biología y Medicina Experimentales. 1987;**20**:117-124

[60] Parra O. La eutroficación de la Laguna Grande de San Pedro, Concepción, Chile: Un caso de estudio. Ambiente y Desarrollo. 1989;**5**:117-136

[61] Parra O, Valdovinos C, Urrutia R, Cisternas M, Habit E, Mardones M. Caracterización y tendencias tróficas de cinco lagos costeros de Chile Central. Limnetica. 2003;**22**:51-83

[62] Quirós R. Empirical relationships between nutrients, phyto and zooplankton and relative fish biomass in lakes and reservoirs of Argentina. Verhandlungen des Internationalen Verein Limnologie. 1991;**24**:1198-1206

[63] Quirós R. La eutrofización de las aguas continentales en Argentina. In: Fernández-Cirelli A, editor. El Agua en Iberoamérica: Acuíferos, lagos y embalses. Buenos Aires: Ciencia y Tecnología para el Desarrollo (CYTED), Subprograma XVII, Aprovechamiento de Recursos Hídricos; 2000. pp. 43-47

[64] Quirós R, Boveri MB, Petracchi CA, Rennella AM, Rosso JJ, Sosnovsky A, et al. Los efectos de la agriculturización del humedal pampeano sobre la eutrofización de sus lagunas. In: Tundisi JG, Matsumura-Tundisi T, Sidagis Galli C, editors. Eutrofização na América do Sul: Causas, Conseqüências e Tecnologias de Gerenciamento e Controle. São Carlos: Instituto Internacional de Ecologia, Instituto Internacional de Ecologia e Gerenciamento Ambiental, Academia Brasileira de Ciências, Conselho Nacional de Desenvolvimento Científico e Tecnológico, InterAcademy Panel on International Issues, InterAmerican Network of Academies of Sciences; 2006. pp. 1-16

[65] Chalar G, Conde D. Antecedentes y estado actual del conocimiento científico de los embalses del Uruguay. In: Fernández-Cirelli A, editor. El Agua en Iberoamérica: Acuíferos, Lagos y Embalses. Buenos Aires: Ciencia y Tecnología para el Desarrollo (CYTED). Subprograma XVII. Aprovechamiento de Recursos Hídricos; 2000. pp. 145-147

[66] De León L, Chalar G. Abundancia y diversidad del fitoplancton en el Embalse de Salto Grande (Argentina-Uruguay). Ciclo estacional y

distribución espacial. Limnetica. 2003;**22**:103-113

[67] Chalar G, De León L, Brugnoli E, Clemente J, Paradiso M. Antecedentes y nuevos aportes al conocimiento de la estructura y dinámica del embalse Salto Grande. In: Fernández-Cirelli A, Chalar-Marquisá G, editors. El agua en Iberoamérica. De la Limnología a la Gestión en Sudamérica. Buenos Aires: Programa Iberoamericano de Ciencia y Tecnología para el Desarrollo, CYTED XVII. Aprovechamiento y Gestión de Recursos Hídricos, Centro de Estudios Transdisciplinarios del Agua, Facultad de Ciencias Veterinarias de la Universidad de Buenos Aires; 2002. pp. 123-141

[68] Chalar G. Dinámica de la eutrofización a diferentes escalas temporales: Embalse Salto Grande (Argentina-Uruguay). In: Tundisi JG, Matsumura-Tundisi T, Sidagis Galli C, editors. Eutrofização na América do Sul: Causas, Conseqüências e Tecnologias de Gerenciamento e Controle. São Carlos: Instituto Internacional de Ecologia, Instituto Internacional de Ecologia e Gerenciamento Ambiental, Academia Brasileira de Ciências, Conselho Nacional de Desenvolvimento Científico e Tecnológico, InterAcademy Panel on International Issues, InterAmerican Network of Academies of Sciences; 2006. pp. 87-101

[69] Chalar G. The use of phytoplankton patterns of diversity for algal bloom management. Limnologica. 2009;**39**:200-208

[70] Vidal L, Kruk C, Aubriot L, Piccini C, Fabre A, Bonilla S. Floraciones de la especie *Cylindrospermopsis raciborskii* en Uruguay. In: Bonilla S, editor. Cianobacterias Planctónicas del Uruguay. Manual para la Identificación y Medidas de Gestión. Montevideo: Programa Hidrológico Internacional de la UNESCO para América Latina y el Caribe; 2009. pp. 81-82

[71] RAP-AL Uruguay. Contaminación y Eutrofización del Agua. Impactos del Modelo de Agricultura Industrial. Red de Acción en Plaguicidas y sus Alternativas para América Latina (RAPAL): Montevideo; 2010. p. 36

[72] López Moreira GA, Hinegk L, Salvadore A, Zolezzi G, Hölker F, Monte Domecq RA, et al. Eutrophication, research and management history of the shallow Ypacaraí Lake (Paraguay). Sustainability. 2018;**10**:1-32. DOI: 10.3390/su10072426

[73] González EJ, Quirós R. Eutrophication of reservoirs in Venezuela: Relationships between nitrogen, phosphorus and phytoplankton biomass. Oecologia Australis. 2011;**15**:458-475. DOI: 10.4257/oeco.2011.1503.03

[74] International Lake Environment Committee (ILEC). Data book of world lake environments. A survey of the state of world lakes. 3. The Americas. Kusatsu: International Lake Environment Committee and United Nations Environment Programme; 1995. p. 868

[75] Pratts J, Morales-Baquero R, Dolz J, Armengol J. Aportaciones de la Limnología a la gestión de embalses. Ingeniería del Agua. 2014;**18**:83-97. DOI: 10.4995/ia.2014.3145

[76] Baho DL, Drakare S, Johnson RK, Allen CR, Angeler DG. Is the impact of eutrophication on phytoplankton diversity dependent on volume/ ecosystem size? Journal of Limnology. 2017;**76**:199-210. DOI: 10.4081/ jlimnol.2016.1562

[77] Downing JA, Watson SB, McCauley E. Predicting cyanobacteria dominance in lakes. Canadian Journal of Fisheries and Aquatic Sciences. 2001;**58**:1905-1908. DOI: 10.1139/ cjfas-58-10-1905

[78] Fernández C, Estrada V, Parodi ER. Factors triggering cyanobacteria dominance and succession during blooms in a hypereutrophic drinking water supply reservoir. Water, Air, and Soil Pollution. 2015;**226**:73-85. DOI: 10.1007/s11270-014-2290-5

[79] Pizzolon L. Importancia de las cianobacterias como factor de toxicidad en las aguas continentales. Interciencia. 1996;**21**:239-245

[80] Vasconcelos V. Eutrophication, toxic cyanobacteria and cyanotoxins: When ecosystems cry for help. Limnetica. 2006;**25**:425-432

[81] Dokulil MT, Teubner K. Cyanobacteria dominance in lakes. Hydrobiologia. 2000;**438**:1-12

[82] Kim S, Chung S, Park H, Cho Y, Lee H. Analysis of environmental factors associated with cyanobacterial dominance after River Weir installation. Water. 2019;**11**:1-24. DOI: 10.3390/w11061163

[83] Reynolds CS, Oliver RL, Walsby AE. Cyanobacterial dominance: The role of buoyancy regulation in dynamic lake environments. New Zealand Journal of Marine and Freshwater Research. 1987;**21**:379-390. DOI: 10.1080/00288330.1987.9516234

[84] Shapiro J. Current beliefs regarding dominance by blue-greens: The case for the importance of CO_2 and pH. Verhandlungen des Internationalen Verein Limnologie. 1990;**24**:38-54

[85] Reynolds CS. The plant life of the pelagic. Verhandlungen des Internationalen Verein Limnologie. 1996;**26**:97-113

[86] Reynolds CS. The Ecology of Phytoplankton. 2nd ed. Cambridge: Cambridge University Press; 2006. p. 535

[87] Reynolds CS, Huszar VLM, Kruk C, Naselli-Flores L, Melo S. Towards a functional classification of the freshwater phytoplankton. Journal of Plankton Research. 2002;**24**:417-428

[88] Brasil J, Huszar VML. O papel dos traços funcionais na ecología do fitoplancton continental. Oecologia Australis. 2011;**15**:799-834. DOI: 10.4257/oeco.2011.1504.04

Chapter 3

Isolation, Characterization, and Biotechnological Potential of Native Microalgae From the Peruvian Amazon

Marianela Cobos, Juan C. Castro, Jae D. Paredes, Sheyla Pérez, J. Dylan Maddox, Segundo L. Estela, Hicler N. Rodríguez, Alvaro B. Tresierra, Juan R. Saldaña, Jorge L. Marapara, Pedro M. Adrianzén and Rosana Gonzales

Abstract

The objective of this chapter is to provide scientific information on basic aspects to be taken into account to achieve the successful isolation, biochemical, and molecular characterization and then to evaluate the biotechnological potential of native microalgae of the Peruvian Amazon. Recent investigations reported by our research team has demonstrated that the isolated native microalgae from the Peruvian Amazon have a great potential for the biotechnological production of biodiesel and nutraceuticals. This biotechnological potential was identified thanks to the application of various protocols that were standardized by the authors over the last 5 years. In conclusion, the native microalgae of the Peruvian Amazon have biotechnological potential and are therefore promising for the production of both biodiesel and nutraceuticals. Various species of microalgae were identified, isolated, cultured, and characterized using biochemical, nutraceutical, and molecular techniques, the isolation stage being the starting point to achieve various biotechnological applications. *Ankistrodesmus* sp. is one of the microalgae with potential for the production of biodiesel and microalgae such as *Haematococcus pluvialis*, *Scenedesmus* sp., and *Chlorella* sp., among others demonstrated a high potential for nutraceutical production. The stress conditions to which microalgae are subjected are being a determining factor for the production of biodiesel and nutraceuticals.

Keywords: biotechnological potential, biodiesel, nutraceuticals, Peruvian Amazon

1. Introduction

From the biotechnological point of view, the term microalgae refer to those microorganisms that contain chlorophyll-a and/or other similar pigments, which allow them to perform oxygenic photosynthesis. In this context, cyanobacteria or green-blue algae, prokaryotes, have traditionally been considered within the group of microalgae. According to this definition, photosynthetic bacteria are excluded, since they do not contain chlorophyll-a and perform anoxygenic photosynthesis. Therefore, the term microalgae have no taxonomic meaning and within it, organisms with two different cell types are included: cyanobacteria that have prokaryotic cell structure and the remaining microalgae with eukaryotic cell structure [1]. Microalgae are characterized by accumulating triglycerides due to their photobiosynthetic capacity [2, 3], can sequester CO_2 from industrial sources [4] and demand less cropping area than traditional oleaginous plants [5]. In addition, microalgae can produce various substances of commercial interest, such as nutrients, food additives, drugs, and other substances [6, 7].

Due to this great biosynthetic diversity, isolates have been made and there are collections of microalgae in several institutions around the world [8]. An estimated 50,000 species have been identified and are kept in collections [9]. These only represent a small fraction of the enormous biodiversity of species that exist. Likewise, it is estimated that less than 10% have been evaluated for their biodiesel production capacity and only of some species their genomes have been sequenced [10, 11].

Therefore, in the Laboratory of Biotechnology and Bioenergetics of the Scientific University of Peru, efforts have been initiated to isolate and increase the diversity of the collections and be more likely to find ideal microalgae strains for the production of biodiesel, nutraceuticals, bioremediation, and other biotechnological applications. It should be noted that there are several methods of isolation, which depend on the dimensions of the microalgae, their mobility, and their morphology. The most commonly used methods are: (a) micropipette isolation, (b) on agar plates, and (c) serial dilutions. However, it is advisable to combine all these techniques to allow isolation and have unialgal cultures [12]. Finally, in this chapter, we will focus on the different isolation techniques, cultivation of freshwater microalgae, biochemical and molecular characterization to evaluate the biotechnological potential of native microalgae of the Peruvian Amazon based on our experience acquired over the years.

2. Isolation and culture techniques of native microalgae

2.1 Isolation techniques

The application of the different techniques of isolation in microalgae has as main objective to obtain a population of microalgae, starting from a single individual or clone (cells, filaments, colonies, and/or cysts) achieving unialgal cultures [3]. The use of the isolation technique depends on the dimensions of the microalgae, its mobility, and its morphology; however, according to our experience in this field, it is advisable to combine the different techniques [8]. The isolation techniques that the authors used are micropipette isolation and agar plate isolation, which are described below:

2.1.1 Isolation of microalgae with micropipette

This technique consists in isolating microalgae with the help of a Pasteur pipette with a reduced tip and/or with a capillary obtained by casting and then sterilized at 121°C for later use. Once the capillary is prepared and sterilized, a drop is removed from the natural collection and placed on a slide sheet, observed with the inverted light microscope or compound microscope to verify the presence of microalgae to be isolated. Under the microscope and with the help of the capillary, the desired microalgae are "trapped" and transferred to a slide sheet having a drop of sterile culture medium. This technique requires constant practice, since microalgal transfers must be done quickly and carefully avoiding causing stress to the microalgae. The microalgae in isolation process can be continued transferring at least five times successively in drops of sterile culture medium, until a single type of microalgal cell is obtained. After obtaining the isolated microalgae, it is transferred to a test tube containing between 200 and 500 μL of sterile culture medium and grown for 7–10 days until microalgal growth is evidenced. Subsequently, the content is transferred to a test tube by adding fresh culture medium in order to continue with the culture. It is advisable to verify if the isolation was successful, that is to say that it is not contaminated with other species of microalgae, otherwise the isolation process must be repeated.

2.1.2 Isolation of microalgae in Petri dishes with agar

This technique is generally used when microalgae are 10 μm in diameter or less and consist of preparing Petri dishes with the appropriate culture medium to which 2% agar is added, and it is autoclaved at 120°C and 15 lb.ft.$^{-2}$ (1.1 kg.cm^{-2}) pressure for 15 min. Subsequently, it is left at room temperature and before it solidifies, they are emptied into the Petri dishes allowing their solidification. In Petri dishes with solid medium, 50–100 mL of the natural collection or microalgal suspension obtained using the capillary technique (to complement the isolation with capillaries) is added on the surface of the medium and with the help of the seeding handle or Drigalsky sterile handle homogenizes the suspension in the middle. It is immediately covered with the lid, sealed with parafilm, inverted, and cultivated for 5–10 days until observing the first colonies. The recommended culture conditions are photoperiod of 12 h light/12 h dark, light intensity of 100 μmol photons.m^2.s^{-1} and temperature of 25°C. Subsequently, observations are made on the inverted or compound microscope, and the microalgal colonies free of other microorganisms are selected with the help of the sowing handle and re-seeded in another Petri dish with culture medium. Repeating this procedure as many times as necessary to achieve a unialgal culture (**Figure 1**).

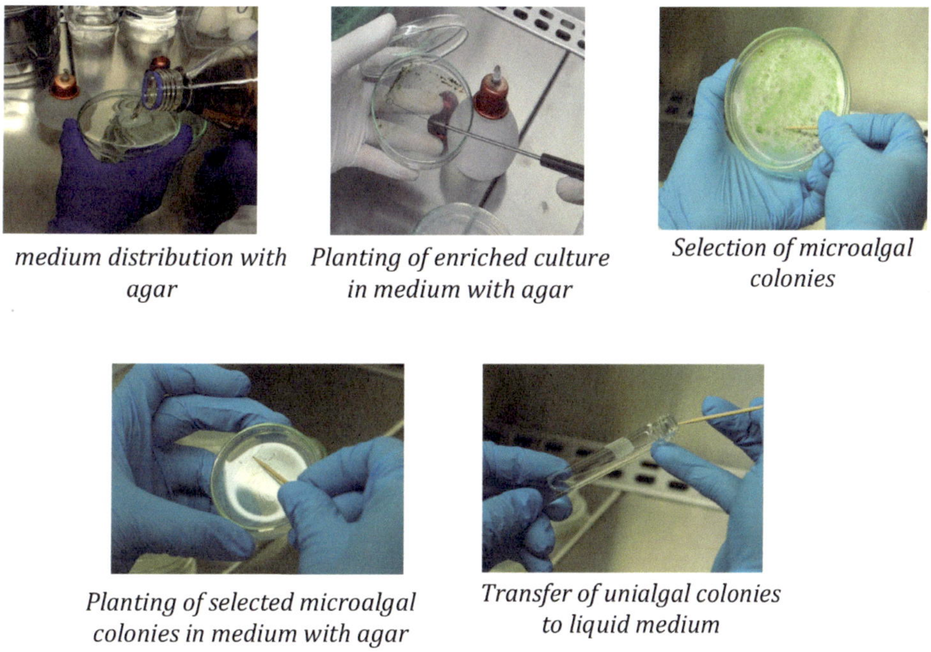

Figure 1.
Isolation of microalgae in Petri dishes with agar.

2.2 Microalgae cultivation techniques

Microalgae are characterized by the ability to synthesize various substances from water, CO_2, and minerals with the help of light energy (photosynthesis) [2]. In order to obtain these substances, culture conditions are required that guarantee a successful microalgal growth and its subsequent direct consumption as live food or nutritional supplements and indirectly for obtaining algal extracts (antibiotics, enzymes, essential fatty acids, among others) [13]. Currently, in the world, a large number of technologies and culture systems are used, especially those that are applied in laboratory conditions, which are addressed in detail in this chapter.

The culture conditions of the microalgae depend on the species to be cultivated and on the planned experimental tests. Various factors such as the composition of the culture medium, temperature, relative humidity, air flow, CO_2 concentration, lighting, among others influence the cultivation of microalgae. Initially, the microalgal suspensions are kept in test tubes of 10 or 15 mL at 25°C with a light intensity of 100 μmol photons.m^{-2}.s^{-1} with a photoperiod 12 h light/12 h dark. As the cell density of the cultures increases, they are transferred to 50, 100, and 250 mL flasks to volumes of 5 L or more (depending on the needs of microalgal biomass) for 4–8 weeks in an orbital shaker at 200 rpm or with constant aeration (**Figure 2**). If you do not have an agitator or air flow, it is recommended to shake manually 2 or 3 times a day. In addition, it is advisable to monitor the cultures daily by microscopic observations.

The amount of inoculum and cell density of the culture are important aspects in the cultivation of microalgae. For example, small inoculums and cultures with low cell densities may be lost due to photooxidation. Also, cultures with high cell densities are affected by the self-name effect. In addition, the inoculum must consist of cells of a single species and preferably in exponential growth phase [7, 13].

Figure 2.
Microalgae culture process under laboratory conditions.

2.2.1 Culture media used

The culture of microalgae under laboratory conditions requires culture medium, this being an aqueous solution that transports the nutrients necessary for its growth, such as water, light, CO_2, and mineral salts; among which are mainly some source of nitrogen and a source of phosphorus. The requirements of certain minerals vary widely between species and type of study. However, the supply of culture medium and nutrient concentrations must be directly related to the production of biomass so that it is necessary to periodically add enough to avoid the decrease in the productivity of the biomass or even some dysfunction of the culture due to photoinhibition.

The use of the culture medium depends on the type of microalgae to be cultivated, since there are different compositions and even with some modifications with which excellent results were achieved in the growth of these microorganisms. However, it is important to take into account certain considerations when preparing the culture medium, such as pH of the medium, hardness and salinity of the water, validity of the reagents, etc. The culture media that the authors generally used were Chu-10, BG-11, Beijerinck, and Bold Basal (see composition in **Table 1**).

For example, to prepare the Chu-10 medium, stock solutions are required, as detailed in **Table 2**.

Each of the stock solutions are added in a ratio of 1:10 v/v. Likewise, before weighing the reagents, it is advisable to perform simple calculations that guarantee the good preparation of the culture medium. For example, to prepare a stock solution of 100 mL of calcium chloride ($CaCl_2.2H_2O$), 3.67 g of this salt is required and flush up to 100 mL with distilled water.

2.2.2 Microalgal growth evaluation

It is recommended that microalgal growth records be interdiary. Commonly, the 0.1 mm deep Neubauer counting chamber is used to count microalgae from 2 to 30 mm and cultures of densities between 5×10^4 and 5×10^7 cells/mL. For this type of record, a sample of the culture (~100 mL) is taken, 50 mL of lugol is added and gently homogenized. About 50 mL of the mixture is placed in the Neubauer counting chamber that has the coverslip, allowed to stand for 5–10 min for the sample to stabilize. The cells are then counted using a microscope using a manual cell counter (**Figure 3**). It is advisable to double count in each of the fields (upper and lower chamber) for each of the samples. Once the average number of microalgal cells is obtained, the calculation is carried out with the following equation (Eq. 01):

$$N \text{ (cells/mL)} = Pnc \times Fd/6 \times 10^6 \quad (1)$$

where N = number of cells/mL, Pnc = average number of cells obtained from the four fields of the Neubauer counting chamber, Fd = dilution factor (250,000 for small microalgae such as *Chlorella* sp. and *Euglena* sp. 10,000 for larger microalgae such as *Scenedesmus* sp., and *Ankistrodesmus* sp.).

A second method to evaluate the microalgal growth used in our laboratory is using the Nanodrop 2000 C UV/visible spectrophotometer. This equipment can measure accurately and reproducibly up to 2 mL of concentrated cultures. The system retains the sample between two optical fibers thanks to the surface tension [14]. The procedure is simple and consists of the following steps: (1) in the software that controls the spectrophotometer, the option to read cells at 680 nm (wavelength absorbed by chlorophylls a and b) is selected, (2) the baseline reading is performed (bleaching) by placing 2 mL of the culture medium in the sensor, and (3) the same volume of the microalgal culture is read.

Chemical components	CHU-10	BG-11	Beijerinck	Bold basal
$NaHCO_2$	12.6 g	—	—	—
$NaNO_3$	85 g	1.5 g	—	1.5 g
KH_2PO_4	—	—	—	1.05 mg
K_2HPO_4	8.7 g	40 mg	1.18 g	0.45 mg
$MgSO_4.7H_2O$	36.9 g	75 mg	20 mg	0.45 mg
$CaCl_2.2H_2O$	36.7 g	36 mg	10 mg	1.2 g
NO_3NH_4	—	—	150 mg	—
PO_4H_2K	—	—	907 mg	—
NaCl	—	—	—	0.15 mg
$NaCO_3$	—	20 mg	—	—
$Na_2SiO_3.9H_2O$	28.4 g	—	—	—
HCl (1 mol/L)	0.05 mL	—	—	—
NaEDTA	50 mg	1.04 g	5 mg	50 mg
KOH	—	—	—	31 mg
Citric acid	—	6 mg	—	—
Ferric ammonium citrate	3.35 g	6 mg	—	—
pH	7.5	7.4	6.8	6.6
Total volume	1 L	1 L	1 L	1 L
H_2BO_3	618 mg	2.86 g	1 mg	11.42 mg
$MnCL_2.4H_2O$	12.6 mg	1.81 g	0.15 mg	1.44 mg
$ZnSO_4.7H_2O$	44 mg	0.22 g	2.2 mg	8.8 mg
$NaMoO_4.2H_2O$	12.6 mg	0.39 g	—	—
$CuMoO_4.2H_2O$	19.6 mg	79 mg	0.15 mg	1.57 mg
$Co(NO_3)_2.6H_2O$	—	49.4 mg	—	0.49 mg
H_2SO_4	—	—	—	1 mg
$FeSO_4.7H_2O$	—	—	—	4.98 mg
MoO_3	—	—	—	0.71 mg
$Mo_7O_{24}(NH_4)_6.4H_2O$	—	—	0.10 mg	—
$CoCl_2$	20 mg	—	—	—
Distilled water	1.0 L	1.0 L	1.0 L	1.0 L

Table 1.
Composition of microalgae culture media.

However, the spectrophotometer absorbance readings are not sufficient, it is necessary to determine their correlation with the number of microalgal cells per milliliter of culture. This is done through a standard curve of absorbances versus cell count for each species under study. To prepare the standard curve, 100 mL of the microalgal culture (in the logarithmic growth phase) is transferred to a microtube, and serial dilutions are made (1/2, 1/4, 1/8, 1/16, 1/32, 1/64, 1/128) with the culture medium. The absorbance at 680 nm in the spectrophotometer is measured from each dilution, and the cell number was determined using a Neubauer counting chamber. The absorbance values and the corresponding microalgae cell number data/mL, are entered in two columns in a Microsoft Excel® spreadsheet. Select both columns, insert the scatter plot, add

Stocks	Components	g/100 mL de H_2O
1	$CaCl_2.2H_2O$	3.67 g
2	$MgSO_4.7H_2O$	3.69 g
3	$NaHCO_3$	1.26 g
4	K_2HPO_4	0.87
5	$NaNO_3$	8.50 g
6	$Na_2SiO_3.9H_2O$	2.84 g
7	Ferric citrate solution	3.35 g/1000 mg water
8	Micronutrient solution	—
	NaEDTA	50.0 mg
	H_2BO_3	618.0 mg
	$CuSO_4.5H_2O$	19.6 mg
	$ZnSO_4.7H_2O$	44.0 mg
	$CaCl_2.6H_2O$	20.0 mg
	$MnCL_2.4H_2O$	12.6 mg
	$NaMoO_4.2H_2O$	12.6 mg

Table 2.
Composition of stock solutions of the Chu-10 medium.

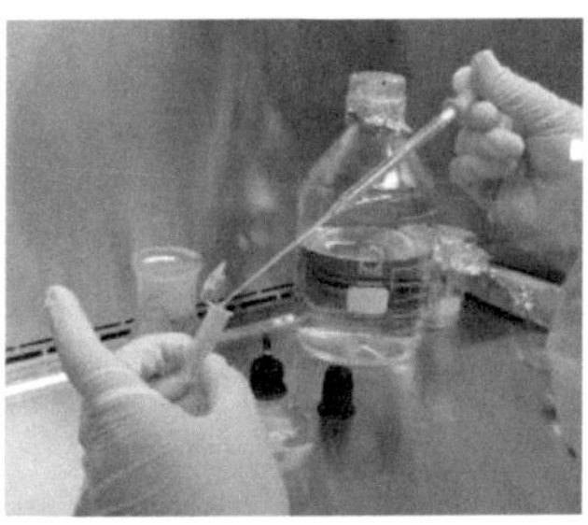

Removing a culture aliquot for reading under a microscope

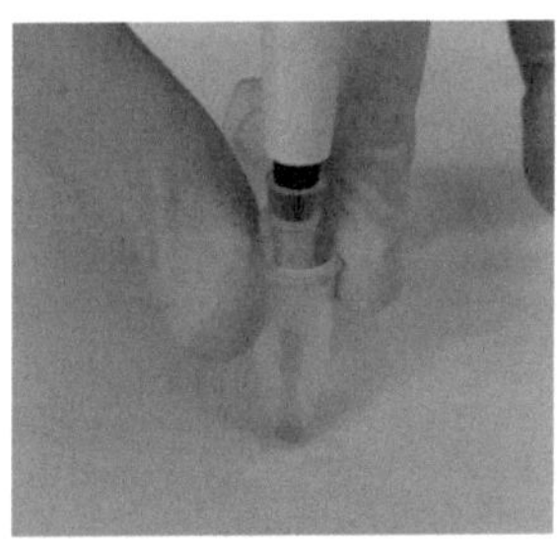

Adding lugol to cell count

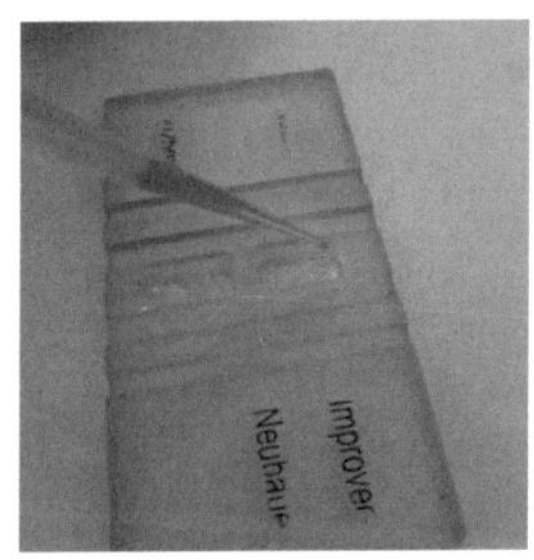

Sliding the sample into the Neubahuer chamber

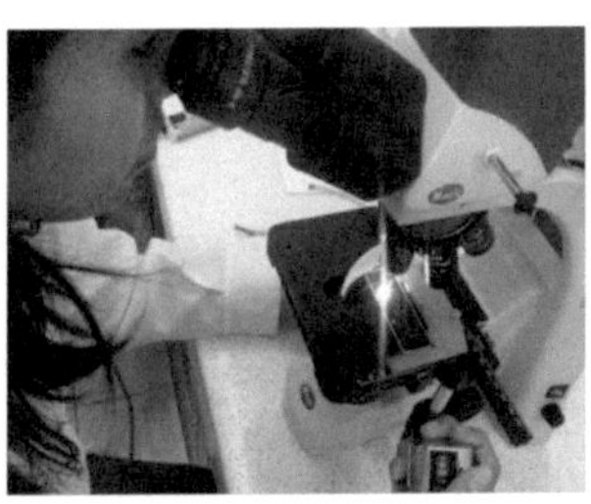

Microalgal growth lectur using a manual contometer

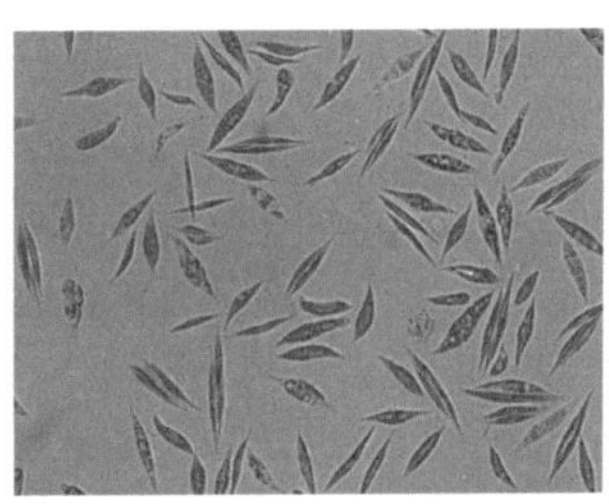

Microalgae observed during cell count by microscopy

Figure 3.
Microalgal growth evaluation process.

the trend line. In the trend line options select "Lineal", activate "present equation in the graph", and "present R^2 value in the graph". This will give us an equation of a straight line and the corresponding R^2 value (should be > 0.98). For example, for a strain of *Chlorella* sp. from our isolated microalgae culture collection, it has been determined that the R2 value was 0.99 and the following equations were obtained (Eq. 02 and 03):

$$A = (3 \times 10^9)\,(n) - 0.0025 \tag{2}$$

$$n = A + 0.0025/(3 \times 10^9) \tag{3}$$

where A is the absorbance at 680 nm and n is the microalgae cell number/mL.

Finally, it is important to consider some aspects that could contribute to the cultivation of microalgae.

- It is important to keep the crops in proper condition because they will be useful for future research. Likewise, it is advisable to use plastic tips with a filter to avoid contamination of a microalgal culture with other strains or species of microalgae.

- The maintenance of the strains can be done in liquid or solid medium.

- Periodically (15–20 days), culture medium (liquid medium) should be added under sterile conditions.

- The plates or flasks must be labeled with the codes and/or names of the microalgal strains, date of inoculation among other relevant data.

3. Techniques for biochemical and molecular characterization

3.1 Techniques for biochemical characterization

The biochemical characterization of the microalgae developed in our laboratory is based on the following analyzes:

3.1.1 Total lipid extraction

Total lipid extraction according to Yu et al. [15]. It consists of transferring the dry biomass to mortars for crushing with 8 mL of a mixture of chloroform:methanol (2:1). The extract obtained is transferred to 2 mL microtubes, and 100 μL of 0.9% NaCl is added for every 1000 μL of extract. The solution is homogenized in vortex for 30 sec and centrifuged at 23,000 × *g* at 4°C for 5 min. The chloroform phase is filtered with 0.45 μm syringe filters and transferred to beakers of known weight. Cellular debris and other components are retained in the intermediate phase (aqueous and chloroform phases) and are treated several times with the lipid extraction solution (chloroform:methanol) after homogenization in the vortex and centrifugation. All extracts with organic solvents are filtered and transferred to the same beaker. The organic solvents are evaporated from the beaker in a hotplate at 50°C for 4 h. Then the lipid components retained in the beaker are dried at 50°C for 4 h. Finally, the beaker is tempered to 25°C, and its weight is determined. The amount of total lipids obtained is determined by weight difference of the beaker with and without the lipids. With the following equation:

$$\text{Total lipid content (\%)} = (P_L/P_M) \times 100 \tag{4}$$

where P_L is the dry weight of total lipids and P_M is the dry weight of microalgae.

3.1.2 Qualitative determination of total lipids

The intracellular triglycerides of the microalgae are detected by the fluorescence they emit when interacting with Nile Red [16]. For this, the cells are stained with 2 mg/mL of Nile Red (dissolved in acetone) for 15 min and photographed using a trinocular microscope of Carl Zeiss-AxioLab.A1 epifluorescence and a real-time AxioCamERc 5 s digital camera. The images are obtained with a magnification of 1000× with visible light and epifluorescence (excitation: 510–560, emission: 590) [17].

3.1.3 Protein and carbohydrate

Protein determination is performed according to Lowry [18], and for carbohydrate evaluation, the Dreywood method [19] prior acid hydrolysis with 2 N HCI.

3.1.4 Humidity and ashes

The moisture content of the sample is determined weighing 0.1 g of the microalgae and dried in a vacuum oven at a temperature of 105°C for 16 h and at a pressure less than 0.1 bar, the result is expressed as a percentage.

The most common method to determine ashes is mucin calcination at temperatures between 500 and 600°C. Water and volatile substances are evaporated, while organic substances are incinerated in the presence of oxygen from the air to produce CO_2 and nitrogen oxide [20]. Most minerals are converted to oxides, sulfate, phosphate, chloride, and silicate.

$$\text{Ash}\ \% = [(P1 - P2) \times 100]/(P - P2) \tag{5}$$

where P is the weight in grams of the capsule plus that of the sample, P1 is the weight in grams of the capsule plus ashes, and P2 is the weight in grams of the empty capsule.

In general, microalgae have variations in the content of their biochemical parameters. For example, *Chlorella lewinii* showed a higher protein content (31.2%), *Ankistrodesmus* sp., a higher total lipid content (39.5%), and *Acutodesmus obliquus* a higher percentage of carbohydrates (49.6%) compared to other freshwater microalgae [21]. These parameters vary even more in conditions of physiological stress to which microalgae are subjected, as evidenced in the species of microalgae that accumulated a greater amount of total lipids (mg/g dry biomass) when grown in media without nitrogen; *Ankistrodesmus nannoselene* (316 mg/g dry biomass), *Ankistrodesmus* sp. (263.6 mg/g dry biomass), and *Scenedesmus* sp. (243.3 mg/g dry biomass), with respect to *Scenedesmus quadricauda* and *Chlorella* sp. which showed lower lipid content. Likewise, *Ankistrodesmus* sp., *A. nannoselene*, and *Scenedesmus* sp. showed statistically significant differences in total lipid content when grown in media with and without nitrogen, while in *S. quadricauda* and *Chlorella* sp., no significant differences were observed. However, the ash and moisture content remain very low [17]. Therefore, it is a fact that microalgae increase their lipid content when subjected to stress conditions in particular under nutrient restrictions [22].

These results suggest that some microalgae species have the ability to modify lipid metabolism in response to changes in environmental conditions, such as mentions Thompson [23] and Guschina and Harwood [24], producing large quantities of microalgal biomass but with relatively low lipid contents [25]. In essence, the production of biomass and microalgal triglycerides compete for photosynthetic assimilation, often requiring reprogramming of physiological pathways to stimulate lipid biosynthesis, which allows microalgae to withstand adverse conditions [26].

3.2 Techniques for molecular characterization

3.2.1 DNA and RNA extraction and purification

The hereditary basis of all living organisms is its genomic DNA, which contains the encoded information that is transmitted from generation to generation [27]. The first step of molecular biology studies and DNA recombination techniques begins with the extraction and purification of nucleic acids (DNA and RNA). The objective of all extraction methods is to obtain purified nucleic acids sufficient for downstream applications. Quantity and quality of extracted nucleic acids are especially important as these factors generally determine whether subsequent molecular techniques are successful. Inadequate methods, therefore, can compromise subsequent procedures for which much labor, time, and money are invested.

The specific procedure of nucleic acid extraction depends largely on the type of sample to be processed but generally consists of three steps: disintegration of cells or tissues (cell lysis), inactivation of intracellular nucleases, and separation of nucleic acids from other cellular components. RNA extraction is not always a simple process; however, since it is less stable than DNA, and the presence of pollutants such as RNAase, proteins, polysaccharides, and genomic DNA can complicate procedures [28]. Additionally, it has been reported that the presence of these contaminants may interfere with the amplification of nucleic acids [29]. Countless protocols now exist for obtaining nucleic acids that range from inexpensive home-brew protocols to commercial kits to complete automation. Each laboratory has generally optimized a few commonly used techniques, and their use is dictated on the time and money available for each research project.

In this chapter, we focus on the extraction and purification of DNA and RNA in freshwater microalgae, which is based on our experiences acquired over the better part of the last decade. Microalgae have attracted world-wide interest in the field of biotechnology due to their current and potential products of commercial interest such as biofuel, nutrients, food additives, and drugs [30].

3.2.1.1 Materials

3.2.1.1.1 Biological material

Approximately, 100–500 mg of microalgal biomass is necessary to achieve the best results. Given the amount of biomass, a large and active microalgae culture is required. Harvest must be carried out at 4°C and stored immediately at −80°C. In the case of RNA extraction, it is advisable to utilize liquid nitrogen to avoid RNA degradation.

3.2.1.1.2 Material and equipment of laboratory

For the extraction and purification of DNA, verify that materials and equipment are available (**Table 3**).

3.2.1.2 DNA extraction and purification

Extraction consists in the isolation of the total dissolved genomic DNA. The first step is the rupture of the plasma membrane in the case of animal cells, and the cell wall in the case of plant cells, managing to release the DNA, and the second is its precipitation. After these extraction steps are finished, agarose gel electrophoresis allows to visualize the genomic DNA bands and quantify approximately the size of the DNA obtained, by direct comparison with a marker whose

Materials	Equipment
Micropipettes of variable volume: 0.5–10, 10–100 and 100–1000 μL Eppendorf of 0.2, 1.5, and 2.0 mL Mortar and pestle Parafilm Magnetic stirrer Wash bottle or Pizetas Plastic tips with and without filter of 1–10, 20–200, 100–1000 μL Graduated Test Tubes of 25, 50, 100, 500, and 1000 mL Beakers of 50, 250, 500, and 1000 mL Flasks of 50, 250, and 500 mL	Autoclave (Yamato SM 510) Stove (Ecocell 111) Water bath (Labnet) Centrifuge (Hettich) Microcentrifuge (Spectrafuge Labnet) Water distiller (Barnstead Fistreem III Glass Still) Analytical balance (Sartorius) Gel Imaging System (BiodocAnalyze Biometra) Power source and horizontal electrophoresis system Spectrophotometer UV/Vis nanodrop 2000c pH-metro (Thermo Scientific) Water purifier (EASY pure RoDi Ultrapure) Vórtex-T Genie 2 (Scientific Industries) Dry block heater (Labnet).

Table 3.
Material and equipment of laboratory for the extraction and purification of DNA.

band size is previously known, and the amount based on the intensity of the band in the gel. Visualization is possible thanks to the use of fluorescence emission markers under UV light. There are various ways to extract DNA. Therefore, depending on the nature of the species, the most suitable total DNA isolation protocol is selected.

3.2.1.2.1 Reagents and solutions

For the visualization of DNA, verify that the following are available (**Table 4**).

3.2.1.2.2 Methodology

Approximately 100 mL of liquid biomass from culture harvested by centrifugation at 1900 × *g* for 10 min was used. Genomic DNA was extracted using a modified version of the CTAB method as described by Doyle and Doyle [31]. Briefly, microalgae cells were completely ground by hand using a mortar and pestle containing 50 mg of sterilized sand and 3 mL of extraction buffer (300 mM Tris-HCl pH 8.0, 50 mM EDTA, 2 M NaCl, 2% cetyltrimethylammonium bromide, 3% polyvinylpyrrolidone (MW

Reagents and solutions
Sterilized sand.
Extraction buffer: Tris-HCl 300 mM, pH 8.0, ethylenediamine tetraacetic acid (EDTA) 50 mM, NaCl 2 M, cetyltrimethylammonium bromide to 2, 3% polyvinylpyrrolidone (MW 40,000), and 2% de β-mercaptoethanol.
Phenol/chloroform/isoamyl alcohol (25:24:1, v/v).
Isopropanol
70% alcohol
RNase treated sterilized water
Absolute ethanol
TE buffer: Tris-HCl 10 mM, pH 8.0, EDTA 1 mM
Agarose gel
Ethidium bromide

Table 4.
Reagents and solutions for the extraction and purification of DNA.

40,000) and 2% β-mercaptoethanol). Homogenized cells were incubated at 70°C for 30 min with gentle inversion every 2 min. An equal volume of phenol/chloroform/isoamyl alcohol (25:24:1, v/v) was then added followed by centrifugation at 20,000 × *g*. The aqueous supernatant was transferred to a new microtube, and an equal volume of chloroform/isoamyl alcohol (24:1, v/v) was added followed by centrifugation at 20,000 × *g*. The aqueous supernatant was transferred to a new microtube and mixed with 0.1 volumes of 3 M sodium acetate (pH 5.2), and an equal volume of chilled isopropanol was added to precipitate the DNA that was pelleted by centrifugation at 15,000 × *g*. The DNA pellet was washed with 70% alcohol, air dried, and then dissolved in 100 μL of sterilized water treated with RNase A at 40°C for 30 min and then extracted with chloroform/isoamyl alcohol and DNA precipitation with absolute ethanol. Finally, the air-dried DNA pellet was dissolved in TE buffer (10 mM Tris-HCl pH 8.0, 1 mM EDTA) and stored at −20°C. Once the purified microalgal DNA has been obtained, the following quality control methods are recommended.

Electrophoretic analysis allows nucleic acid molecules to be separated by size that is dependent on the density of the matrix used (19–21). High molecular weight DNA should produce a single, bright band with little to no smearing. For microalgae, samples were resolved by standard gel electrophoresis using 1.2% agarose gels [32] stained with ethidium bromide and visualized under ultraviolet light.

Spectrophotometric analysis allows the determination of the concentration and purity of extracted DNA [33]. It is important to identify low quality or contaminated samples early in the process as common contaminates such as proteins, phenols, polyphenols, and carbohydrates, can negatively affect downstream procedures. Absorbance measurements are commonly conducted at 230, 260, and 280 nm, because carbohydrates and polyphenols absorb at 230 nm, nucleic acids at 260 nm, and proteins at 280 nm. To verify the quality and quantity of genomic DNA were evaluated by standard OD measurement [32] using a Nanodrop 2000 UV-Vis spectrophotometer (Thermo Fisher Scientific, Wilmington, DE, USA).

3.2.2 RNA extraction and purification

To perform studies at the molecular level, fast, simple, economical, reproducible, and high-performance protocols must be available for the extraction and purification of high-quality total RNA [34]. To cover these gaps in this chapter of the book, a standardized protocol is described to purify the total RNA of Amazonian oilseed microalgae, considering that the purity and integrity of the total RNA is essential to study the genetic expression at the level of individual genes or at the transcriptomic level.

3.2.2.1 Reagents and solutions

For the extraction and purification of RNA, verify that the following materials and equipment are available (**Table 5**).

3.2.2.2 Methodology

The most important process is to break the cell wall of the microalgae and for this you need to freeze the cells with liquid nitrogen and crush it with mortar and pestle. If liquid nitrogen is not available, crushing can be done with sterile sand, which is the case in our experience. Subsequently, the integrity of RNA must be ensured, by removing the other cellular components present in the cell, with the help of extraction buffers. As to our experience with the standardization of the RNA extraction protocol, the procedure was as follows.

Reagents and solutions
Hexadecyltrimethylammonium bromide (CTAB)
Tris base
Ethylenediaminetetraacetic acid (EDTA)
Sodium chloride (NaCl)
Polyvinylpyrrolidone (PVP)
Sodium dodecyl sulfate (SDS)
Sarkosyl, β-mercaptoetanol
Activated carbon
Absolute ethanol
Ethanol 70%
Lithium chloride (LiCl)
Sodium hydroxide (NaOH)
Hydrochloric acid (HCl)
Diethyl pyrocarbonate (DEPC)
Proteinase K
Chloroform
Isoamyl alcohol
Ethidium bromide
Boric acid
Formamide
Agarose
Bromophenol blue
H_2O ultrapure
Hydrogen peroxide
Sodium hypochlorite

Table 5.
Reagents and solutions for the extraction and purification of RNA.

In a mortar previously cooled to −20°C for 30 min, 0.25 g of microalgal biomass is added, add 2.5 mL of the extraction buffer [Tris-HCl 300 mM pH 8,0, EDTA 100 mM, NaCl 2 M, CTAB 2.25%, SDS 0.75%, Sarkosyl 0.13% and PVP 3%, water treated with DEPC (up to ~ 90% of the total volume), measure and adjust the pH (NaOH or HCl) and autoclave] and 100 μL of 2-mercaptoethanol, 20 μL of proteinase K (10 mg/mL), 50 μL of activated carbon (10 mg/mL), and 100 mg of sterile sand. Thoroughly crush the samples for ~8 min and transfer it to a 2 mL microtube, add chloroform:isoamyl alcohol (24:1) in a ratio of 1:1 v/v, homogenize in the vortex and centrifuge at 21,000 × *g* per 10 min at 4°C. Put the supernatant (700 μL) in a microtube, add 0.3 volumes of absolute ethanol, gently homogenize by inversion 8–10 times. Then, add equal volume of chloroform:isoamyl alcohol (24:1), homogenize in the vortex, and centrifuge at 21,000 × *g* for 10 min at 4°C. Transfer the supernatant to a microtube and repeat the previous step. To precipitate the RNA, transfer the supernatant to a microtube, add lithium chloride (LiCl) until a final concentration of 2.5 M is achieved, incubate at −20°C for 2 h or −80°C for 1 h, and centrifuge at 23,000 × *g* for 30 min at 4°C. Discard the supernatant, resuspend the RNA precipitate with 500 μL of 2 M LiCl, incubate at −20°C for 10 min, and centrifuge at 23,000 × *g* for 10 min at 4°C, repeat this step once more. Then wash the precipitated RNA by resuspending it with 500 μL of 70% ethanol and then with 500 μL of absolute ethanol, interspersing with centrifugation steps at 21,000 × *g* for 8 min at 4°C. Finally, discard the absolute ethanol, dry the precipitated RNA in a Thermo block at 45°C for 5 min, resuspend with 30–60 μL of TE buffer (10 mM Tris pH 8.0, 1 mM EDTA), and to stock at −80°C. Once the purified microalgal RNA is obtained, it is recommended to perform complementary analyzes:

- Electrophoretic analysis: it is a method that allows to determine the integrity of the purified RNA, which consisted in putting in a 0.2 mL microtube 5 μL of the total purified RNA and 5 μL of the seeding buffer (contains 6% SDS, 0.5% bromophenol blue, 0.5% ethidium bromide, 10 μM EDTA, and 95% formamide), homogenize them by pipetting 3–5 times and then incubate in a water bath at 70°C for 10 min and cool on ice for 5 min, centrifuge for 5 s at room temperature, sow the treated sample on the agarose gel and perform the electrophoretic run for 30–60 min at 100 volts, and finally, observe the RNA bands in the photo documentation system.

- Spectrophotometric analysis: this analysis allows us to determine the concentration and purity of the purified total RNA [33]. The Nanodrop 2000c was used to perform the spectrophotometric analysis, the first step is to select the nucleic acid reading (RNA) option, subsequently, the blank reading (bleaching) is carried out by placing 2 μL of the TE buffer on the sensor and the same volume of the purified RNA is read, subsequently, the absorbance results are recorded at 260, 280, the quality ratios (A_{260}/A_{230} and A_{260}/A_{280}) and the RNA concentration in ng/μL.

3.2.3 De novo assembly and annotation of the microalgae transcriptome

Transcriptomics studies the level of expression of all transcribed genes (mRNAs, non-coding RNAs, small RNAs, etc.) in a cell or tissue. Transcriptomic analysis is often focused on the study of mRNA populations, which are molecules with encoded information for the synthesis of all proteins. Therefore, with this type of analysis, you can reconstruct the multiple metabolic pathways and physiological processes that are active, because these proteins are those that participate in fundamental processes as diverse and complex as transport, intracellular signaling, defense, enzymatic catalysis, among others.

There is a great variety of techniques used in transcriptomics, which allow quantifying millions of RNA molecules at the same time; this thanks to the recent development of massive sequencing technologies and the implementation of different bioinformatics tools, and it is now possible to analyze the transcriptome of any organism, even species that do not have sequenced genomes. Massive sequencing technologies differ in their details, but typically consist of three similar stages. These stages are: (1) template preparation, (2) clone amplification, and (3) cyclic rounds of parallel and massive sequencing.

However, Illumina sequencers are the ones that generate the most data at low costs, so these machines currently dominate the market [35]. The sequencing process with these platforms consists in the clonal amplification of DNA fragments linked to adapters on the surface of a glass sheet. In all Illumina models, the overall error rate is less than 1% and the most common type of error is substitution [36, 37]. Once the data are obtained, which are millions of short sequences (100–200 bp), they must be pre-processed, assembled, and finally their functional annotation.

Preprocessing or "cleaning" consists in eliminating erroneous sequences, low quality sequences, and technical sequences (adapters, primers, etc.). This is a process where bioinformatics programs are used, for example, the frequently used bioinformatics program for this cleaning process is the Trimmomatic and CutAdapt [38, 39] are two commonly used programs.

Subsequently, the sequences are assembled, that is to say, correctly join the short DNA fragments (100–200 bp) to assemble the thousands of transcripts that commonly have sizes from 1000 to 3000 bp. To achieve this purpose, there are different bioinformatics programs such as ABySS, ALLPATHS, SHORTY, Velvet, Oases, SOAPdenovo, Trinity, CAP3, among others [40, 41].

Finally, after assembling the transcripts, they proceed to their functional annotation, that is, based on their homology with sequences, previously reported and stored in the biological databases, the molecular and cellular function of each of the transcripts is deduced. For this bioinformatic analysis, the scientific community frequently uses the Blast2GO program [42]. This program also allows us to rebuild all metabolic pathways.

4. Biotechnological potential for biofuel and nutraceutical production

Climate change, the growing global energy demand, the increase in the cost and depletion of fossil fuels, high rates of child malnutrition, among others are problems that require medium- and long-term solutions. To mitigate these problems, in the different countries, several legal devices are available with the purpose of promoting and diversifying the energy matrix based on renewable sources such as biofuels and are considered alternatives to venturing into the production of functional (nutraceutical) foods, as it happens in our country. However, many times due to low domestic production volumes (approx. 20% of demand), these measures are forcing the importation of both biofuels and various food products from other countries. In the case of biofuels in our country, its production depends mainly on crops of large areas of oil plants such as oil palm, white pine nut, and other species, thus contributing to a loss of biodiversity and a decrease in ecosystem services [43]. In the case of functional or nutraceutical foods, it is common to find them under different commercial presentations (syrups, capsules, powders, etc.) imported from abroad that do not allow increasing the supply in each country. Therefore, to minimize these negative impacts, it is urgently necessary to have other alternative sources of biofuel and nutraceutical production.

An excellent alternative is offered by microalgae, which are organisms that play a key role in aquatic ecosystems because of their photosynthetic capacity, and that in aerobic conditions can fix carbon dioxide just like plants [44]. Among the compounds of most interest obtained from microalgae, carotenoids, biodiesel, phycobiliproteins, lipids, polysaccharides, and compounds with biological activity stand out. Currently, researchers from around the world have demonstrated the potential of several microalgae species (*Chlorella minutissima*, *Thalassiosira fluviatilis*, and *T. pseudonana*) [45]. Our team has carried out studies with *Ankistrodesmus* sp., using different culture media, registering a higher total lipid production of 263.6 mg/g dry biomass with 68.56 ± 2.35 palmitic acid under physiological stress conditions [46], as raw material to produce biodiesel.

Likewise, it is demonstrated that the determination of the fatty acid profile is important to assess the potential for the quality and production of biodiesel in microalgae, since the quality of biodiesel depends on the type of fatty acid present in each microalgal cell. As also, it is necessary to know the number of cetones, oxidative stability and could-flow, which depend on the length of the hydrocarbon chains of saturated and unsaturated fatty acids (UFA) [47]. Studies carried under conditions of nitrogen limitation in the culture, the microalgae showed variation in the composition of saturated fatty acids (SFA) with monounsaturated fatty acids (MUFA). And that 20–30% are SFA and 69–80% are UFA, the range of SFA/UFA being in the microalgae evaluated between 0.25 and 0.45 [17]. Also, it was reported that palmitic (C16:0), oleic (C18:1n-9), linoleic (C18:2n-6), and γ-linolenic acids (C18:3n-3) were the most abundant fatty acids and showed greater variations within and between the species studied [46].

Regarding the nutraceutical potential of microalgae, it can be mentioned that in the early 50s, humans began using microalgae in their diets, mainly as dietary supplements (protein and vitamin, in the form of powder, capsules, pills or tablets). These are usually incorporated into foods such as pasta, cookies, bread, candies, yogurts, soft drinks, among others. It is currently estimated that approximately 30% of the microalgae produced in the world is used in human nutrition due to its high protein content [48]. Therefore, *Arthrospira*-based nutraceutical compounds with anti-cancer protection properties are being commercialized due to their high content of β-carotene [12], the prevention of malnutrition in vitamin A and blood sugar levels, the stimulation of certain prostaglandins, prevention of degenerative diseases, and accelerated wound healing [48].

Recent research has shown that human consumption of microalgae is limited to few species due to strict control over food safety, commercial factors, market demand, and specific preparation for consumption, since the most prominent genera are *Chlorella*, *Arthrospira,* and *Dunaliella*, which are marketed as a food supplement [49, 50]. Studies on nutraceutical substances have been carried out almost exclusively in algae and microalgae, identifying more than 600 naturally occurring carotenoids in plants, animals, and fungi, of which 400 have been isolated and characterized [51], but of these, only a small number are commercially used among them B-carotene and astaxanthin, being only two species of marine microalgae recognized commercial sources of carotenoids: *Dunaliella salina* and *Haematococcus pluvialis* [52]. Recently, our team has registered that *Spirulina maxima* presented 269.54 ± 0.021 μg/g dry mass of β-carotene and *Scenedesmus* sp. presented 15.29 ± 0.01 μg/g dry mass.

In addition, microalgae have the ability to increase the nutritional content of traditional foods and even positively affect both human health and animal health. This is due to its original chemical composition, since the high protein content in some species is one of the main reasons to consider them as a source of unconventional protein. In turn, the amino acid profile of almost all microalgae is more favorable than compared to conventional sources. The carbohydrates in the microalgae can be found in the form of starch, glucose, sugars or other polysaccharides. Its digestibility is high, so there is no limitation of use in food preparations for both animals and humans [53, 54].

Many microalgal species are marketed for their medicinal value, as *Chlorella* protects against kidney failure and promotes the growth of *Lactobacillus*. In addition, it favors the decrease in blood cholesterol concentration [49, 55], increases the level of antioxidants in the body [56], and stimulates collagen synthesis, supporting tissue regeneration and wrinkle reduction [50]. *Dunaliella* is marketed for its high content of β-carotene [49, 55]. *Haematococcus pluvialis* is the only microalgae that have been commercially exploited for the production of astaxanthin [57, 58]. Today, microalgae are used to feed livestock, and human consumption is increasing, particularly in food supplements [59]. Algae dietary supplements can be particularly useful for supporting some diets.

Therefore, the microalgae native to the Peruvian Amazon have been characterized in order to determine their application and use from a biotechnological point of view, due to the abundant source of antioxidants that microalgae possess, which makes them candidates for use in biological processes (aging), as well as in the degenerative changes of different organs [60]. In addition, the beneficial effects of nutraceuticals have been attributed to polyphenols, polyunsaturated fatty acids, terpenes, chlorophyll, and accessory pigments of the photosynthetic apparatus of microalgae and are therefore considered excellent sources of proteins, small peptides, and amino acids that contribute to providing high amount of phosphorus to fight hypertension and lower cholesterol levels, help in the formation and regeneration of blood cells together with iron and stimulate the release of substances that control satiety.

In that sense, nutraceuticals currently play an important role in the daily life of a large number of people, being more prevalent in those with chronic diseases and high overall mortality. The various reasons for promoting the consumption of these products vary according to age, sex, nationality, customs, and existence of comorbidities [61]. Due to these interesting qualities, in recent years, research has been carried out with various species of algae for various biotechnological applications. As a result, today the biotechnology of microalgae has gained relevance due to the wide range of applications derived from its use, from biomass production for food, its use in aquaculture, as biofertilizer, to obtaining products of therapeutic or industrial value. Therefore, the success of microalgal biotechnology lies in choosing the correct species with relevant properties that, under specific culture conditions, produce the compounds of interest [48]. The applications range from the production of simple biomass for animal feed or for the production of valuable products for ecological applications. Due to the enormous biodiversity of microalgae and recent developments, this group of organisms represents one of the most promising sources for new products and applications. With the investigation of sophisticated crops, strain search and microalgal biotechnology, the demands of the food and pharmaceutical industries can be met [53, 54].

5. Conclusions

The native microalgae of the Peruvian Amazon have biotechnological potential and are therefore promising for the production of both biodiesel and nutraceuticals. Various species of microalgae were identified, isolated, cultured, and characterized using biochemical, nutraceutical, and molecular techniques, the isolation stage being the starting point to achieve various biotechnological applications. *Ankistrodesmus* sp. is one of the microalgae with potential for the production of biodiesel and microalgae such as *Haematococcus pluvialis*, *Scenedesmus* sp., and *Chlorella* sp., among others demonstrated a high potential for nutraceutical production. The stress conditions to which microalgae are subjected are being a determining factor for the production of biodiesel and nutraceuticals.

Acknowledgments

This work was supported by the Scientific University of Peru (UCP) and the National Fund for Technological Scientific Development and Technological Innovation (FONDECYT) with grant contract No. 213-2015-FONDECYT.

Conflict of interest

The authors declare no conflict of interest.

Author details

Marianela Cobos[1*], Juan C. Castro[2*], Jae D. Paredes[1], Sheyla Pérez[3], J. Dylan Maddox[4,5], Segundo L. Estela[1], Hicler N. Rodríguez[2], Alvaro B. Tresierra[3], Juan R. Saldaña[3], Jorge L. Marapara[2], Pedro M. Adrianzén[2] and Rosana Gonzales[3]

1 Laboratory of Biotechnology and Bioenergetics, Scientific University of Peru (UCP), Iquitos, Peru

2 Specialized Unit of Biotechnology, Research Center of Natural Resources of the Amazon (CIRNA), National University of the Peruvian Amazon (UNAP), Iquitos, Peru

3 Laboratory of Basic Sciences, Scientific University of Peru (UCP), Iquitos, Peru

4 Pritzker Laboratory for Molecular Systematics and Evolution, The Field Museum of Natural History, Chicago, IL, USA

5 Environmental Sciences, American Public University System, Charles Town, WV, USA

*Address all correspondence to: mcobos@ucp.edu.pe and juan.castro@unapiquitos.edu.pe

IntechOpen

© 2019 The Author(s). Licensee IntechOpen. This chapter is distributed under the terms of the Creative Commons Attribution License (http://creativecommons.org/licenses/by/3.0), which permits unrestricted use, distribution, and reproduction in any medium, provided the original work is properly cited.

References

[1] Lee RE. Phycology. 4th ed. Cambridge: Cambridge University Press; 2008. p. 615

[2] Menetrez MY. An overview of algae biofuel production and potential environmental impact. Environmental Science & Technology. 2012;**46**: 7073-7085. DOI: 10.1021/es300917r

[3] Lam MK, Lee KT. Microalgae biofuels: A critical review of issues, problems and the way forward. Biotechnology Advances. 2012;**30**:673-690. DOI: 10.1016/j.biotechadv.2011.11.008

[4] Ho S-H, Chen C-Y, Lee D-J, Chang J-S. Perspectives on microalgal CO_2-emission mitigation systems—A review. Biotechnology Advances. 2011;**29**:189-198. DOI: 10.1016/j.biotechadv.2010.11.001

[5] Smith VH, Sturm BSM, deNoyelles FJ, Billings SA. The ecology of algal biodiesel production. Trends in Ecology & Evolution. 2010;**25**:301-309. DOI: 10.1016/j.tree.2009.11.007

[6] Li Y, Horsman M, Wu N, Lan C, Dubois N. Biocatalysts and bioreactor design. Biotechnology Progress. 2008;**24**:815-820

[7] Stephens E, Ross I, Mussgnug JH, Wagner LD, Borowitzka MA, Posten C, et al. Future prospects of microalgal biofuel production systems. Trends in Plant Science. 2010;**15**:554-564

[8] Duong VT, Li Y, Nowak E, Schenk PM. Microalgae isolation and selection for prospective biodiesel production. Energies. 2012;**5**:1835-1849. DOI: 10.3390/en5061835

[9] Richmond A, Qiang H. Handbook of Microalgal Culture: Applied Phycology and Biotechnology. 2nd ed. Hoboken, NJ, USA: John Wiley & Sons, Ltd; 2013. pp. 719. DOI: 10.1002/9781118567166

[10] Derelle E, Ferraz C, Rombauts S, Rouzé P, Worden AZ, Robbens S, et al. Genome analysis of the smallest free-living eukaryote *Ostreococcus tauri* unveils many unique features. Proceedings of the National Academy of Sciences of the United States of America. 2006;**103**:11647-11652

[11] Jinkerson RE, Radakovits R, Posewitz MC. Genomic insights from the oleaginous model alga *Nannochloropsis gaditana*. Bioengineered. 2013;**4**:37-43. DOI: 10.4161/bioe.21880

[12] Richmond A. Handbook of Microalgal Culture: Biotechnology and Applied Phycology. 1st ed. Hoboken, NJ, USA: Blackwell Publishing Ltd; 2004. p. 556. DOI: 10.1002/9780470995280

[13] Rosenberg JN, Oyler GA, Wilkinson L, Betenbaugh MJ. A green light for engineered algae: Redirecting metabolism to fuel a biotechnology revolution. Current Opinion in Biotechnology. 2008;**19**:430-436. DOI: 10.1016/j.copbio.2008.07.008

[14] NanoDrop 2000/2000c Spectrophotometer. V1.0 User Manual. Available from: http://rai.unam.mx/manuales/lgen_ND2000_user-manual.pdf

[15] Yu X, Zhao P, He C, Li J, Tang X, Zhou J, et al. Isolation of a novel strain of *Monoraphidium* sp. and characterization of its potential application as biodiesel feedstock. Bioresource Technology. 2012;**121**:256-262. DOI: 10.1016/j.biortech.2012.07.002

[16] Greenspan P, Mayer EP, Fowler SD. Nile red: A selective fluorescent stain for intracellular lipid droplets. The Journal of Cell Biology. 1985;**100**:965-973. DOI: 10.1083/jcb.100.3.965

[17] Cobos M, Paredes JD, Castro JC. Inducción de la producción de lípidos totales en microalgas sometidas a estrés nutritivo. Acta Biológica Colombiana. 2016;**21**:17-26. DOI: 10.15446/abc.v21n1.47439

[18] Hartree EF. Determination of protein: A modification of the Lowry method that gives a linear photometric response. Analytical Biochemistry. 1972;**48**:422-427. DOI: 10.1016/0003-2697(72)90094-2

[19] DuBois M, Gilles KA, Hamilton JK, Rebers PA, Smith F. Colorimetric method for determination of sugars and related substances. Analytical Chemistry. 1956;**28**:350-356. DOI: 10.1021/ac60111a017

[20] Horwitz W, Chichilo P, Reynolds H. Official Methods of Analysis of the Association of Official Analytical Chemists. 11th ed. Washington, DC, USA: Association of Official Analytical Chemists; 1970

[21] Cobos M, Paredes JD, Maddox JD, Vargas-Arana G, Flores L, Aguilar CP, et al. Isolation and characterization of native microalgae from the Peruvian Amazon with potential for biodiesel production. Energies. 2017;**10**:224. DOI: 10.3390/en10020224

[22] Gonzales A, Magallanes J. Efecto del estrés nutritivo (Nitrógeno y Fierro) sobre la producción de lípidos en tres cepas de microalgas clorófitas. In: Encuentro de Jóvenes Investigadores. México D.F: CONACYT; 2012. pp. 23-25

[23] Thompson GA. Lipids and membrane function in green algae. Biochimica et Biophysica Acta (BBA)—Lipids and Lipid Metabolism. 1996;**1302**:17-45. DOI: 10.1016/0005-2760(96)00045-8

[24] Guschina IA, Harwood JL. Lipids and lipid metabolism in eukaryotic algae. Progress in Lipid Research. 2006;**45**:160-186. DOI: 10.1016/j.plipres.2006.01.001

[25] Khozin-Goldberg I, Cohen Z. The effect of phosphate starvation on the lipid and fatty acid composition of the fresh water eustigmatophyte Monodus subterraneus. Phytochemistry. 2006;**67**:696-701. DOI: 10.1016/j.phytochem.2006.01.010

[26] Bigogno C, Khozin-Goldberg I, Cohen Z. Accumulation of arachidonic acid-rich triacylglycerols in the microalga *Parietochloris incisa* (Trebuxiophyceae, Chlorophyta). Phytochemistry. 2002;**60**:135-143. DOI: 10.1016/S0031-9422(02)00037-7

[27] Lewin B, Krebs J, Kilpatrick ST, Goldstein ES. Lewin's GENES X. Sudbury, MA, USA: Jones & Bartlett Publishers; 2011. p. 930

[28] Zakaria Z, Umi SH, Mokhtar SS, Mokhtar U, Zaiharina MZ, Aziz ATA, et al. Methodology an alternate method for DNA and RNA extraction from clotted blood. Genetics and Molecular Research. 2013;**12**:302-311. DOI: 10.4238/2013

[29] Vermeulen J, De Preter K, Lefever S, Nuytens J, De Vloed F, Derveaux S, et al. Measurable impact of RNA quality on gene expression results from quantitative PCR. Nucleic Acids Research. 2011;**39**:e63-e63. DOI: 10.1093/nar/gkr065

[30] Bowler C, Allen AE, Badger JH, Grimwood J, Jabbari K, Kuo A, et al. The *Phaeodactylum* genome reveals the evolutionary history of diatom genomes. Nature. 2008;**456**:239-244. DOI: 10.1038/nature07410

[31] Doyle JJ, Doyle JL. A rapid DNA isolation procedure from small quantities of fresh leaf tissue. Phytochemical Bulletin. 1987;**19**:11-15

[32] Cornejo A, Serrato A, Rendón B, Rocha MG. Herramientas Moleculares

Aplicadas en Ecología: Aspectos Teóricos y Prácticos. 1st ed. Tlalpan, México, DF: Secretaría de Medio Ambiente y Recursos Naturales; 2014. p. 274

[33] Campbell NA, Mitchell LG, Reece JB. Biología: Conceptos y Relaciones. 3rd ed. San Francisco, CA, EUA: Pearson Educación; 2001

[34] Castro JC, Navarro GA, Cerdeira LA, Cobos M. Purificación de ADN genómico a partir de hojas, raíces y neumatóforos de Mauritia flexuosa "aguaje". Revista de la Sociedad Química del Perú. 2013;**79**:209-217

[35] Cloonan N, Forrest ARR, Kolle G, Gardiner BBA, Faulkner GJ, Brown MK, et al. Stem cell transcriptome profiling via massive-scale mRNA sequencing. Nature Methods. 2008;**5**:613-619. DOI: 10.1038/nmeth.1223

[36] Buermans HPJ, den Dunnen JT. Next generation sequencing technology: Advances and applications. Biochimica et Biophysica Acta (BBA)—Molecular Basis of Disease. 2014;**1842**:1932-1941. DOI: 10.1016/j.bbadis.2014.06.015

[37] Ansorge WJ. Next-generation DNA sequencing techniques. New Biotechnology. 2009;**25**:195-203. DOI: 10.1016/j.nbt.2008.12.009

[38] Bolger AM, Lohse M, Usadel B. Trimmomatic: A flexible trimmer for illumina sequence data. Bioinformatics. 2014;**30**:2114-2120. DOI: 10.1093/bioinformatics/btu170

[39] Martin M. Cutadapt removes adapter sequences from high-throughput sequencing reads. EMBnet.journal. 2011;**17**:10-12. DOI: 10.14806/ej.17.1.200

[40] Miller JR, Koren S, Sutton G. Assembly algorithms for next-generation sequencing data. Genomics. 2010;**95**:315-327. DOI: 10.1016/j.ygeno.2010.03.001

[41] Nagarajan N, Pop M. Sequence assembly demystified. Nature Reviews Genetics. 2013;**14**:157-167. DOI: 10.1038/nrg3367

[42] Conesa A, Gotz S, Garcia-Gomez JM, Terol J, Talon M, Robles M. Blast2GO: A universal tool for annotation, visualization and analysis in functional genomics research. Bioinformatics. 2005;**21**:3674-3676. DOI: 10.1093/bioinformatics/bti610

[43] Cobos M, Castro JC, Del Águila J, Bosmediano M. Identificación de microalgas oleaginosas en el área de concesión para conservación, cuenca alta del río itaya. Loreto-Perú. Ciencia Amazónica. 2012;**2**:162-168. DOI: 10.22386/ca.v2i2.42

[44] Kumar K, Dasgupta CN, Nayak B, Lindblad P, Das D. Development of suitable photobioreactors for CO_2 sequestration addressing global warming using green algae and cyanobacteria. Bioresource Technology. 2011;**102**:4945-4953. DOI: 10.1016/j.biortech.2011.01.054

[45] Neto CJD, Letti LAJ, Karp SG, Vítola FMD, Soccol CR. Chapter 18—Production of biofuels from algae biomass by fast pyrolysis. In: Pandey A, Chang J-S, Soccol CR, Lee D-J, Chisti Y, editors. Biofuels from Algae. 2nd ed. Cambridge, MA, USA: Elsevier; 2019. pp. 461-473. DOI: 10.1016/B978-0-444-64192-2.00018-4

[46] Cobos M, López LC, Paredes JD, Maddox JD, Vargas-Arana GE, Flores L, et al. Media composition affects the fatty acids profiles of three oleaginous microalgae from the Peruvian Amazon. Asian Journal of Microbiology and Biotechnology. 2016;**1**:14-24

[47] Knothe G. Dependence of biodiesel fuel properties on the structure of fatty acid alkyl esters. Fuel Processing Technology. 2005;**86**:1059-1070. DOI: 10.1016/j.fuproc.2004.11.002

[48] Pulz O, Gross W. Valuable products from biotechnology of microalgae. Applied Microbiology and Biotechnology. 2004;**65**:635-648. DOI: 10.1007/s00253-004-1647-x

[49] Brennan L, Owende P. Biofuels from microalgae—A review of technologies for production, processing, and extractions of biofuels and co-products. Renewable and Sustainable Energy Reviews. 2010;**14**:557-577. DOI: 10.1016/j.rser.2009.10.009

[50] Spoloare P, Joannis-Cassan C, Duran E, Isambert A. Commercial applications of microalgae. Journal of Bioscience and Bioengineering. 2006;**101**:87-96. DOI: 10.1263/jbb.101.87

[51] Adarme-Vega TC, Lim DKY, Timmins M, Vernen F, Li Y, Schenk PM. Microalgal biofactories: A promising approach towards sustainable omega-3 fatty acid production. Microbial Cell Factories. 2012;**11**:96. DOI: 10.1186/1475-2859-11-96

[52] Quintero E, Arredondo BO, Flores TE, et al. Selección de un medio de cultivo a nivel laboratorio para el cultivo de la cianobacteria spirulina (*Arthrospira maxima*) y cuantificación de los nutracéuticos. Investigación y Desarrollo en Ciencia y Tecnología de Alimentos. 2016;**1**:532-537

[53] Becker EW. Micro-algae as a source of protein. Biotechnology Advances. 2007;**25**:207-210. DOI: 10.1016/j.biotechadv.2006.11.002

[54] Sousa I, Gouveia L, Batista AP, Raymundo A, Bandarra NM. Chapter 2.Microalgae in novel food products. In: Papadopoulos KN, editor. Food Chemistry Research Developments. New York, USA: Nova Science Publishers, Inc; 2008. pp. 1-37

[55] Segovia-Mendoza M, Díaz L, González-González ME, Martínez-Reza I, García-Quiroz J, Prado-Garcia H, et al. Calcitriol and its analogues enhance the antiproliferative activity of gefitinib in breast cancer cells. The Journal of Steroid Biochemistry and Molecular Biology. 2015;**148**:122-131. DOI: 10.1016/j.jsbmb.2014.12.006

[56] Lee SH, Kang HJ, Lee H-J, Kang M-H, Park YK. Six-week supplementation with chlorella has favorable impact on antioxidant status in Korean male smokers. Nutrition. 2010;**26**:175-183. DOI: 10.1016/j.nut.2009.03.010

[57] Del Campo JA, García-González M, Guerrero MG. Outdoor cultivation of microalgae for carotenoid production: Current state and perspectives. Applied Microbiology and Biotechnology. 2007;**74**:1163-1174. DOI: 10.1007/s00253-007-0844-9

[58] Guerin M, Huntley ME, Olaizola M. Haematococcus astaxanthin: Applications for human health and nutrition. Trends in Biotechnology. 2003;**21**:210-216. DOI: 10.1016/S0167-7799(03)00078-7

[59] Rodolfi L, Zittelli GC, Bassi N, Padovani G, Biondi N, Bonini G, et al. Microalgae for oil: Strain selection, induction of lipid synthesis and outdoor mass cultivation in a low-cost photobioreactor. Biotechnology and Bioengineering. 2009;**102**:100-112. DOI: 10.1002/bit.22033

[60] Fernández-RojasB,Hernández-JuárezJ, Pedraza-Chaverria J. Nutraceutical properties of phycocyanin. Journal of Functional Foods. 2014;**11**:375-392. DOI: 10.1016/j.jff.2014.10.011

[61] Raposo MF, de Morais RMSC, de Morais AMMB. Health applications of bioactive compounds from marine microalgae. Life Sciences. 2013;**93**: 479-486. DOI: 10.1016/j.lfs.2013.08.002

Section 2

Microalgae as Nutrition

Chapter 4

Drying and Quality of Microalgal Powders for Human Alimentation

Fábio de Farias Neves, Mariana Demarco and Giustino Tribuzi

Abstract

The demand for natural foods with high protein content and functional properties is constantly growing in the last years. In this context, microalgae as *Spirulina* (*Arthrospira* spp.), *Chlorella* spp., *Haematococcus pluvialis*, *Dunaliella salina*, and others, assume a key role to diversify the offer of nutritious and functional ingredients and supplements. Microalgae are commercialized, mostly, as dried powders to facilitate their use as food ingredients and to allow easy transportation and long-term stability. Microalgal powder quality and storage stability depend mainly on drying method, packaging, and storage conditions. Most of the studies that approach the subject of microalgal drying evaluate the efficiency of the process and suitability for this raw material. However, studies that assess the effect of traditional and innovative drying methods on quality of microalgal powder for human consumption are rare in literature. In this chapter, the state of the art of drying processing technology for microalgae was reviewed, discussing the effect of dehydration on quality and stability of microalgal powders with potential use in human alimentation.

Keywords: microalgae, dry biomass, biomass quality, microalgal powder, functional supplements

1. Introduction

Microalgae are photosynthetic microscopic organisms that convert CO2 and water in biomass and O2. This group of organisms is very diverse and abundant around the globe. They occur most in freshwater and saltwater aquatic ecosystems but also at other environments [1]. The main groups are Cyanophyta, Chlorophyta, Ochrophyta, Dinophyta, Rhodophyta, Euglenophyta, Haptophyta, and Prymnesiophyta [2]. It is estimated that there are around 300,000 of microalgal species around the world [1].

Microalgae have a great ecological importance as they are primary producers contributing to a lot of food chains; they produce around 40–60% of the oxygen available on Earth atmosphere, convert inorganic nutrients in organic matter, and for millions of years have produced the oil that today economy is still dependent [1].

Also, microalgae have been used in different industries for decades. These microscopic organisms are produced in ponds or photobioreactors to be used directly as live feeds for aquaculture hatcheries or to be used in food industry as food supplements or source of nutrients and vitamins, in agriculture as biofertilizer,

IntechOpen

in pharmaceutical and cosmetic industry as raw material to extract specific molecules, and in other different biotechnology applications [3]. Also, microalgae have been used for environmental purposes as in tertiary wastewater treatment, as system for carbon fixation, and as raw materials to produce biofuels [4].

Despite all economic potential applications of microalgae and the great diversity of species, the microalgal industry is based mainly in few applications and species. The majority of microalgal biomass production is destined for food industry as food supplement. The main species belongs to the genus *Spirulina* also known as *Arthrospira*, *Chlorella*, *Haematococcus*, *Dunaliella*, and few others [5].

Spirulina (*Arthrospira*) is a cyanobacteria and the same as the green algae *Chlorella* (chlorophyte); both usually are sold as dry biomass. Meanwhile, the microalgae *Haematococcus pluvialis* and *Dunaliella salina* are usually used as source of the carotenoids astaxanthin and beta-carotene, respectively. However, some companies also sell the entire biomass and, even when extracting the pigments, need to dry the biomass previously [5].

To dry the microalgal biomass is the way microalgae are most commercialized because this method increases the product stability and durability, also allowing easy storage and transportation.

There are different dryers being used in microalgal industry. The main dryer used is the spray dryer [6], and in small spirulina farms, the utilization of ovens with forced ventilation is very common. However, freeze-dryer, drum dryer, natural sun dryer, and other processes to dry the biomass can be also used [6–8].

It is well known that, depending on the drying process method and conditions, there is a potential to lose quality of the microalgal biomass. For example, the microalgal properties change with the dry process, and for the food industry, the nutritional quality can decrease as proteins, lipids, pigments, and other nutrients are lost. In particular, functional components (i.e., phycocyanin from spirulina or astaxanthin from *H. pluvialis*) are very sensible to drying conditions, that is, time, temperature, and oxygen, among others.

Therefore, the purpose of this chapter is to provide a review of the common and innovative dry process technologies available for microalgal biomass, discuss the effect of dehydration on quality and stability of microalgal powders used in human alimentation, and then support the microalgal industry and researchers to choose the most suitable drying method for each different use of dried microalgae.

2. Preprocess in algae drying

Microalgae have been studied largely because they have an industrial importance with their bioproducts, such as lipids, carotenoids, etc. Also, their lipid content is considered a potential feedstock for biodiesel production [9, 10]. Some adversities found in preprocessing are the presence of rigid cell walls surrounding the algae cells and the biomass moisture that interferes in some extraction solvent performance [11]. It can be solved with chemical, mechanical, and biological means of cell wall disruption and can be used alone or in combined forms [9]. Also, in these cases, the step of dewatering is important, that is, with flocculation and centrifuge [12]. The proportions of microalgal biomass in cultivation are generally relatively low, being around 0.02–0.05% of dry biomass in raceway tanks and between 0.1 and 0.5% of dry biomass in tubular photobioreactors. This aspect, together with the size of microalgae, turns microalgal biomass separation very complex [13]. Ref. [14] presents that separation costs can be around 20 and 30% of the total production cost.

There are different ways to perform biomass separation. The most common processes are flocculation followed by sedimentation, centrifugation, and filtration. Sedimentation, considered the simplest option, can retain 85% of biomass, with the percentage of dry biomass around 3%, depending on the species used. However, this process requires significant additional space. The most efficient method for separating biomass is centrifugation, but it represents a significant increase in production costs, so it is widely used when the extracted product has high value. Filtration can be performed mainly to separate biomass from filamentous microalgae, but it is a slow process and in large-scale systems requires a large infrastructure [13].

Cell disruption is an alternative to cellular disintegration of many microorganisms, like bacteria, yeasts, and microalgae, and can be classified as mechanical and nonmechanical manner [15, 16]. Among the mechanical methods, there are high-pressure homogenization, ultrasonication, bead milling, autoclaving, lyophilization, and microwaving, being the first and second ones the widely used methods for laboratory-scale microalgal cell disruption. The nonmechanical methods involve lysing the cell wall with acids, alkalis, enzymes, or osmotic shocks [15, 17].

Some mechanisms involved to the cellular disruption are achieved: impingement of the cells on the hard surface of the valve seat and their impact on each other during collision, turbulence, viscous and high-pressure shear, pressure-drop-induced shear passing from the valve to the chamber, and sudden pressure drop caused by rapid release of gas bubbles within the cells [10, 15, 18].

Concerning low-energy input, chemical treatments have advantages, in addition to showing good scalability. However, they should be carefully selected and applied considering their bio-toxicity and reactivity to some compounds [9].

Physicochemical extraction process can cause thermal and/or chemical stresses inducing structural changes and denaturation/degradations of compounds, like astaxanthin isomers, and affecting significantly the product qualities, such as antioxidative activity, bioavailability, and purity [9]. To guarantee the extraction efficiency of astaxanthin, some operating conditions should be properly considered, like temperature and the use and minimization of less toxic chemicals [19].

3. Drying and microalgal quality

Drying of foods can be defined as a unit operation of water removal aiming to reduce moisture content and water activity and consequently stabilize foods by inhibiting the microbial grow and enzymatic activity and slowing chemical reactions [20]. Dried foods present advantages which are easy to store and transport and have long shelf life.

After separation and concentration, microalgal biomass has a high water content, presenting high perishability once it represents a good substrate to microbial grow and enzymatic activity if commercialized without any stabilization treatment. Stabilization of moist biomass by pasteurization is possible, but the prolongation of the shelf life is limited, refrigerate storage is needed [21], and degradation of functional components may occur if high temperatures are used. Thus, to extend shelf life and allow storage at room temperature, drying of microalgae biomass is generally considered an effective alternative. Many drying technologies can be potentially used to dry microalgae biomass as well as other foods with high viscosity. Different factors should be taken in consideration to choose the best drying method. In most cases the main factors that influence this choice are energy efficiency and installation and operation costs. However, if the dried microalgal biomass is produced to human alimentation, the preservation of nutritional and functional components of the biomass must be also considered [22, 23].

In the literature many drying methods applied to microalgae biomass are discussed. Most of them are related to the stabilization of biomass for nonhuman use such as oil extraction for biofuel industry or feed production. These methods are in general very effective on the point of view of energy efficiency and processing time; among them the following methods can be cited: rotary drying, solar drying, cross-flow and vacuum shelf drying and flash drying. The main challenge in these cases is the processing cost and energy requirement, but not much attention is given to degradation on functional and nutritional components [6, 24].

On the other hand, studies on the assessment of drying methods applied to microalgae to human use are not so common in literature. In **Table 1**, a summary of the different drying methods used to produce dried biomass with potential application in human alimentation is presented in which, together with the engineering of the drying process, the impact on nutritional and/or functional components was assessed.

One of the conditions that affect the choice of method and the drying performance is the initial moisture that the microalgal biomass presents. The algal biomass starts the drying process with initial moisture values between 55 and 88% (wet basis) in different dehydration processes [25–30].

Apart from the color change, the compound degradation, and the drying kinetics, the final moisture content can be one parameter to compare different drying methods and demonstrates how these methods can affect the sample. Based on the initial moisture and the chosen method, the drying can show high or low drying rates, can influence the velocity heat and water mass transfer through the samples, and in some conditions not allow the diffusion between the interior and the surface [26, 28]. In short, apart from the initial moisture that can influence the chosen method, some physical parameters can influence in velocity and water outlet, impacting in final moisture and in quality of the product. Further, some storage conditions, like light, temperature, water activity, oxygen concentration, relative humidity, existence of coating matrix, etc., are important parameters to study compound degradation and product shelf life [31, 32]. Assessing sorption isotherms, stability studies and DSC, the sample behavior during storage, and how the environmental factors influence these parameters is important to evaluate the effect of storage on quality retention of the final products [28, 32–34].

Many different methods were presented in **Table 1**; however, similar drying technologies have the same principle with few modifications of processing parameters or equipment design but are named with different denominations, according to the authors, in papers. **Figure 1** exemplifies this segregation, based on the same principle. The principle depends on the conditions during the drying, physical apparat, and intrinsic processes and interactions that occur with the sample and the drying environment, that is, mass transference process and water outlet [38]. As it can be seen from **Figure 1** in most of the drying methods used for microalgal drying, water is removed by an airflow. Different heat transfer principles can be found, that is, convection, conduction (e.g., cast-tape drying), and radiation. The moisture and the viscosity of the sample are also variable of these processes.

To enrich the discussion about drying methods and applicability to microalgal powder production, the main, traditional, and innovative drying methods are briefly presented in their principle and applications to microalgae.

3.1 Spray drying

Spray drying is the most common drying method applied to microalgae biomass for human uses [6] and, more general, is one of the most widely diffused drying technologies when dehydration of liquid foodstuff is required.

Algae species*	Dry method	Dry specifications/ variables	Quality assessment	Findings/conclusions
Chlorella vulgaris [26]	Freeze-drying (FD)	Temperature: −30°C Pressure: 3 mbar Time: 4.5 h Final moisture content: 0.88 ± 0.05% (w.b.)	• Color characterization • Total carotenoid • Chlorophyll content • Characterization of carotenoid-rich extracts • Protein content • Determination of antiradical activity	• FD powder shows intense green color, while HAD powder shows dark brown color • Carotenoid degradation was of 57.12 ± 3.74% for FD and 91.06 ± 2.37% for HAD • Protein content was not significantly influenced by drying method Freeze-drying is the most suitable drying method to maintain the nutrient and bioactive compounds
	Hot-air drying (HAD)	Temperature: 60°C Time: 4.5 h Final moisture content: 3.58 ± 0.19% (w.b.)		
Spirulina sp. [29]	Heat pump drying	Temperature: 30, 40, and 50°C Sample thickness: 1, 3, and 5 mm Drying time: 85–560min Final moisture content: 10.4 ± 1.2% (w.b.)	• Color measure • Phycocyanin content • Total activity antioxidant (DPPH) determination	• Color difference (ΔE): 4.22–13.51 • Phycocyanin content loss: 15–83% • TAA loss: 11–87% The optimal condition for lower color and phycocyanin degradation was air temperature of 50°C and sample thickness of 5 mm
Spirulina sp. [33]	Convective drying in thin layer	Temperature: from 40 to 60°C Air velocity: 1.9–3.8 m/s	• Drying kinetic • Effect of drying conditions and temperature on sorption isotherm	• Spirulina is very hygroscopic in the 25–40°C temperature range • Equilibrium moisture content is not dependent on the storage temperature

Algae species*	Dry method	Dry specifications/ variables	Quality assessment	Findings/conclusions
Spirulina sp. [35]	Convective drying (CD)	Temperature: 40, 50, and 60°C Air velocity: 0.15 m/s Drying time: 2–3 h	• Microscope characterization • Protein analysis • Total sugar analysis	• Protein loss: 10% in FD; 10–15% in SD; 10–25% in ID • Total sugar loss: 30% at 40°C and higher temperatures (mean) FD showed the highest retention of proteins and sugars Structure damage is caused by the air-drying temperatures
	Infrared drying (ID)	Temperature: 40, 50, and 60°C Radiative flux: 2.71 kW/m^2		
	Spray drying (SD)	Temperature: 130–150°C Feed rate: 0.09 l/h		
	Freeze-drying (FD)	Temperature: - 20°C Pressure: 8 Pa Drying time: 18 h		
Spirulina sp. [36]	Convective drying	Average temperature: 50°C Relative humidity: 12%	• Shrinkage coefficient and isotropicity • Porosity and apparent density	• Weak and anisotropic shrinkage • Final porosity approaching 80%

Algae species*	Dry method	Dry specifications/variables	Quality assessment	Findings/conclusions
Arthrospira spirulina LEB-18 [28]	Vacuum drying (laboratory scale)	Temperature: 40, 50, and 60°C Pressure: 13.3 kPa Final moisture content: 0.10 g/g (w.b.)	• Phycocyanin content • Total phenolic compounds (TPC) • Lipid oxidation • Color parameters • Rehydration analysis • FTIR spectroscopy • DSC analysis • Morphologic analysis	• Phycocyanin content: higher losses (80.5%) in OD at 55 °C than the VD at 60°C drying (71.7%) • TPC: VD samples showed increase of values (from 18 to 48% (w/w)) in relation to in natura sample • The lipid oxidation increases with the increase in drying temperature • Sample drying at 60°C presented the highest peak of temperature (143.8°C) and the lowest enthalpy value (189 J/g) • Morphologic analysis: rigid shape of irregular and compact particles for both samples Vacuum drying at 40°C causes the lowest losses of phycocyanin, phenolic compounds, the minor lipid oxidation, the good rehydration capacity, and the highest thermal stability
Spirulina LEB-18 [37]	Spouted bed dryer (SBD)	Temperature: 80, 90, 100, and 110°C Air velocity: 0.33 ± 0.01 m.s^{-1} Drying time: 210 min	• Centesimal composition • Total antioxidant activity (TAA)—DPPH • Total phenolic compounds • Protein solubility • Phycocyanin content • Lipid oxidation (TBA) • Color analysis • Thermal analysis • FTIR spectroscopy • Scanning electron microscopy	• TAA values: 34.6 ± 1.1% for CTD sample; TAA increases with the inlet drying air temperature increase for SBD samples • Phycocyanin content, the color difference for samples, and TBA values were affected by the temperature in SBD • SB drying at 80°C and CT drying obtain minor lipid oxidation and phycocyanin degradation SBD drying at 100°C reached greater thermal stability
	Conventional tray drying (CTD)	Temperature: 55°C Air velocity: 2.5 m.s^{-1} Tray thickness: 4 mm Final moisture: 0.10 kg.kg^{-1} (w.b.)		

Algae species*	Dry method	Dry specifications/ variables	Quality assessment	Findings/conclusions
Spirulina platensis LEB-52 [30]	Perpendicular airflow drying	Temperature: 50 and 60°C Air velocity: 1.5 m/s Relative humidity: 7–10%	• Experimental design for protein solubility response • Centesimal composition • Phycocyanin content	• Protein content: 74% (d.b.) • Protein solubility in acid medium: 42.6–79.1% • Higher solubility results occurred at 60°C • Phycocyanin determination: 12.6% (w.b.)
Spirulina LEB-18 [38]	Discontinuous tray drying	Temperature: 50, 60, and 70°C Sample thickness: 3, 5, and 7 mm Hot-air velocity: 2.5 $m.s^{-1}$	• Phycocyanin content • Lipid oxidation (TBA) • Fatty acid profiles	Oil obtained from spirulina: important source of monounsaturated and polyunsaturated fatty acids The best drying condition, 55°C and 3.7 mm, showed 37% phycocyanin loss and 1.5 $mg_{MDA}kg^{-1}$ TBA value, and the fatty acid composition did not show significance difference in relation to fresh biomass
Spirulina sp. [22]	Tray drying	Temperature: 50, 60, and 70°C Air velocity: 2.2 $m.s^{-1}$ Foaming agent: glair/albumin	• Color • Texture • Beta-carotene	Beta-carotene: 140.0 mg/100 g in dried sample at 60°C Improve the drying rate: 2.5% with the foaming agent The quality of spirulina dried (color, texture, and beta-carotene content) by foam mat drying is higher than that of produced by industry
Spirulina maxima [31]	Convective drying at atmospheric pressure	Benchtop chamber Temperature: 30, 50, 70, and 80°C Relative humidity: 13, 20, 50, and 60% Air velocity: 2.0 m/s	• Phycocyanin content • Total phenolic content (TPC) • Antioxidant capacity—ABTS • Phycocyanin denaturation kinetics	Significant loss of phycocyanin at drying temperatures above 70°C Significant influence in TPC values at drying temperature of 80°C Phycocyanin and total phenolic contents were largely dependent on the drying temperature rather than on humidity

Algae species*	Dry method	Dry specifications/variables	Quality assessment	Findings/conclusions
Arthrospira platensis [39]	Convective drying at atmospheric pressure	Layer thickness: 1 and 4 mm Temperature: 45°C	• Photography and SEM view • True density • Volume shrinkage characterization	Cylinders: initial porosity, 20%; final porosity, 65–78% Layers do not show macroporosity; the product is homogeneous without any pores Microporosity: present in cylinder and layer forms Porosity can be linked to the shrinkage phase durations, an improvement of organoleptic taste of dried spirulina Cylinders for drying indicate optimum drying conditions
		Cylinder diameters: 2, 3, 4, and 6 mm Temperature: 45°V		
Haematococcus pluvialis [40]	Spray drying	Not informed	Economic feasibility and the return for astaxanthin production	The results have proven the economic feasibility of the production for different astaxanthin market prices Evaporative rate: 26.125 kg/h
Aphanothece microscopica Nugeli [41]	Tray drying with air circulation	Temperature: 40, 50, and 60°C Constant speed: 1.5 m/s Sample thickness: 5 and 7 mm	• Total protein determination • Total carbohydrate determination • Total lipid determination • Fatty acid determination	• Protein content: 0.413–0.493 g/g (dry weight) • Carbohydrate fraction: 0.134–0.176 g/g (dry weight) • Lipid fraction: 0.071–0.079 g/g (dry weight) • Fatty acids: chain lengths with 14 and 24 C The drying conditions were shown to affect the macronutrient composition (protein, carbohydrate, and lipid contents), but did not influence the polyunsaturated/saturated ratio of the biomass.
Dunaliella salina [32]	Spray drying	Inlet temperature: 130°C Outlet temperature: 85°C Sprayed rate: 200 mL.h^{-1} Prevent degradation agent: antioxidants	• Stability studies • Solid determination • Carotenoid analysis	Spray drying with TBHQ and α-tocopherol was efficient to preserve algal carotenoid and minimize degradation of beta-carotene Between the two antioxidants, α-tocopherol had a small protective effect on beta-carotene degradation

Algae species*	**Dry method**	**Dry specifications/ variables**	**Quality assessment**	**Findings/conclusions**
Dunaliella salina [34]	Fluid-bed drying with alginate cells	Temperature: 70°C Airflow: 3.5 $m.s^{-1}$ Time: 10 min	• Beta-carotene analysis • Stability of total carotenoid during storage of the beads	• Total carotenoid losses: 13–20% during fluid-bed drying • Spray dry microencapsulation can reduce degradation of carotenes *D. salina* cells in alginate followed by fluid-bed drying have the potential in producing a carotene-rich nutraceutical product with good carotenoid stability characteristics
Tetraselmis chuii [42]	Spray drying	Temperature: 110/130/150°C Pressure: 40 bar and 2.5 mL/min Microencapsulation with maltodextrin	• Carotenoid analysis • Beta-carotene estimation • Antioxidant activity analysis	• Preservation of 80–92% of beta-carotene and 46–81% of phenolic compounds in microencapsulated microalgae and dried in spray dryer

Species name used in the paper.

Table 1.
Different drying methods applied in some microalgal species with an interest in evaluating quality characteristics.

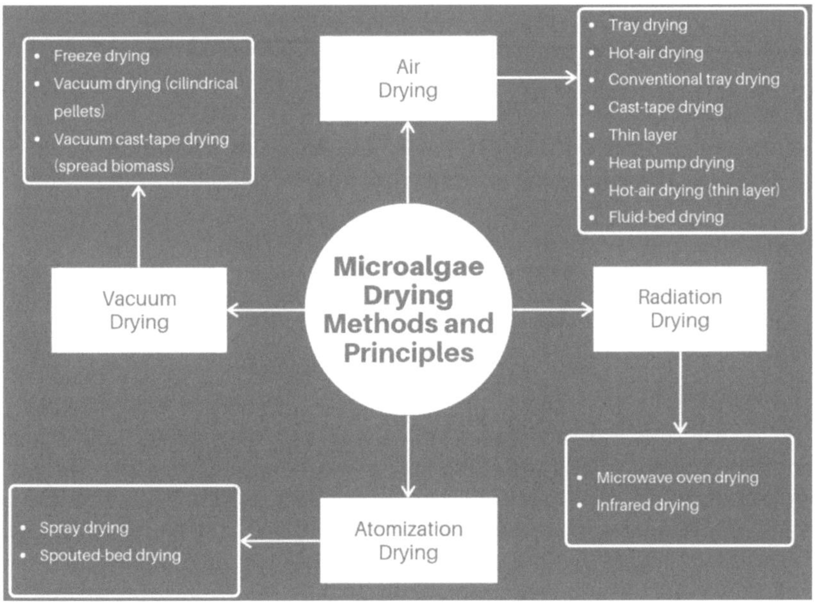

Figure 1.
Drying methods of microalgal biomass. Filled shapes, general method denomination; empty shapes, method denominations used in articles.

Spray drying uses the atomization of a liquid food to create droplets which are dried as individual particles while moving through a heated gas (hot air) [20]. Drying of single droplet provides a large surface area per unit volume of liquid, which favors rapid drying [43] and also causes a very short exposition of food to a very high temperature causing moderate degradation of product quality (high-temperature exposition for short time). The main steps of the spray drying process are atomization of the liquid, mixing of the droplets with the heated air, and separation of the dried powder in a cyclone [44]. Size of the droplet, air temperature, and liquid flow are the main factors that influence the quality of the dried product. Other factors that should be taken into account for the optimization of quality of spray-dried products are related to the biomass characteristics such as glass transition temperature, surface tension, liquid density, viscosity, and composition. The presence of high content of sugars, for example, impacts negatively the yield of this process. This problem is especially present when fruit pulps are dried; on the other hand for microalgae that present mostly long-chain carbohydrates, it does not represent an important issue [43, 44].

Among the advantage of this technology, it can be cited the high versatility, the possibility of pack directly, the powder produced without any milling process, and the easiness of the processing control allowing quality of the product remain constant (uniform) during processing [44]. On the other hand, this technology has a high installation and energy/operation costs, volatile compounds can be lost, and products that present high sensibility to high temperature could lose quality. It can cause rupture of cells, due to the high pressure generated during the atomization process, causing, in some cases, degradation in product quality [6], that is, promoting oxidation. However, spray drying is the only drying technology used in large-scale microalgal biomass drying for human consumption [6] mainly due to the equilibrium between high productivity and quality of the dried product. Although this drying method is the most used by the industry, few papers approach the effect of processing parameters, such as air temperature, viscosity of the moist biomass, and size of the biomass droplet on microalgal powder quality loss [32, 35, 40, 42].

Spray drying of *D. salina* biomass allowed production of powder with very low degradation of β-carotene and its isomers. On the other hand, during 5 days of storage, it degraded to less than 10% of retention. Damages to the membrane cell caused by the very fast water vaporization facilitate the oxidation and degradation of this functional compound, due to oxygen and light exposition [32]. To avoid this problem, a possible strategy is mixing the microalgal biomass with some encapsulating agent (e.g., maltodextrin, gum arabic, etc.) producing microcapsules by spray drying; in this case the high retention of functional compounds can be maintained during storage [42].

3.2 Drum drying

Another technology widely diffused in the food industry to produce dried product, from viscous foodstuff, is the drum drying. Drum drying consists in cylindrical metallic heated rollers or drums rotating at a variable speed. The material to be dried comes into contact with the surface of the drum in a thin layer of film, and heat is transferred through the metal. A slide is arranged in the apparatus to remove the dry thin film layer from the drum surface [8]; after that, the dried material is commonly milled to produce a uniform powder. This technology presents low operation costs and can be easily managed by small producers. On the other hand, it presents some limitation such as the processing time/temperature binomial which the sample must be submitted to be dried [8]. Although this method is widely used in microalgae biomass drying, the high temperature of the drum causes degradation of quality of the dried product; for this reason, this method is used to produce raw materials for biofuel industry but presents no particular interest for dried biomass for human alimentation. Alternatives have been developed to overcome the problem of high degradation of nutritional and functional components and allow the use of this technology to produce algal powders for human use. One example that can be cited is the use of an inert bed to increase the surface contact between a hot-air flow inside of the drum and the moist spirulina biomass, increasing the drying rate and the processing yield; this system also allows to overcome problems such as the bed agglomeration [25]; on the other hand, no assessment on biomass quality was done with this method, and further studies should be done to improve quality of products dried by this method.

3.3 Freeze-drying

Freeze-drying is a well-known drying process that allows production of dried food with high added value and high quality. Freeze-drying consists in two main steps; firstly the product is frozen and is transferred in a vacuum chamber, and water is sublimated [45] providing heat (latent heat of sublimation) by radiation or conduction (hot plates). Freeze-drying is particularly indicated to dry products with high sensibility to high temperature and oxygen exposition and with high added value. On the other hand, it presents high installation and operational cost, especially for industrial-scale equipment and requires long drying time (commonly up to 12 h). This method is highly recommended when the conservation of the nutritional and functional components of the raw material is desired. On the other hand, stability of freeze-dried foods could be compromised by their very high porosity that facilitates the contact with oxygen and air humidity promoting oxidation during storage [46]. In general, freeze-drying of microalgal biomass is considered an ideal method because it causes no degradation of biomass quality [21]. On the other hand, the very low water activity reached in freeze-dried powder could promote, combined with the high porosity, the oxidation of lipids and pigments;

thus, vacuum packaging should be considered in freeze-dried powder storage. The effect of drying method on stability of functional components of microalgae was assessed by [7]. Different drying conditions and storage methods were studied assessing their effect on the astaxanthin concentration in *Haematococcus pluvialis*-dehydrated powder. As expected, freeze-drying resulted in products with higher (approximately 30%) astaxanthin retention than spray-dried biomass. On the other hand, both powders present similar pattern of degradation during storage under different temperatures (from −20 to 37°C) and packaging (vacuum and air). A higher degradation of functional component was found in samples stored at 20 and 37°C in normal packaging (without vacuum); the highest astaxanthin degradation for freeze-dried powder was >80% at 37°C and > 60% at 20°C, both after 20 weeks of storage. Vacuum packaging efficiency was confirmed avoiding degradation of functional components of microalgae also when storage was performed at room temperature [7, 47]. However, due to the higher initial retention of function components, freeze-drying was considered a good option for high-quality dried biomass production. Stability of dehydrated product during storage plays a key role for the drying method chosen. For freeze-dried products, to maintain the high quality of the product obtained by this expensive method, specifically storage strategies must be used, that is, vacuum, light barrier, and low temperature.

3.4 Solar drying

Solar drying is a traditional drying method used for hundreds of years to stabilize the moist algal biomass. In this method the heat for water evaporation is provided by the solar radiation and the moisture removal by the natural airflow. Although it presents the obvious advantage of the low processing cost both in direct solar radiation method or in solar dryers, the efficiency of the method is directly dependent on weather condition and only applicable in few producing locations. Moreover, the long processing time and the exposition to the open environment increase the risk of spoilage or production of off-odors [6]. Strategies and dryers have been developed to overcome these problems optimizing the efficiency of the drying equipment and allowing drying of microalgal biomass with acceptable quality retention [48].

3.5 Convective drying and thin layer drying

Convective drying of a thin layer of spread biomass or extruded biomass cylinder is a method widely used especially by small-scale producers. Ref. [41] produced an *Aphanothece microscopica* Nägeli powder by spreading, in a convective oven, layers of biomass with thickness of 5 and 7 mm and testing the effect of drying temperature (40–60°C) on protein, carbohydrates, lipid content, and fatty acid profile, finding only small differences among treatments. On the other hand, the same research group found that in the same drying condition, chlorophyll a content and hue angle (related with sample color) were strongly influenced by the process temperature and chlorophyll concentration decreases intensely at temperature up to 40°C of drying [49], demonstrating the importance of the optimization of the drying temperature for producing high-quality dried microalgal powder. The effect of temperature on stability of functional components was also proven with spirulina. Many papers approach this issue [33, 30, among others] showing the importance of temperature and thickness optimization during convective drying (air-drying) to maximize the conservation of phycocyanin. This fact was confirmed by [50] that assessed the effect of drying temperature on spirulina functional components showing that temperatures above 45°C cause degradation, reducing its health benefits.

Other technology that can be used to dehydrate viscous foodstuff is the refractance window drying [51] or the cast-tape drying [52–55]. In both methods the liquid food is spread on surface, and the heat transfer occurs by radiation or by conduction. These methods allow producing high-quality fruit pulp powder with very low processing time. On the other hand, the temperature used in these processes could be above the limit for functional component preservation in microalgal biomass; thus, a vacuum chamber can be coupled to the cast-tape drier, and lower processing temperature can be used [56]. The vacuum drying removes the sample moisture thru low atmospheric pressure, showing many advantages comparing the conventional drying methods, that is, oxidation reducing. The low pressure in the drying chamber substitutes the hot-air flow, avoiding significantly compound degradations that lead to low product quality [57, 58]. This technology has been studied recently to spirulina biomass with interesting results in terms of quality and processing time [59, 60]. The cast-tape drying method used in preliminary studies for spirulina biomass drying has proved to be a very effective method in terms of drying time and efficiency. Using the same principle (thin layer of sample on a heated surface), vacuum cast-tape drying allows drying at lower temperatures, thus avoiding the degradation of important compounds such as phycocyanin. Preliminary studies conducted by our group showed phycocyanin preservation values greater than 60% in the method using vacuum and milder temperatures. The authors showed and reported that this technology is a promising method, which can achieve excellent moisture and water activity values, better performance, and low energy costs compared to conventional and/or expensive drying processes.

4. Conclusions

Microalgae are dried to allow easy storage and transportation as well as to facilitate their use in biorefinery and food and feed industry. In this chapter a concise overview of the state of the art about drying of microalgal biomass with potential use for human alimentation was presented. In literature, drying of algal biomass is approached, in most cases, focusing on energy efficiency and engineering of processes. Not much attention is given to the effect of dehydration on functional and nutritional components of the final product. Among the methods that are studied and applied to produce algal biomass for human use, spray drying is the most widely used; it is a very efficient method especially adequate for large-scale producers. Although, this method allows producing powders with relatively high retention of functional components, it was demonstrated that the due to cell structure degradation that occurs during drying, these components can be lost during storage under inadequate conditions. The same problem was found in freeze-dried powders; however, freeze-drying allows higher retention of functional components immediately after drying; thus, with adequate storage condition, this can be considered the best method to maintain quality of biomass. On the other hand, this method is very expensive and energy costing and thus is adequate only to produce high added value products. Air-drying is one of the most studied methods, and it was proposed, when performed adequately, as a good method to allow quality retention in dehydrated products. For example, spreading of the moist biomass in a thin layer increases the evaporation rate allowing lower drying time and the use of lower temperature, thus allowing higher retention of functional components in the final products. Alternatives to this technology have been developed combining thin layer drying with a vacuum chamber allowing reducing drying temperature and time and obtaining higher-quality products. Air-drying or thin layer drier (air or vacuum) is, in general, less expensive than spray driers or freeze-driers, in terms of installation and operational costs, and thus is a better option for small-scale producers. Finally, it can be concluded that more studies are necessary to improve not only the drying methods but also to understand the degradative phenomena that occur during storage in particular with regard to the high sensitivity to light, heat, and oxygen of dried microalgal biomass, to allow providing the consumers high-quality products.

Author details

Fábio de Farias Neves[1*], Mariana Demarco[2] and Giustino Tribuzi[2]

1 Santa Catarina State University, Laguna, Brazil

2 Federal University of Santa Catarina, Florianópolis, Brazil

*Address all correspondence to: fabio.neves@udesc.br

IntechOpen

© 2019 The Author(s). Licensee IntechOpen. This chapter is distributed under the terms of the Creative Commons Attribution License (http://creativecommons.org/licenses/by/3.0), which permits unrestricted use, distribution, and reproduction in any medium, provided the original work is properly cited.

References

[1] Andersen RA. Algal Culturing Techniques. Amsterdam, The Netherlands: Elsevier Academic Press; 2005. 578 p

[2] Graham LE, Wilcox LW, Graham JM. Algae. 2nd ed. San Francisco, CA: Pearson/Benjamin Cummings; 2009. ISBN 9780321559654

[3] Van Harmelen T, Oonk H. Microalgae Biofixation Processes: Application and Potential Contribution to Greenhouse Gas Mitigation Options. Apeldoorn, The Netherlands: TNO Built Environment and Geosciences; 2006. 45 p

[4] Sheehan J, Dunahay T, Benemann J, Roessler PA. Look Back at the US Department of Energy's Aquatic Species Program: Biodiesel from Algae. Golden, CO: US Department of Energy's, Office of Fuels development. NREL/TP-580-24190; 1998. 328 p

[5] Richmond A. Handbook of Microalgal Culture. Biotechnology and Applied Phycology. Cornwall: Blackwell Science Ltd.; 2004. 566 p. ISBN 0-632-05953-2

[6] Show K, Lee D, Tay J, Lee T, Chang J. Microalgal drying and cell disruption – Recent advances. Bioresource Technology. 2015;**184**:258-266. DOI: 10.1016/j.biortech.2014.10.139

[7] Ahmed F, L Y, Fanning K, Netzel M, Schenk PM. Effect of drying, storage temperature and air exposure on astaxanthin stability from *Haematococcus pluvialis*. Food Research International. 2015;**74**:231-236. DOI: 10.1016/j.foodres.2015.05.021

[8] Barbosa-Cánovas GV, Ortega-Rivas E, Juliano P, Yan H. Food Powders – Physical Properties, Processing and Functionality. New York, NY: Kluwer Academic/Plenum Publishers; 2005

[9] Kim D, Vijayan D, Praveenkumar R, Han J, Lee K, Park J, et al. Cell-wall disruption and lipid/astaxanthin extraction from microalgae: Chlorella and Haematococcus. Bioresource Technology. 2015;**199**:300-310. DOI: 10.1016/j.biortech.2015.08.107

[10] Lee AK, Lewis DM, Ashman PJ. Disruption of microalgal cells for the extraction of lipids for biofuels: Processes and specific energy requirements. Biomass and Bioenergy. 2012;**46**:89-101. DOI: 10.1016/j.biombioe.2012.06.034

[11] Samarasinghe N, Fernando S, Lacey R, Faulkner WB. Algal cell rupture using high pressure homogenization as a prelude to oil extraction. Renewable Energy. 2012;**48**:300-308; 9 p. DOI: 10.1016/j.renene.2012.04.039

[12] Belarbi E, Molina E, Chisti Y. A process for high yield and scaleable recovery of high-purity eicosapentaenoic acid esters from microalgae and fish oil. Enzyme and Microbial Technology. 2000;**26**(7):516-529. PII S0141-0229(00)00191-X

[13] FAO. Algae-Based Biofuels - a Review of Challenges and Opportunities for Developing Countries. Roma, Italy: FAO; 2009

[14] Molina-Grima E, Belarbi EH, Acien-Fernandez FG, Medina AR, Chisti Y. Recovery of microalgal biomass and metabolites: Process options and economics. Biotechnology Advances. 2003;**20**(7-8):491-515

[15] Chisti Y, Moo-Young M. Disruption of microbial cells for intracellular products. Enzyme Microbiology Technology. 1986;**8**:194-204. DOI: 10.1016/0141-0229(86)90087-6

[16] Harrison RG, Todd P, Rudge SR, Petrides DP. Cell Lysis and flocculation.

In: Bioseparations Sciences and Engineering. London: Oxford University Press; 2003. pp. 83-93

[17] Halim R, Rupasinghe T, Tull DL, Webley PA. Mechanical cell disruption for lipid extraction from microalgal biomass. Bioresource Technology. 2013;**140**:53-63. DOI: 10.1016/j.biortech.2013.04.067

[18] Middelberg APJ. Process-scale disruption of microorganisms. Biotechnology Advances. 1995;**13**:491-551. DOI: 10.1016/0734-9750(95)02007-P

[19] Kaczor A, Baranska M. Structural changes of carotenoid astaxanthin in a single algal cell monitored in situ by Raman spectroscopy. Analytical Chemistry. 2011;**83**:7763-7770. DOI: 10.1021/ac201302f

[20] Fellows P. Food Processing Technology Principles and Practice. 2nd ed. Vol. 591. Boca Raton: CRC Press LLC; 2000

[21] Morist A, Montesinos JL, Cusidó JA, Gòdia F. Recovery and treatment of *Spirulina platensis* cells cultured in a continuous photobioreactor to be used as food. Process Biochemistry. 2011;**37**:535-547. DOI: 10.1016/S0032-9592(01)00230-8

[22] Prasetyaningrum A, Djaeni M. Drying spirulina with foam mat drying at medium temperature. International Journal of Food Science and Engineering. 2012;**3**(2):1-3. DOI: 10.12777/ijse.3.2.1-3

[23] Show KY, Lee DJ, Chang JS. Algal biomass dehydration. Bioresource Technology. 2013;**135**:720-729. DOI: 10.1016/j.biortech.2012.08.021

[24] Chen C, Chang J, Lee D. Dewatering and drying methods for microalgae. Drying Technology. 2015;**33**(4):443-454. DOI: 10.1080/07373937.2014.997881

[25] Silva NC, Machado MVC, Brandão RJ, Duarte CR, Barrozo MAS. Dehydration of microalgae *Spirulina platensis* in a rotary drum with inert bed. Powder Technology. 2019;**351**:178-185. DOI: 10.1016/j.powtec.2019.04.025

[26] Stramarkou M, Papadaki S, Kyriakopoulou K, Krokida M. Effect of drying and extraction conditions on the recovery of bioactive compounds from *Chlorella vulgaris*. Journal of Applied Phycology. 2017;**29**:2947-2960. DOI: 10.1007/s10811-017-1181-8

[27] Hosseinizand H, Lim CJ, Webb E, Sokhansanj S. Economic analysis of drying microalgae chlorella in a conveyor belt dryer with recycled heat from a power plant. Applied Thermal Engineering. 2017;**124**:525-532. DOI: 10.1016/j.applthermaleng.2017.06.047

[28] Larrosa APQ, Comitre AA, Vaz LB, Pinto LAA. Influence of air temperature on physical characteristics and bioactive compounds in vacuum drying of *Arthrospira spirulina*. Journal of Food Process Engineering. 2016;**40**:e12359. DOI: 10.1111/jfpe.12359

[29] Costa BR, Rodrigues MCK, Rocha SF, Pohndorf RS, Larrosa APQ, Pinto LAA. Optimization of *Spirulina sp.* drying in heat pump: Effects on the physicochemical properties and color parameters. Journal of Food Processing & Preservation. 2016;**40**:934-942. DOI: 10.1111/jfpp.12672

[30] Oliveira EG, Rosa GS, Moraes MA, Pinto LAA. Characterization of thin layer drying of *Spirulina platensis* utilizing perpendicular air flow. Bioresource Technology. 2009;**100**:1298-1303. DOI: 10.1016/j.biortech.2008.05.052

[31] Nakagawa K, Ritcharoen W, Sri-Uam P, Pavasant P, Adachi S. Antioxidant properties of

convective-air-dried *Spirulina maxima*: Evaluation of phycocyanin retention by a simple mathematical model of air-drying. Food and Bioproducts Processing. 2016;**100**:292-302. DOI: 10.1016/j.fbp.2016.07.014

[32] Orset S, Leach GC, Morais R, Young AJ. Spray-drying of the microalga *Dunaliella salina*: Effects on beta-carotene content and isomer composition. Journal of Agricultural and Food Chemistry. 1999;**47**:4782-4790. DOI: 10.1021/jf990571e

[33] Desmorieux H, Decaen N. Convective drying of spirulina in thin layer. Journal of Food Engineering. 2005;**66**:497-503. DOI: 10.1016/j.jfoodeng.2004.04.021

[34] Leach G, Oliveira G, Morais R. Production of a carotenoid-rich product by alginate entrapment and fluid-bed drying od *Dunaliella salina*. Journal of the Science of Food and Agriculture. 1998;**76**:298-302. DOI: 10.1002/(SICI)1097-0010(199802)76:2<298::AID-JSFA953>3.0.CO;2-0

[35] Desmorieux H, Hernandez F. Biochemical and physical criteria of spirulina after different drying processes. In: Proceedings of the 14th International Drying Symposium (IDS). 2004. pp. 900-907

[36] Dissa AO, Desmorieux H, Savadogo PW, Segda BG, Koulidiati J. Shrinkage, porosity and density behaviour during convective drying of spirulina. Journal of Food Engineering. 2010;**97**:410-418. DOI: 10.1016/j.jfoodeng.2009.10.036

[37] Larrosa APQ, Camara AS, Pohndorf RS, Rocha SF, Pinto LAA. Physicochemical, biochemical and thermal properties of *Arthrospira* (Spirulina) biomass dried in spouted bed at different conditions. Journal of Applied Phycology. 2017;**30**:1019-1029. DOI: 10.1007/s10811-017-1265-5

[38] Oliveira EG, Duarte JH, Moraes K, Crexi VT, Pinto LAA. Optimization of *Spirulina platensis* convective drying: Evaluation on phycocyanin loss and lipid oxidation. International Journal of Food Science and Technology. 2010;**45**:1572-1578. DOI: 10.1111/j.1365-2621.2010.02299.x

[39] Desmorieux H, Madiouli J, Herraud C, Mouaziz H. Effects of size and form of *Arthrospira Spirulina* biomass on the shrinkage and porosity during drying. Journal of Food Engineering. 2010;**100**(4):585-595. DOI: 10.1016/j.jfoodeng.2010.03.021

[40] Zgheib N, Saade R, Khallouf R, Takache H. Extraction of astaxanthin from microalgae: Process design and economic feasibility study. In: IOP Conference Series: Materials Science and Engineering. Vol. 323. 2018. DOI: 10.1088/1757-899X/323/1/012011

[41] Zepka LQ, Jacob-Lopes E, Goldbeck R, Queiroz MI. Production and biochemical profile of the microalgae *Aphanothece microscopica Nägeli submitted* to different drying conditions. Chemical Engineering and Processing Process Intensification. 2008;**47**(8):1305-1310. DOI: 10.1016/j.cep.2007.04.013

[42] Bonilla-Ahumada FJ, Khandual S, Lugo-Cervantes EC. Microencapsulation of algal biomass (*Tetraselmis chuii*) by spray-drying using different encapsulation materials for better preservation of beta-carotene and antioxidant compounds. Algal Research. 2018;**36**:229-238. DOI: 10.1016/j.algal.2018.10.006

[43] Chen XD, Özkan N. Stickiness, functionality and microstructure of food powders. Drying Technology. 2007;**25**:969-979. DOI: 10.1080/07373930701397400

[44] Bhandari B, Bansal N, Zhang M, Schuck P. Handbook of Food Powders.

1st ed. Sawston: Woodhead Publishing; 2013. 688 p 9780857095138

[45] Liapis AI, Bruttini R. Freeze Drying. In: Mujumdar AS, editor. Handbook of Industrial Drying. Boca Raton, FL: CRC Press; 2006

[46] Ochoa-Martínez CI, Quintero AA, Ayala MJ, Ortiz-Jerez M. Drying characteristics of mango slices using the refractance window technique. Journal of Food Engineering. 2012;**109**:69-75. DOI: 10.1016/j.jfoodeng.2011.09.032

[47] Gouveia L, Empis J. Relative stabilities of microalgal carotenoids in microalgal extracts, biomass and fish feed: Effect of storage conditions. Innovative Food Science & Emerging Technologies. 2003;**4**:227-233. DOI: 10.1016/S1466-8564(03)00002-X

[48] Prakash J, Pushparaj B, Carlozzi P, Torzillo G, Montaini E, Materassi R. Microalgal biomass drying by a simple solar device. International Journal of Solar Energy. 1997;**18**(4):303-311. DOI: 10.1080/01425919708914325

[49] Lopes EJ, Zepka LQ, Pinto LAA, Queiroz MI. Characteristics of thin-layer drying of the cyanobacterium *Aphanothece microscopica Nägeli*. Chemical Engineering and Processing Process Intensification. 2007;**46**:63-69. DOI: 10.1016/j.cep.2006.04.004

[50] Sarada R, Pillai MG, Ravishankar GA. Phycocyanin from Spirulina sp: Influence of processing of biomass on phycocyanin yield, analysis of efficacy of extraction methods and stability studies on phycocyanin. Process Biochemistry. 1999;**34**:795-801. DOI: 10.1016/S0032-9592(98)00153-8

[51] Zotarelli MF, Carciofi BAM, Laurindo JB. Effect of process variables on the drying rate of mango pulp by refractance window. Food Research International. 2015;**1**:410-417. DOI: 10.1016/j.foodres.2015.01.013

[52] Durigon A, Souza PG, Carciofi BAM, Laurindo JB. Cast-tape drying of tomato juice for the production of powdered tomato. Food and Bioproducts Processing. 2016;**100**:145-155. DOI: 10.1016/j.fbp.2016.06.019

[53] Zotarelli MF, Da Silva VM, Durigon A, Hubinger MD, Laurindo JB. Production of mango powder by spray drying and cast-tape drying. Powder Technology. 2017;**305**:447-454. DOI: 10.1016/j.powtec.2016.10.027

[54] Simão RS, De Moraes JO, De Souza PG, Carciofi BAM, Laurindo JB. Production of mango leathers by cast-tape drying: Product characteristics and sensory evaluation. LWT- Food Science and Technology. 2019;**99**:445-452. DOI: 10.1016/j.lwt.2018.10.013

[55] Frabetti ACC, Durigon A, Laurindo JB. Effect of process variables on the drying of guava pulp by cast-tape drying. LWT- Food Science and Technology. 2018;**96**:620-626. DOI: 10.1016/j.lwt.2018.06.021

[56] Costa JD. Secagem de iogurte natural por Cast-tape sob vácuo [dissertation]. Federal University of Santa Catarina; 2017. URI: 123456789/182602

[57] Comitre AA, Vaz LB, Silva LA, Larrosa APQ, Pinto LAA. Secagem a vácuo de *Spirulina sp. Leb 18*: análise das características dos compostos bioativos do produto desidratado. In: XI Brasilian Congress of Chemistry Engineering in Scientific Initiation, Unicamp, Campinas – SP. 2015. DOI: 10.5151/chemeng-cobeqic2015-255-33034-259977

[58] Zhang M, Tang J, Mujumdar AS, Wang S. Trends in microwave-related drying of fruits and vegetables. Trends

in Food Science and Technology. 2006;**17**:524-534. DOI: 10.1016/j.tifs.2006.04.011

[59] Demarco M, Ferrari MC, Canto FQ, Moraes JO, Neves FF, Laurindo JB, et al. Estudo da evolução temporal da umidade e da atividade de água durante a secagem de biomassa de spirulina. Florianópolis, Santa Catarina, Brazil: XXI Nacional Meeting and VII American-Latin Congress of Food Analysts; 2019

[60] Bonetti A, Moraes JO, Demarco M, Neves FF, Laurindo JB, Tribuzi G. Efeito de diferentes processos de desidratação na qualidade de Spirulina (*Arthrospira platensis*). Belém, Pará, Brazil: XXVI Brazilian Congress of Food Science and Technology; 2018

Chapter 5

Microalgae and Its Use in Nutraceuticals and Food Supplements

Joshi Nilesh Hemantkumar and Mor Ilza Rahimbhai

Abstract

Microalgae are a large diverse group of microorganisms comprising photoautotrophic protists and prokaryotic cyanobacteria—also called as blue-green algae. These microalgae form the source of the food chain for more than 70% of the world's biomass. It contains higher nutritional values, with rapid growth characteristics. Microalgae are autotrophic organisms and extensively desired for use in nutraceuticals and as supplement in diet. Many microalgal species are documented for health benefits, by strengthening immune system and by increasing the nutritional constitution of body. In this chapter the major economically important species like *Spirulina*, *Chlorella*, *Haematococcus*, and *Aphanizomenon* are described with reference to its importance as nutraceuticals and food.

Keywords: microalgae, *Spirulina*, *Chlorella*, nutraceuticals, food supplements

1. Introduction

Microalgae are a large diverse group of microorganisms comprising photoautotrophic protists and prokaryotic cyanobacteria—also called as blue-green algae. These microalgae form the source of the food chain for more than 70% of the world's biomass [1]. Microalgae are single-celled, microscopic photosynthetic organisms, found in freshwater and marine environment. They produce compounds such as protein, carbohydrates, and lipids. Mostly, microalgae are photosynthetic microorganisms; it does not contain cell organelles unlike land plants. They use the carbon from air for energy production.

Microalgae can be cultivated photosynthetically using CO_2, solar energy, and water. It can be cultivated in shallow lagoons, marginal ponds, raceway ponds, or artificial tanks. The use of plastic tubes/reactors in pond system can achieve up to seven times the production efficiency compared to open culture system [2].

There are more than 300,000 species of microalgae, out of which around 30,000 are documented. They live in complex natural habitats and can adapt rapidly in extreme conditions (in variation of extreme weather conditions). This ability makes them capable to produce secondary metabolites, with novel structure and biologically active functions.

Microalgae produce some useful bio-products including β-carotene, astaxanthin, docosahexaenoic acid (DHA), eicosapentaenoic acid (EPA), bioactive and functional pigments, natural dyes, polysaccharides, antioxidants, and algal extracts.

IntechOpen

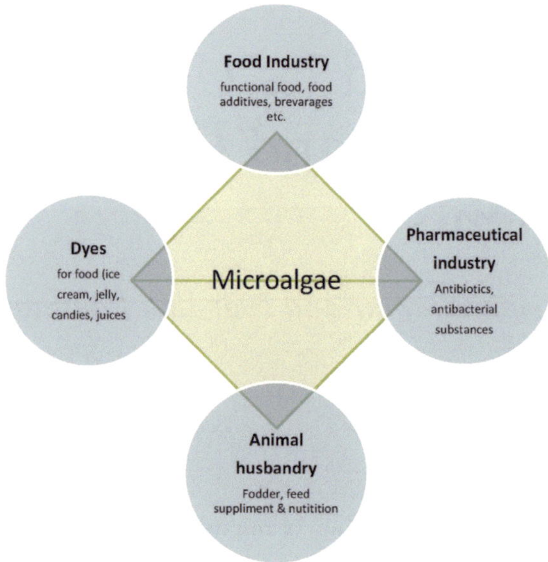

Figure 1.
Application of microalgae in various fields.

The first commercial cultivation of *Chlorella* was started in 1960 in Japan for nutraceuticals. Microalgae grow fast and produce large biomass with high protein content and consist the source of "single cell protein" [3] (**Figure 1**).

Algae are classified into many major groups, based on pigment composition, storage compound, and diversity in features of its ultrastructure. However, advance molecular biology-based techniques are nowadays used to check the relation between taxonomic groups and families of the specific class.

The global market value of microalgae is estimated to be around US$ 6.5 billion, out of which about US$ 2.5 billion is generated by the health food sector, US$ 1.5 billion by the production of DHA, and US$ 700 million by aquaculture. The annual production of microalgae is approximately 7.5 million tons.

2. Microalgal diversity

The diversity of microalgae is vast and represents an intact resource. The scientific literature indicates the existence of 200,000 to several million species of microalgae when compared to about 250,000 species of higher plants [4].

Green microalgae usually grow in freshwater and seawater, whereas several other species of microalgae grow in extremely saline environments, such as the Great Salt Lake in UT, USA, and the Dead Sea in Israel. Within these aqueous habitats, some algae grow inside the deeper waters, others populate the subsurface water column, and a few grow at the limits of the photic zone, 200–300 m below the water surface [5–8]. The microalgae are small in size (mostly 5–50 μm) and characterized by a simple morphology, usually unicellular. Accordingly, most of the species are not observed as an individual cell/specimen but become noticeable only when it generates large colony, specially in the form of green, black, red, or brown patches on the water surface. Coastlines between 45 and 30°N are suitable regions for algal farming, in particular in those territories at the south of the Mediterranean that experience warmer climates and whose temperature does not go too much below 15°C throughout the year [2]. This type of warmer climate of the Mediterranean region can facilitate the algal growth in the open or closed pond system.

The environmental parameters favorable for mass scale culture are explored by many counties. For example, Israel and few Mediterranean countries have explored specific parameters for each economically important alga and cultivating it by maintaining them artificially for mass scale production. Few advance countries have started culturing several microalgal strains of biofuel production, while countries like Libya, Cyprus, and Turkey also have plenty of marginal lands to harvest algae. For these countries a limited water resource is not a constraint as they are using recycled brackish or saline water. With the high temperatures in the Mediterranean region, the open or closed pond system would probably be the most efficient and suitable to grow algae.

Terrestrial microalgae belong primarily to three diverse evolutionary pedigrees: the blue-green algae (Cyanobacteria), the green algae (Chlorophyta and Streptophyta), and the diatoms (Bacillariophyceae, Ochrophyta) [9]. However, the species of green and blue-green algal group are the majorly studied group, taxonomically as well as with economical perspective. Nevertheless, the understanding of the patterns of geographical distribution in terrestrial algae is inadequate, mainly due to poor understanding of the diversity of these organisms [10].

3. Microalgae: uses as nutraceuticals and food

Microalgae have a wide range of industrial applications, in food industries, wastewater purification, and pharmaceutical formulations [11]. Microalgae can also be used for high-value food, health food for human, polysaccharides, food and fodder additives, cosmetics, antioxidants, anti-inflammatory objects, dyes and feed for aquaculture, and preparation of biofilms [3, 12, 13].

The most widely used microalgae include Cyanophyceae (blue-green algae), Chlorophyceae (green algae), Bacillariophyceae (including diatoms), and Chrysophyceae (including golden algae). **Table 1** highlights some major microalgal species, products, and their application.

Since the last 20 years, biotechnological and nutraceutical application of microalgae has focused specifically on four major microalgae: (a) *Spirulina* (*Arthrospira*), (b) *Chlorella,* (c) *Dunaliella salina*, and (d) *Haematococcus pluvialis*.

3.1 *Spirulina*

Spirulina is a prokaryotic cyanobacterium that has been commercially produced for over 30 years for uses including fish food, vitamin supplements, food dyes, aquaculture, pharmaceuticals, and nutraceuticals [15, 16]. *Spirulina* is manufactured by many pharmaceutical companies. This alga is thought of as a super food and is widely cultured, primarily in specifically designed raceway ponds and photobioreactors, to meet the current demand.

Spirulina is one of the algae studied for large-scale commercial culture. It grows best at a high pH (9–11) and high bicarbonate concentrations. Generally raceway ponds are used to culture *Spirulina*. The water depth in the pond is generally 300–500 mm depending on the physical–chemical parameters and density of microalgae. The depth is also dependent on pond size, water flow velocity, and light absorption by the algal culture. Water temperature and pH have a large effect on the productivity of this species. It grows well between the temperatures of 35 and 37°C [17]. *Spirulina* is a filamentous microalga; hence, its harvesting is relatively easy.

Spirulina contains 60–70% protein by weight (including many amino acids) and contains up to 10 times more beta-carotene than carrots per unit mass [18]. *Spirulina* is rich in nutrients such as B vitamins, phycocyanin, chlorophyll, vitamin E,

Group/species	Extract	Use/application
Arthrospira (Spirulina) platensis	Phycocyanin, biomass	Health food, cosmetics
Arthrospira (Spirulina)	Protein, vitamin B12	Antioxidant capsule, immune system
Aphanizomenon flos-aquae	Protein, essential fatty acids, β-carotene	Health food, food supplement
Chlorella spp.	Biomass, carbohydrate extract	Animal nutrition, health drinks, food supplement
Dunaliella salina	Carotenoids, β-carotene	Health food, food supplement, feeds
Haematococcus pluvialis	Carotenoids, astaxanthin	Health food, food supplement, feeds
Odontella aurita	Fatty acids, EPA	Pharmaceuticals, cosmetics, anti-inflammatory
Porphyridium cruentum	Polysaccharides	Pharmaceuticals, cosmetics
Isochrysis galbana	Fatty acids	Animal nutrition
Phaeodactylum triconutum	Lipids, fatty acids	Nutrition, fuel production
Lyngbya majuscule	Immune modulators	Pharmaceuticals, nutrition
Schizochytrium sp.	DHA and EPA	Food, beverage, and food supplement
Crypthecodinium cohnii	DHA	Brain development, infant health and nutrition
Nannochloropsis oculata	Biomass	Food for larval and juvenile marine fish

Adopted from [14].

Table 1.
The major microalgal species, products, and application.

omega 6 fatty acids, and many minerals [19]. *Spirulina* is used for weight loss [20], diabetes [21], high blood pressure, and hypertension [22]. It has documented antiviral [23, 24] and anticancer properties [25]. *Spirulina* positively affects cholesterol metabolism by increasing HDL levels, which can lead to healthy cardiovascular functions [26]. Romay et al. [27] described the antioxidant and anti-inflammatory properties of C-phycocyanin, which is a prevalent pigment in *Spirulina*. Essential amino acids like leucine, isoleucine, and valine are significantly present in *Spirulina*. *Spirulina* contains fatty acids like linolenic and γ-linolenic acid and ω-3 and ω-6 polyunsaturated fatty acids. *Spirulina platensis* is a natural source of DHA accounting up to 9.1% of the total fatty acids. The mineral content of *Spirulina* depends on the water in which it was grown, and its content of iron, calcium, and magnesium provides high nutritional value [16]. Belay et al. [18] reported that *Spirulina* powder contains provitamin A (2.330×10^3 IU/kg), β-carotene (140 mg 100/g), vitamin E (100 mg 100/g), thiamin B1 (3.5 mg 100/g), riboflavin B2 (4.0 mg 100/g), niacin B3 (14.0 mg 100/g), vitamin B6 (0.8 mg 100/g), inositol (64 mg 100/g), vitamin B12 (0.32 mg 100/g), biotin (0.005 mg 100/g), folic acid (0.01 mg 100/g), pantothenic acid (0.1 mg 100/g), and vitamin K (2.2 mg 100/g). The amino acid composition of *Spirulina* is given in **Table 2**.

3.2 *Chlorella*

Chlorella is a single-cell, spherical shaped (2–10 μm in diameter), and photoautotrophic green microalga with no flagella. It multiplies rapidly requiring only CO_2, water, sunlight, and a small amount of minerals. *Chlorella* has been grown commercially cultured in photobioreactors [28] and harvested by centrifugation or

Amino acids (g kg^{-1})	*Spirulina*	*Chlorella*
Alanine	47	48
Arginine	43	36
Aspartic acid	61	52
Cysteine	6	4
Glutamic acid	91	63
Glycine	32	34
Histidine	10	13
Isoleucine	35	26
Leucine	54	53
Lysine	29	35
Methionine	14	15
Phenylalanine	28	31
Proline	27	29
Serine	32	28
Threonine	32	27
Tryptophan	9	6
Tyrosine	30	21
Valine	40	36

Adapted from [16].

Table 2.
Amino acid profile of Spirulina *and* Chlorella *microalgae (in mg kg^{-1}).*

autoflocculation. After harvesting the biomass is spray-dried, and the cell powder is sold directly. *Chlorella* contains 11–58% protein, 12–28% carbohydrate, and 2–46% lipids of its dry weight [29]. *It also* contains various vitamins such as β-carotene (180 mg 100/g), provitamin A (55,500 IU/kg), thiamin B1 (1.5 mg 100/g), vitamin E (<1 mg 100/g), riboflavin B2 (4.8 mg 100/g), niacin B3 (23.8 mg 100/g), vitamin B6 (1.7 mg 100/g), inositol (165.0 mg 100/g), vitamin B12 (125.9 mg 100/g), biotin (191.6 mg 100/g), folic acid (26.9 mg 100/g), and pantothenic acid (1.3 mg 100/g) [18, 30]. The amino acid composition of *Chlorella* is shown in **Table 2.**

Chlorella is able to decrease blood pressure, lower cholesterol levels, and enhance the immune system [16]. It also has the potential to relieve fibromyalgia, hypertension, or ulcerative colitis [31, 32]. The presence of aortic atheromatous lesions was significantly inhibited, and low-density lipoprotein (LDL) cholesterol levels were greatly suppressed upon consumption of *Chlorella* [33]. Some *Chlorella* consumers have mentioned a potential correlation between some brands of *Chlorella* tablets and nausea, vomiting, and other gastrointestinal issues. *Chlorella* has been labeled as a weak allergen and may be of clinical significance to certain types of people [34].

3.3 *Dunaliella*

Dunaliella (*D. salina)* is a unicellular green alga which contains large amounts of β-carotene, glycerol, and protein that can easily be extracted through its thin cell wall. *Dunaliella* does not required waters appropriate for agricultural and domestic uses and can be cultured in brackish water, marine water, and highly saline water. Global production of *Dunaliella* is estimated to be 1200 tons dry weight per year [16]. The dominant companies that produce *Dunaliella*, mainly for beta-carotene production, are located in Israel, China, the USA, and Australia and include Betatene, Western Biotechnology, AquaCarotene Ltd., Cyanotech Corp., and Nature Beta Technologies [35].

Dunaliella produces many carotenoid pigments with the dominant being beta-carotene and smaller amounts of lutein and lycopene [36]. Some strains of *Dunaliella* contain up to 14% of beta-carotene on dry weight basis. The total carotenoid content of *Dunaliella* varies with the physicochemical parameters and growth conditions. In optimal environmental condition, it can yield around 400 mg beta-carotene/m^2 of cultivation area [37]. Carotenoids from *Dunaliella* are potent free radical scavengers that reduce levels of lipid peroxidation and enzyme inactivation, thereby restoring enzyme activity. Research has shown beta-carotene to prevent cancer of various organs like the lungs, cervix, pancreas, colon, rectum, breast, prostate, and ovary by means of antioxidant activity [36]. It has also been shown to promote regression of certain types of cancer. Supplements of *Dunaliella* have also shown excellent hepatoprotective effects and reduced the occurrence of liver lesions [38].

3.4 *Haematococcus pluvialis*

Haematococcus pluvialis (*H. pluvialis*) is unicellular biflagellate freshwater green microalga. This species is known for its ability to accumulate large quantities of

Composition content (% of DW)	Green stage	Red stage
Proteins	29–45	17–25
Lipids (% of total)	20–25	32–37
Neutral lipids	59	51.9–53.5
Phospholipids	23.7	20.6–21.1
Glycolipids	11.5	25.7–26.5
Carbohydrates	15–17	36–40
Carotenoids (% of total)	0.5	2–5
Neoxanthin	8.3	n.d
Violaxanthin	12.5	n.d
β-carotene	16.7	1
Lutein	56.3	0.5
Zeaxanthin	6.3	n.d
Astaxanthin (including esters)	n.d	81.2
Adonixanthin	n.d	0.4
Adonirubin	n.d	0.6
Canthaxanthin	n.d	5.1
Echinenone	n.d	0.2
Chlorophylls	1.5	2 0

Adapted from [16].

Table 3.
Composition of H. pluvialis *biomass in green and red cultivation stages.*

strong antioxidant astaxanthin (up to 2–3% on dry weight) under any conditions. The principal commercial astaxanthin-producing microalga is *H. pluvialis* [9, 26]. Astaxanthin is used as a nutritional supplement and anti-inflammatory and anticancer agent for cardiovascular diseases and is recently recorded to prevent diabetes and neurodegenerative disorders and stimulates immunization. It also has anti-inflammatory properties and is used for various commercial applications in the dosage forms as biomass, capsules, creams, granulated powders, oils, soft gels, syrups, and tablets [39].

Photoautotrophic culture of *H. pluvialis* is mainly carried out in open raceway ponds or closed photobioreactors. The accumulation of astaxanthin is affected by environmental factors such as light, temperature, pH, salt concentration, and nutritional stresses. The cellular composition of *H. pluvialis* varies notably between its "green" and "red" stages of cultivation [40]. Specific biochemical characters of green and red stage of *H. pluvialis* are described in **Table 3**. The table shows that carbohydrate content in the green stage is approximately half of the red stage. *H. pluvialis* can accumulate approximately 5% DW of astaxanthin which is considered as a natural source of this high-value carotenoid protein [37].

3.5 *Aphanizomenon*

Aphanizomenon is a prokaryotic cyanobacterium commonly found in freshwater systems. There are approximately 500 tons of dried *Aphanizomenon* produced annually for use in food and pharmaceutical industries [41]. The dominant production source of *Aphanizomenon* in North America is Upper Klamath Lake and Klamath Falls, Oregon, and currently constitutes a significant part of the health food supplement industry throughout North America. *Aphanizomenon* contains a significant amount of C-phycocyanin, a light-harvesting pigment. It has antioxidant and anti-inflammatory properties [42]. *Aphanizomenon* also exhibits high

Component	*Spirulina*	*Dunaliella*	*Haematococcus*	*Chlorella*	*Aphanizomenon*
Protein	63	7.4	23.6	64.5	1.0
Fat	4.3	7.0	13.8	10.0	3.0
Carbohydrate	17.8	29.7	38.0	15.0	23.0
Chlorophyll	1.15	2.2	0.4 (red) 1.1 (green)	5.0	1.8
Magnesium	0.319	4.59	1.14	0.264	0.2
B-carotene	0.12	1.6	0.054	0.086	0.42
Vitamin B1 (thiamin)	0.001	0.0009	0.00047	0.0023	0.004
Vitamin B2 (riboflavin)	0.0045	0.0009	0.0017	0.005	0.0006
Vitamin B3 (niacin)	0.0149	0.001	0.0066	0.025	0.025
Vitamin B5 (pantothenic acid)	0.0013	0.0005	0.0014	0.0019	0.0008
Vitamin B6 (pyridoxine)	0.00096	0.0004	0.00036	0.0025	0.0013
Vitamin B9 (folic acid)	0.000027	0.00004	0.00029	0.0006	0.0001
Vitamin B12 (cobalamine)	0.00016	0.000004	0.00012	0.000008	0.0006

Adopted from [16].

Table 4.
Summary of referenced biochemical constitutions of average nutritional compositions (g per 100 g DW).

hypo-cholesterolemic activity, significantly greater than soybean oil, which decreases blood cholesterol and triglyceride levels [43–45]. It also produces poly-unsaturated fatty acids (i.e., omega 3 and omega 6), a deficiency of which has been linked to immunosuppression, arthritis, cardiovascular diseases, mental health issues, and dermatological problems [16]. A summary on biochemical characters of all these economically important species is described in **Table 4**.

4. Summary

As the human population continues to increase, the demand for nutritive food and health products increases concomitantly. The sources of nutritive biomass that can meet this demand are pursued rampantly. Their wide diversity, fast growth, and diverse uses make them easily accepted for commercial culture. Microalgae require much fewer resources as compared to other crops. The role of algae in human health and nutrition will continually increase with additional research in the areas of health benefits and culturing. The usage of currently produced algae primarily includes food, food additives, aquaculture, colorants, cosmetics, pharmaceuticals, and nutraceuticals. Very few algal species are being cultivated for human use. There are likely more species of algae that have not been identified than ones that have and those still numbers in the thousands. Therefore, the potential for algal use in the realms of food consumption, health supplements, energy production, and many more is likely to intensify in the years to come.

Author details

Joshi Nilesh Hemantkumar[1*] and Mor Ilza Rahimbhai[2]

1 College of Fisheries Science, Junagadh Agricultural University, Veraval, Gujarat, India

2 Department of Botany, H & H. B College of Science, Saurashtra University, Rajkot, Gujarat, India

*Address all correspondence to: nileshhj@gmail.com

IntechOpen

© 2019 The Author(s). Licensee IntechOpen. This chapter is distributed under the terms of the Creative Commons Attribution License (http://creativecommons.org/licenses/by/3.0), which permits unrestricted use, distribution, and reproduction in any medium, provided the original work is properly cited.

References

[1] Wiessner W, Schnepf E, Starr RC. Algae, Environment and Human Affairs. Bristol: Biopress Ltd.; 1995

[2] Singh J, Gu S. Commercialization potential of microalgae for production of biofuels. Renewable and Sustainable Energy Reviews. 2010;**14**:2596-2610

[3] Varfolomeev SD, Wasserman LA. Microalgae as source of biofuel, food, fodder and medicine. Applied Biochemistry and Microbiology. 2011; **47**:789-807

[4] Norton TA, Melkonian M, Andersen RA. Algal biodiversity. Phycologia. 1996;**35**:308-326

[5] Rindi F, McIvor L, Sherwood AR, Friedl T, Guiry MD, Sheath RG. Molecular phylogeny of the green algal order Prasiolales (Trebouxiophyceae, Chlorophyta). Journal of Phycology. 2007;**43**:811-822

[6] Lewis LA, Flechtner VR. Green algae (*Chlorophyta*) of desert microbiotic crusts: Diversity of North American taxa. Taxon. 2002;**51**:443-451

[7] Broady PA. Diversity, distribution and dispersal of Antarctic terrestrial algae. Biodiversity and Conservation. 1996;**5**:1307-1335

[8] Sharma NK, Rai AK, Singh S, Brown RM. Airborne algae: Their present status and relevance. Journal of Phycology. 2007;**43**:615-627

[9] Fritsch F. The subaerial and freshwater algal flora of the tropics. A phyto-geographical and ecological study. Annals of Botany. 1907;**21**: 235-275

[10] Lewis LA, McCourt RM. Green algae and the origin of land plants. American Journal of Botany. 2004;**91**: 1535-1556

[11] Mobin S, Alam F. Biofuel production from algae utilizing wastewater. In: Proceedings of 19th Australasian Fluid Mechanics Conference; Article No 27; 2014

[12] Borowitzka MA. Algae as food. In: BJB W, editor. Microbiology of Fermented Foods. Thomson Science; 1998. pp. 585-602

[13] Hudek K, Davis LC, Ibbini J, Erickson L. Commercial products from algae. In: Bajpai et al., editors. Algal Biorefineries. Dordrecht: Springer Science+Business Media; 2014. pp. 275-295

[14] Mobin S, Alam F. Some promising microalgal species for commercial applications: A review. Energy Procedia. 2017;**110**:510-517

[15] Ciferri O, Tiboni O. The biochemistry and industrial potential of *Spirulina*. Annual Review of Microbiology. 1985;**39**: 503-526

[16] Bishop WM, Zubeck HM. Evaluation of microalgae for use as nutraceuticals and nutritional supplements. Journal of Nutrition & Food Sciences. 2012;**2**:147

[17] Richmond A. Spirulina. In: Borowitzka MA, Borowitzka LJ, editors. Micro-Algal Biotechnology. Cambridge: Cambridge University Press; 1988. pp. 85-121

[18] Belay A, Ota Y, Miyakawa K, Shimamatsu H. Current knowledge on potential health benefits of *Spirulina*. Journal of Applied Phycology. 1993;**5**: 235-241

[19] Gershwin ME, Belay A. *Spirulina* in Human Nutrition and Health. Boca Raton, FL, USA: CRC Press; 2008

[20] Becker EW, Jakover B, Luft D, Schmuelling RM. Clinical and biochemical evaluations of the alga *Spirulina* with regard to its application in the treatment of obesity: A double-blind cross-over study. Nutrition Reports International. 1986; **33**:565-574

[21] Takai Y, Hosoyamada Y, Kato T. Effect of water-soluble and water insoluble fractions of *Spirulina* over serum lipids and glucose resistance of rats. Journal of Japan Society of Nutrition and Food Sciences. 1991;**44**: 273-277

[22] Iwata K, Inayama T, Kato T. Effects of *Spirulina platensis* on plasma lipoprotein lipase activity in fructose-induced hyperlipidemic rats. Journal of Nutritional Science and Vitaminology (Tokyo). 1990;**36**:165-171

[23] Kromhout D. Fish oil consumption and coronary heart disease. In: Dietary LW3 and LW6 Fatty Acids: Biological Effects and Nutritional Effects and Nutritional Essentiality. New York, USA: Plenum Publishing; 1989

[24] Simopoulos AP. Omega-3 fatty acids in health and disease and in growth and development. The American Journal of Clinical Nutrition. 1991;**54**:438-463

[25] Tang G, Suter PM. Vitamin A, nutrition, and health values of algae: *Spirulina*, *Chlorella*, and *Dunaliella*. Journal of Pharmacy and Nutrition Sciences. 2011;**1**:111-118

[26] de Caire GZ, de Cano MS, de Mule CZ, Steyerthal N, Piantanida M. Effect of *Spirulina platensis* on glucose, uric acid and cholesterol levels in the blood of rodents. International Journal of Experimental Botany. 1995;**57**:93-96

[27] Romay C, Armesto J, Remirez D, González R, Ledon N. Antioxidant and anti-inflammatory properties of C-phycocyanin from blue-green algae. Inflammation Research. 1998;**47**:36-41

[28] Borowitzka MA. Commercial production of microalgae: Ponds, tanks, tubes and fermenters. Journal of Biotechnology. 1999;**70**:313-321

[29] Zhu LD, Hiltunen E, Antila E, Zhong JJ, Yuan ZH, Wang ZM. Microalgal biofuels: Flexible bioenergies for sustainable development. Renewable and Sustainable Energy Reviews. 2014; **30**:1035-1046

[30] Belay A, Ota Y, Miyakawa K, Shimamatsu H. Production of high quality *Spirulina* at Earthrise Farms. In: Phang SM, Lee K, Borowitzka MA, Whitton B, editors. Algal Biotechnology in the Asia-Pacific Region. University of Kuala Lumpur; 1994. pp. 92-102

[31] Merchant RE, Andre CA. A review of recent clinical trials of the nutritional supplement *Chlorella pyrenoidosa* in the treatment of fibromyalgia, hypertension, and ulcerative colitis. Alternative Therapies in Health and Medicine. 2001;**7**:79-91

[32] Sansawa H, Takahashi M, Tsuchikura S, Endo H. Effect of *Chlorella* and its fractions on blood pressure, cerebral stroke lesions, and life-span in strokeprone spontaneously hypertensive rats. Journal of Nutritional Science and Vitaminology (Tokyo). 2006;**52**:457-466

[33] Sano T, Tanaka Y. Effect of dried, powdered *Chlorella vulgaris* on experimental atherosclerosis and alimentary hypercholesterolemia in cholesterol-fed rabbits. Artery. 1987;**14**: 76-84

[34] Tiberg E, Dreborg S, Björkstén B. Allergy to green algae (*Chlorella*) among children. The Journal of Allergy and Clinical Immunology. 1995;**96**:257-259

[35] Del Campo JA, García-González M, Guerrero MG. Outdoor cultivation of microalgae for carotenoid production: Current state and perspectives. Applied Microbiology and Biotechnology. 2007; **74**:1163-1174

[36] Chidambara Murthy KN, Vanitha A, Rajesha J, Mahadeva Swamy M, Sowmya PR. In vivo antioxidant activity of carotenoids from *Dunaliella salina*—A green microalga. Life Sciences. 2005; **76**:1381-1390

[37] Finney KF, Pomeranz Y, Bruinsma BL. Use of algae *Dunaliella* as a protein supplement in bread. Cereal Chemistry. 1984;**61**:402-406

[38] Hsu YW, Tsai CF, Chang WH, Ho YC, Chen WK. Protective effects of *Dunaliella salina*—A carotenoids-rich alga, against carbon tetrachlorideinduced hepatotoxicity in mice. Food and Chemical Toxicology. 2008;**46**:3311-3317

[39] Guerin M, Huntley ME, Olaizola M. *Haematococcus* astaxanthin: Applications for human health and nutrition. Trends in Biotechnology. 2003;**21**: 210-216

[40] Shah MMR, Liang Y, Cheng JJ, Daroch M. Astaxanthin-producing green microalga *Haematococcus pluvialis*: From single cell to high value commercial products. Frontiers in Plant Science. 2016;**7**:531

[41] Dashwood R, Guo D. Protective properties of chlorophylls against the covalent binding of heterocyclic amines to DNA in vitro and in vivo. Princess Takamatsu Symposia. 1995;**23**:181-189

[42] Kushak RI, Drapeau C, Van Cott EM, Winter HH. Favorable effect of blue-green algae *Aphanizomenon flos-aquae* on rat plasma lipids. Journal of the American Nutraceutical Association. 2000;**2**:59-65

[43] Miyamoto E, Tanioka Y, Nakao T, Barla F, Inui H. Purification and characterization of a corrinoid-compound in an edible cyanobacterium *Aphanizomenon flos-aquae* as a nutritional supplementary food. Journal of Agricultural and Food Chemistry. 2006;**54**:9604-9607

[44] Liang S, Liu X, Chen F, Chen Z. Current microalgal health food R & D activities in China. Hydrobiologia. 2004; **512**:45-48

[45] Pulz O, Gross W. Valuable products from biotechnology of microalgae. Applied Microbiology and Biotechnology. 2004;**65**:635-648

Section 3

Microalgae Applications

Chapter 6

The Red Microalga *Galdieria* as a Promising Organism for Applications in Biotechnology

Mária Čížková, Milada Vítová and Vilém Zachleder

Abstract

The genus *Galdieria* refers to red algae and includes microscopic inhabitants of highly acidic (pH 1–2), often volcanic habitats. They are thermophilic or thermo-tolerant organisms, some of them surviving temperatures up to 56°C. As other extremophilic microorganisms, they exhibit unique features derived from their modified metabolisms. In this chapter, we will review the special abilities of *Galdieria* species such as metabolic flexibility to grow photoautotrophically, heterotrophically or mixotrophically, ability to utilize a whole range of unusual carbon sources, capability of surviving extreme environments or their extremely high resistance to metals. We will discuss the potential of *Galdieria* for applications in biotechnology, for example, phycocyanin production, nutrient removal from urban wastewaters, bio-mining, treatment of acidic mine drainage, selective metal precipitation, bioremediation of acidic metal-contaminated areas or recovery of critical and scarce metals from secondary sources.

Keywords: *Galdieria*, red algae, extremophiles, metals, biotechnology, recovery

1. Introduction

The alga *Galdieria* belongs to the class *Cyanidiophyceae*, which are unicellular red algae classified into three genera *Cyanidium*, *Cyanidioschyzon*, and *Galdieria*. They have a relatively simple morphology consisting of spherical thick-walled cells containing one chloroplast, 1–3 mitochondria, a nucleus, a vacuole, and storage products. They inhabit diverse environments including hot sulfur springs, streams, mud, rock walls, endolithic habitats, etc. *Galdieria* is an extremophilic organism surviving in acidic geothermal environments with very low pH and high temperatures. It exhibits a unique metabolism enabling auto-, mixo-, and heterotrophic growth. Versatility in the utilization of carbon sources for heterotrophic growth is enormous including a large range of sugars and alcohols. Besides low pH and high temperature, the *Galdieria* have several unique mechanisms for metal tolerance, enabling them to grow in environments rich in toxic metals [1, 2], where other organisms cannot grow at all or with difficulties.

For its unique extremophilic properties, the alga *Galdieria* is an unrivaled organism for biotechnological applications in surroundings with high temperatures, low pH, and high concentrations of heavy metals and organic compounds, particularly sugars. It can be cultivated under conditions where other microalgae often become contaminated with microorganisms, which is a major problem for their application in biotechnology [3]. The autotrophic cultivation of *Galdieria* follows predominant research trends in microalgal pigments, ß carotene, astaxanthin, and phycocyanin used in feed, in foods, in health applications or biofuels production. The composition of storage glycogen and lipids for biofuels can be largely changed depending on the growth conditions. *Galdieria* biomass has potential for use as food ingredients, both for protein-rich or insoluble dietary fiber-rich diets and for its low concentration of lipids.

2. Taxonomy and biodiversity

The order Cyanidiales from the class Cyanidiophyceae are a group of asexual, unicellular organisms that diverged from ancestral red algae around 1.3 billion years ago [4, 5]. These unicellular red algae were classified into three genera, *Cyanidium*, *Cyanidioschyzon*, and *Galdieria*. *Cyanidioschyzon merolae* is clearly recognizable thanks to its characteristic size and shape, but the other two algae are morphologically very similar. Until 1981 *Cyanidium caldarium* was used as a synonym for *Galdieria sulphuraria*, however, this species can grow only autotrophically while *G. sulphuraria* is also able to grow heterotrophically [6, 7]. Based on morphological, ultrastructural and ecophysiological studies the class Cyanidiophyceae was therefore instituted, containing the family Cyanidiaceae, including *Cyanidium caldarium* and the family Galdieriaceae, including *Galdieria sulphuraria* [8].

Different Cyanidiophycean species, including *Galdieria*, are found all over the world; however, their distribution is discontinuous as they are restricted to hot springs and geothermal habitats. This is related to the discontinuity of geothermal environments. Two decades ago, little was known about their biodiversity, their population structures, and the phylogenetic relationships of Cyanidiales.

Research based on environmental PCR studies revealed an unexpected level of genetic diversity among Cyanidiales. It was demonstrated that the Cyanidiales comprise a species-rich branch of red algae [9]. The high divergence rates in the Cyanidiales could be possibly explained by an elevated mutation rate in these taxa, resulting potentially from DNA damage in their extreme environments. The analyses also reject the putative mesophilic origin of Cyanidiales and suggest ancestral thermo-acidotolerancy of this lineage [9].

Sequencing of the *rbcL* gene with high sequence divergence within the genus has contributed to the taxonomy of *Galdieria* [10–12]. A cladogram defining molecular relationships among these algae shows that *Cyanidium caldarium* and *Cyanidioschyzon merolae* form a sister group relationship with *Galdieria* [11]. The genus *Galdieria* is divided into two clades, one of which includes *G. sulphuraria* accessions from Naples (Italy), California, and Yellowstone and the other one includes *G. sulphuraria* accessions from Java (Indonesia) and Russian species [11].

Based on molecular phylogenetics, three well supported *Galdieria* species exist: *G. maxima* Sentsova, *G. sulphuraria* (Galdieri) Merola (including two Russian species *G. partita* Sentsova and *G. daedala* Sentsova, described based on morphological differences) and *G. phlegrea* Pinto, Ciniglia, Cascone et Pollio [9, 13, 14].

Consequently, the main lineages were identified: ***G. phlegrea*** [14] comprising strains thriving in acidic non-thermophilic Italian sites; ***G. sulphuraria***, a group that is geographically dispersed worldwide, including *G. sulphuraria* strains as well as *G. partita* and *G. daedala*, isolated from acid-thermal springs in Russia [15]. A final lineage contains the cosmopolitan species ***G. maxima*** [15], which is clustered independently from the *Galdieria* clade (*G. sulphuraria* + *G. phlegrea*) and shows an unexplained sister group relationship with the morphologically distinct *Cyanidioschyzon merolae* [5, 9, 10, 12, 16].

Generally, *Galdieria* (Cyanidiales) is well known from Italy [5, 9, 14, 17] Yellowstone National Park, USA [18, 19], New Zealand [20], Iceland [10], and recently from the Czech Republic [21, 22].

Phylogenetic analyses of the *rbc*L gene also showed that *Galdieria* from the coal mining site at Ostrava, Czech Republic, belongs to the cosmopolitan species *G. sulphuraria*, for now the only eukaryotic organism forming visible biomass on a burning coal-waste heap [10]. This was the first evidence of this species growing in central Europe, and isolates were closely related to the Italian strains, together forming the continental European lineage of *G. sulphuraria* [21]. Another, non-thermophilic strain of *Galdieria*, also found in the Czech Republic [22], referred to as CCALA 965 (Culture Collection of Autotrophic Organisms, Institute of Botany, CAS, Třeboň, Czech Republic) was found to belong to the species *G. phlegrea*, so far known only from Italy [13, 14].

3. Morphology and extremophile properties

Morphology of the unicellular taxa Cyanidiales is relatively simple. Thick-walled cells are of a spherical shape and usually contain one chloroplast, 1–3 mitochondria, a nucleus, a vacuole, and energy reserve products [8, 12, 15, 23, 24].

Representatives of the order Cyanidiales are unparalleled among phototrophic microorganisms (eukaryotes) in their ability to thrive in acidic (pH 0.5–3.5) and high temperature (38–56°C) geothermal environments. Soils, sediments, and endolithic habitats around hot springs, boiling mud pools, and steaming fumaroles are typical for these extremophiles, which are dominant in local microbial communities [5, 9, 10, 14, 17–20, 25–30]. They are the principal photosynthetic organisms found in hot acidic waters [31], where even photosynthetic prokaryotes, such as the cyanobacteria, are completely absent [32–34].

G. sulphuraria is a unicellular, spherical, spore-forming red alga. In addition to acidophilic and thermophilic properties, it has the ability to grow phototrophically, mixotrophically and moreover heterotrophically while utilizing sugars, alcohols or amino acids [7, 35, 36]. *G. sulphuraria* cells are morphologically indiscernible from cells of *Cyanidium caldarium*, however, are well recognizable thanks to their ready ability to grow heterotrophically in the dark [16, 18]. In its natural environment, *G. sulphuraria* has a yellow to green color; however, when heterotrophically grown in liquid medium, it looks like yellow-green to dark blue-green. The cell size of *G. sulphuraria* is larger than that of *C. caldarium*. It reproduces by endospore formation ranging from four to thirty-two. As originally described, *G. sulphuraria* has a single, cup-shaped, parietally localized chloroplast [8] and includes a vacuole and mitochondria [11].

Morphological similarities between *G. sulphuraria* and *G. phlegrea* are so high that methods for recognition of these species, their habitats, and growth requirements, together with molecular analyses, are used.

G. phlegrea (**Figure 1**) prefers a relatively low temperature (25–38°C) for growth and inhabits rather dry endolithic sites with high acidity (pH 0.5–1.5) [9, 14, 22]. The Latin word *phlegreus* means volcanic, what is consistent with the specific locality, (Campi Flegrei, Naples, Italy), where the alga was found [14]. The locality provides diverse environmental conditions in the form of hot springs, streams, mud, rock walls, with different pHs and temperature ranges, producing the different microhabitats occurring in that site. [14].

Strain DB01 of *G. phlegrea* from the Tinto River (Spain) has typical coccoid cells with thick smooth cell walls. Mature cells reach the average size of 6.4 μm. The cell possesses a blue-green chloroplast without pyrenoids. Typical for the alga is asexual

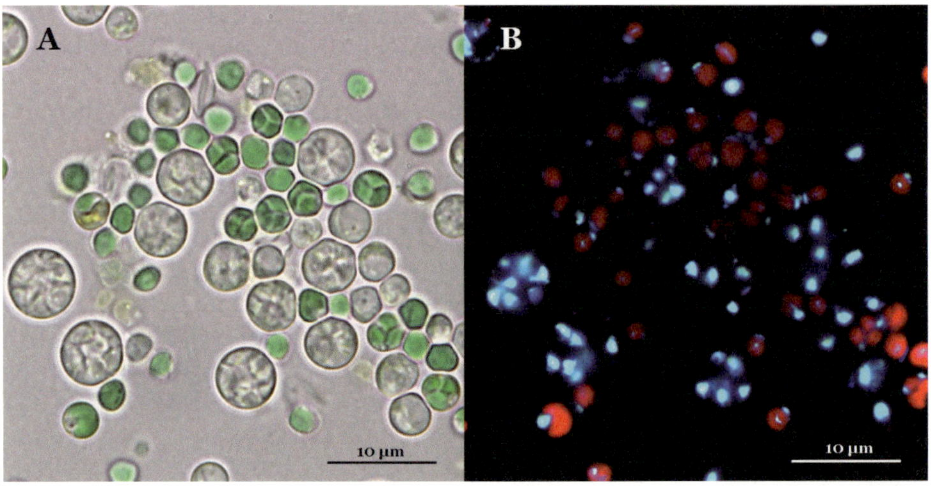

Figure 1.
Microphotographs of Galdieria phlegrea *in bright field (A) and fluorescence (B). Nuclei in the panel B are in blue (stained by DAPI) and chloroplasts are in red (autofluorescence). The chloroplasts of large mother dividing cells are not visible. The bar is 10 µm.*

reproduction by autospores originating in the parental cell and resulting in autosporangia with 2–8 daughter cells. Testing the culture conditions of the isolate DB01 showed that the algae were not strictly thermophilic [13].

G. maxima is characterized by facultative heterotrophy; however, *G. maxima* strains grow very poorly when cultivated under dark conditions. Spherical cells are significantly larger (10–16 µm diameter) compared with other thermoacidophilic algal species [18, 23]. Cell size is thus used as the main character to distinguish one from the other. Inside the cell are at least two parietal plastids, lobe or oval shaped [37].

4. Genomes

Genetic information for the red algae *Galdieria* (*G. maxima, G. partita*, and *G. sulphuraria*) is located in the nucleus, in two small chromosomes, which differ in length. The smaller chromosome ranges from 0.8 to 1.8 µm and the larger one from 1.2 to 2.3 µm. The genome is characterized by an unusually high gene density, small or absent introns, and very few repetitive sequences. A genome size of 10.8 Mbp was estimated for *G. sulphuraria* [38]. In other strains of *G. sulphuraria*, genome sizes were found between 9.8 and 14.2 Mbp [39]. These genome characteristics refer to the smallest known genomes of all free living eukaryotes [38, 40].

The mitochondrial genome is extremely small in size with a very low genetic content. It is characterized by the highest guanine-cytosine content among all red algae.

The plastid genome contains a large number of intergenic stem-loop structures but is otherwise rather typical in size, structure, and content in comparison with other red algae. It is assumed that the unique genomic characteristic resulted from both the harsh conditions in which *Galdieria* lives and its unusual ability to grow mixotrophically, heterotrophically, and endolithically. The authors [41] suggested that "these conditions place additional mutational pressures on the mitogenome due to the increased reliance on the mitochondrion for energy production, whereas the decreased reliance on photosynthesis and the presence of numerous stem-loop structures may shield the plastome from similar genomic stress."

5. Biotechnological applications

Galdieria with its extensive extremophilic properties, which are unique not only among all eukaryotic organisms but even in extremophilic prokaryotes, has a broad utilization in biotechnology. It is the only algae that can grow photo-, mixo-, and heterotrophically to biomass concentrations above 100 g/L dry weight [42]. For heterotrophic growth, it can use over 27 different kinds of sugars and polyols to produce a huge biomass and beneficial compounds [7, 43, 44]. It tolerates concentrations of glucose and fructose up to 166 g/L, salt concentrations up to 1–2 M, and pH values below pH 1 [42, 45] and can attain 80–110 g L^{-1} biomass in continuous flow cultures [43]. The ability of *Galdieria* to grow under conditions intolerable for other organisms, even prokaryotic ones, predetermines its biotechnical applications in such surroundings as different, often toxic, wastewaters, treatment of acid mine drainage, selective metal precipitation, bioremediation of acidic metal-contaminated areas, or recovery of critical and scarce metals from secondary sources.

5.1 Wastewaters

Recycling of valuable components and nutrients from wastewaters using algae has recently been studied extensively. But only limited types of wastewaters can be treated because wastewaters are generally acidic and most algal species grow with difficulty at low pH, and absorption rates that can be achieved by bioaccumulation decrease substantially [46–49]. The acidophilic alga *G. sulphuraria* is the only alga that has commercial potential for remediation of these wastewaters [50, 51]. Nutrient removal from municipal wastewater by the alga *G. sulphurea* was found to be very efficient for ammoniacal-nitrogen (88.3%) and phosphate (95.5%) in large scale outdoor bioreactors [51]. Additionally, many crucial elements, including phosphate and rare earth elements from wastewater were successfully bio-sorbed [51, 52]. It can be concluded that *G. sulphuraria* can be applied as a preferred strain for energy-efficient nutrient removal from urban wastewaters [51], achieving higher nutrient removal efficiencies and removal rates than other strains. This alga can also be used for bio-sorption of precious metals from metal-containing wastewaters. The great advantage is that precious metals can be efficiently bio-sorbed by *Galdieria* cells even if they are present in very low concentrations. Over 90% of gold and palladium were recovered from aqua-regia-based metallic wastewater where metal concentrations were so low that they could not be recycled chemically or pyro-metallurgically. Because the entire process could be completed within 1 h, the use of *G. sulphuraria* has promising applications in metal recovery [53], particularly where Pt and Au could be selectively re-eluted from cells into a solution containing 0.2 M ammonium salts without other contaminating metals [54].

5.2 Rare earth elements

Lanthanides [Rare Earth Elements (REEs)] have unique magnetic and catalytic properties and are, up to now, irreplaceable materials in numerous technologies, for example, wind turbines, solar panels, batteries, fluorescent lamps, computer and mobile monitors, TV screens etc. They are also used as fertilizers in agriculture, in aquaculture, or as animal growth enhancers.

Methods for extraction of lanthanides from ores, including pyro-metallurgy and hydro-metallurgy, have severe negative environmental impacts, as well as being expensive. Currently, industrial extraction of lanthanides from monazite involves either a basic process that uses concentrated sodium hydroxide or an acidic process that uses concentrated sulfuric acid. These processes generate large amounts of hazardous waste containing thorium and uranium [55]. Moreover, requirements for REEs are continually increasing, becoming critical due to risks of reduced availability of resources and their possible exhaustion.

One way to solve the problem would be efficient, sustainable, and cheap recycling of REE-containing wastewaters and others industrial wastes.

Considerable research efforts have been directed toward the development of efficient biological methods for recovering small amounts of these materials from wastewater systems [48, 49]. Research has recently focused on environmentally friendly technologies of metal recovery, including REEs, from secondary resources [56, 57] including bio-sorption by algae or cyanobacteria [52, 58]; for review, see [59, 60].

However, if REEs were present in an aquatic environment, together with other metals, most algae could not accumulate high concentrations of REEs [61, 62] due to metal-inhibited growth. The extraction of REEs or other metals have now been simplified by the use of *Galdieria* cells, which were effective in the recovery of many crucial elements, including phosphates and REEs [51, 52].

Similar to its relative *Cyanidium caldarium, G. sulphuraria* is resistant to high concentrations of metals in solution, including Al^{3+}. Moreover, it could be used to selectively recover lanthanides and Cu^{2+} ions from water containing various kinds of metals at a pH of 2.5. The concentration of soluble metals in solution remained unchanged at pH values within the range 0.5–5.0 [52]. In contrast, this process is usually difficult to achieve by bacterial bio-sorption. Lowering of pH to 1.0–1.5 enabled the recovery of lanthanides from cells whereas Cu^{2+} ions remained dissolved in aqueous acid. The use of *G. sulphuraria* also allowed recovery of over 90% of low levels of metals (0.5 ppm) from solution by cell fractionation at pH values in the range of 1.5–2.5. This system did not require any genetic manipulation or treatment of the cells for the efficient recovery of lanthanides [52].

Recycling from different mineral ores and electronic wastes (luminophores) often meets difficulties in that REEs are not suitable for bio-sorption because they are present in solid forms and are almost insoluble in nutrient solutions for algal cultivation. The material can be readily dissolved in aqueous acid, but the efficiency of metal bio-sorption for most algae is usually decreased under acidic conditions or the algae cannot grow at a low pH. Application of extremophilic red alga *Galdieria* would therefore be an advantageous solution to this problem and seems to be the aim of future research. The species *Galdieria phlegrea* has already been used to test the bio-accumulation of REEs from luminophores added into the medium in the form of a powder. Algal cells were cultured mixotrophically in a liquid medium with the addition of glycerol as a source of carbon. Luminophores from two different sources (fluorescence lamps and energy saving light bulbs) were tested. In spite of the low solubility of luminophores, *G. phlegrea* could grow in the presence of luminophores and accumulate REEs [63, 64].

Another rich source of lanthanides is bauxite residue, called red mud, which is a by-product of the production of alumina (aluminum oxide) from bauxite. However, less than 2% of the residue produced annually is currently being reused [65], due to difficulties related to high pH, salinity, low solid content, size of fine particles, and the leaching of metals [66]. The ability to grow in the presence of red mud and accumulate REEs was successfully tested with *G. phlegrea* (**Figure 2**) [67].

To conclude, the alga *G. sulphuraria* offers great potential for the direct recovery of REEs from metal-containing wastewaters (even if present at very low concentrations) or from solid waste material (luminophores) as well as for bio-mining from bauxite ore residue (red mud).

Figure 2.
Electron microphotograph of the dense freeze-dried culture of Galdieria phlegrea *used for REEs recovery. The bar is 10 µm. (provided by Dr. Jens Hartmann).*

5.3 Phycocyanin

Phycocyanin is an important compound that can be obtained from microalgal and cyanobacterial cultures. It is used as a fluorescent marker in diagnostic histochemistry [68, 69] and as a dye in foods and cosmetics [70] or as a therapeutic agent [71, 72]. Production of phycocyanin as a photosynthetic pigment in most microalgae grown heterotrophically is low and not suitable for biotechnological applications [73]. In contrast, phycocyanin as a major pigment of *G. sulphuraria* can be produced even under heterotrophic conditions in darkness [7, 74]. Due to superior biomass productivity, the productivity of phycocyanin in cultures of *G. sulphuraria* was 1–2 orders of magnitude greater than in *Arthrospira* (*Spirulina*) *platensis*, which was used recently for commercial production of phycocyanin [43] and was dependent on sunlight and climatic conditions. Besides light independence, *G. sulphuraria* tolerates a concentrated medium and can metabolize many different compounds as a source of carbon and energy [7, 75].

For example, *G. sulphuraria* could grow in restaurant and bakery waste hydrolysates in which sugars and free amino acids were utilized as substrates. Ammonium and inorganic nutrients were, however, needed in order to maximize phycocyanin synthesis [76].

The feasibility of utilizing crude glycerol (a major waste by-product of biofuel production from oilseed rape) as a carbon source for heterotrophic growth of green microalgae [54] was confirmed for *Chlorella* and also for *G. sulphuraria*.

G. sulphuraria has also been grown on sugar beet molasses [42]. Under heterotrophic conditions, phycocyanin synthesis depends mostly on available ammonium ions [77]. Ammonium sulphate was tolerated in higher molar concentrations than glucose, fructose, or sodium chloride [22, 23].

G. sulphuraria is well suited for heterotrophic growth to an extremely high cell density, which is among the highest biomass concentrations ever reported for microalgal cultures. Nearly 5% of sugar is employed for biomass yield, which is comparable to the biomass yields in industrially important heterotrophic microorganisms [78].

The high tolerance of *Galdieria* species to concentrated substances is probably an adaptation to the high concentrations of sulfuric acid and other salts present in acidic springs. *G. sulphuraria* tolerated and grew well concentrations of glucose and fructose of up to 166 g/L (0.9 M) and an ammonium sulphate concentration of 22 g/L (0.17 M) without negative effects on specific growth rate. In carbon-limited fed-batch cultures, biomass dry weight concentrations of 80–120 g/L were obtained while phycocyanin accumulated to concentrations between 250 and 400 mg/L [42].

The ability of *G. sulphuraria* to accumulate high levels of phycocyanin in heterotrophic or mixotrophic cultures compete with or at least represents an alternative to the cyanobacterium, *Arthrospira (Spirulina) platensis* that is currently used for synthesis of phycocyanin [77].

Since a number of positive health effects have been associated with phycocyanin [79], and phycocyanin from *A. platensis* has been approved for food use in the USA and EU in 2013 and 2014, respectively, interests in applications of phycocyanin have increased substantially over recent years [80].

In addition to phycocyanin, *G. sulphuraria* could also provide floridosides suggested as a commercial products [81–83]. Its biomass was also tested and found to be a suitable and safe component in foods, as well as a dietary supplement [84].

5.4 Biofuels

The world-wide and continuous increase in fossil fuel consumption, leading probably in the relatively near future to an exhaustion of resources, has led to increased research for alternative energy sources. Production of biofuels by algae might provide a viable alternative to fossil fuels; however, this technology must overcome a number of serious obstacles before it could compete in the fuel market and be broadly deployed. Application of remarkably extremophilic *G. sulphuraria* could overcome at least some of these.

Microalgae often become contaminated with other microorganisms in large-scale outdoor cultivations, which is a major problem that inhibits algal growth and decreases the quality of biofuel and high-value products. A lack of resistance to these factors could be catastrophic for future algae farmers. The red alga *G. sulphuraria* and other species of the same genus have great potential to produce large quantities of biofuel [53] and other beneficial compounds without becoming contaminated with other microorganisms, under both mixotrophic and heterotrophic conditions. Furthermore, the algae are tolerant of pH and temperature extremes that offer a reliable means of controlling the composition of large-scale cultures.

5.5 Glycogen

The extremophilic red algae, similarly to other Rhodophyta, produce glycogen as energy and carbon reserves, instead of starch, which is characteristic of other microalgae and higher plants [81]. Glycogen, in contrast to starch, is readily soluble in cold water and more accessible by enzymes. In red algae, glycogen accumulates in a lower molecular weight form than glycogen from other microalgae and is a highly branched (higher than any other glycogen) glucose polymer [81]. Amylopectin, as a highly branched glucose polymer in starch, is used in various products such as peritoneal dialysis solutions and sports drinks. However, it is costly to prepare because of its insoluble, granular nature. The application of glycogen offers a cheaper alternative.

The alga *G. sulphuraria* can grow to a very high biomass concentration [42], accumulating glycogen up to 50% of the dry cell weight. Another advantage of this alga is that it can grow heterotrophically using many organic sources and also very cheap waste glycerol [84].

Apart from *G. sulphuraria*, the production of glycogen by most other microorganisms is too low for biotechnological applications because it is produced only under growth limiting conditions [85]. Thus, *G. sulphuraria* can be used as a cheap and efficient producer of glycogen, which could be applied as an alternative to starch in several fields [83].

Large amounts of glycogen were obtained in mixotrophic cultures, [53] where the maximum glycogen content per mL of culture was almost 10- and 2-fold greater than those of autotrophic and heterotrophic cultures, respectively. The accumulation of glycogen was enhanced by the addition of glucose, and the amount and composition of glycogen were determined by growth conditions. It is assumed that in addition to glycogen, other forms of carbon may be stored, although pathways are, as yet, unknown [75, 86].

5.6 Nutritional applications

Because of their high content of protein, algal biomass, in general, and green algae particularly, have been used in many foods, mostly in the form of dried biomass.

Difficulties in introducing microalgal-based ingredients into foods are technological and include sensorial obstacles such as its unattractive green-brownish color and unpleasant fishy smell increasing after longer storing [87]. Another problem is bacterial contamination, which decreases the commercial quality of algal biomass. Such disadvantages of green algae are not encountered using *Galdieria* species. *G. sulphuraria* can grow heterotrophically even in large-scale bioreactors under so extreme conditions that contamination by other organisms is not likely. Biomass is colorless, has a low lipid content, mainly of monounsaturated fatty acids, and oxidation during shelf life is negligible. Heterotrophic growth enables high cell densities to be achieved using cheap glycerol as a source of carbon. Consequently, in addition to other specific applications, red algal biomass can be used as a source of protein and other macronutrients. *G. sulphuraria* is rich in proteins (26–32%) and polysaccharides (63–69%), and poor in lipids. Under heterotrophic cultivation conditions, the lipid moiety mainly contained monounsaturated fatty acids. Nutritional applications of red algae were firstly suggested by Bailey and Staehelin [87], who found very high levels of protein in their cell walls.

G. sulphuraria proteins are strictly associated with polysaccharide components and therefore not digestible. However, a commercial enzyme preparation containing a mixture of polysaccharidases was developed, and *G. sulphuraria* proteins were good substrates for human gastrointestinal enzymes. *G. sulphuraria* biomass therefore has the potential to be used either for protein-rich or for insoluble dietary fiber-rich applications. Among micronutrients, some B group vitamins and pigments are present. Carotenoids are minor pigments in *G. sulphuraria*, detected only in the autotrophic algae, the main ones being astaxanthin and lutein. The absence of carotenoids under heterotrophic growth conditions is due to the lack of photosynthesis. Phycobiliproteins are present under heterotrophic and mixotrophic cultivation conditions. The cells grown on organic source of carbon frequently lose their photosynthetic antenna undermining the accumulation of the phycobilins. In *G. sulphuraria*, allophycocyanin is the dominant form in the autotrophic algae, while phycoerithrin was the main phycobiliprotein in the heterotrophic algae.

G. sulphuraria can therefore be used to develop new food ingredients, including preparations that are rich in bioavailable proteins and dietary fiber [84].

6. Conclusions

The unicellular red alga *Galdieria* is an amazing organism, not only because of its ability to live in extreme habitats, but also because of its metabolic flexibility to grow photo-autotrophically, heterotrophically or mixotrophically on diverse carbon sources. These properties make it a suitable candidate for biotechnological applications, including those inappropriate for other organisms. The biotechnological potential of *Galdieria* lies in the tremendous variability of cultivation conditions under which it is able to grow, including the ability to accumulate toxic heavy metals as well as rare ones such as Au, Pt or lanthanides. This alga can also produce biotechnologically attractive products such as phycocyanin, glycogen, protein-rich or insoluble dietary fiber-rich additives for nutritional applications. This opens up fields for its use in a wide range of industries such as the food industry, pharmaceuticals, healthcare, waste management, the metal recycling industry, bioremediation, wastewater management, etc. *Galdieria* is certainly worth the effort to research it for future uses.

Acknowledgements

We thank prof. J. D. Brooker for critical reading and language corrections of the manuscript. We are grateful to Dr. Jens Hartmann, Donau University, Krems, Austria, for the electron-microscopy photograph.

This work was supported by the National Program of Sustainability I, ID: LO1416 and by the Program Interreg V-A Austria-Czech Republic, project ATCZ172 REEgain.

Author details

Mária Čížková*, Milada Vítová and Vilém Zachleder
Institute of Microbiology, The Czech Academy of Sciences, Centre Algatech, Třeboň, The Czech Republic

*Address all correspondence to: cizkova@alga

IntechOpen

© 2019 The Author(s). Licensee IntechOpen. This chapter is distributed under the terms of the Creative Commons Attribution License (http://creativecommons.org/licenses/by/3.0), which permits unrestricted use, distribution, and reproduction in any medium, provided the original work is properly cited.

References

[1] Yoshimura E, Nagasaka S, Sato Y, Satake K, Mori S. Extraordinary high aluminium tolerance of the acidic thermophilic alga, *Cyanidium caldarium*. Soil Science & Plant Nutrition. 1999;**45**:721-724. DOI: 10.1080/00380768.1999.10415835

[2] Nagasaka S, Nishizawa NK, Watanabe T, Mori S, Yoshimura E. Evidence that electron-dense bodies in *Cyanidium caldarium* have an iron-storage role. Biometals. 2003;**16**:465-470. DOI: 10.1023/A:1022563600525

[3] Varshney P, Mikulic P, Vonshak A, Beardall J, Wangikar PP. Extremophilic micro-algae and their potential contribution in biotechnology. Bioresource Technolology. 2015;**184**:363-372. DOI: 10.1016/j.biortech.2014.11.040

[4] Yoon HS, Hackett JD, Ciniglia C, Pinto G, Bhattacharya D. A molecular timeline for the origin of photosynthetic eukaryotes. Molecular Biology and Evolution. 2004;**21**(5):809-818. DOI: 10.1093/molbev/msh075

[5] Yoon HS, Ciniglia C, Wu M, Comeron JM, Pinto G, Pollio A, et al. Establishment of endolithic populations of extremophilic Cyanidiales (Rhodophyta). BMC Evolutionary Biology. 2006;**6**:78. DOI: 10.1186/1471-2148-6-78

[6] De Luca P, Taddei R, Varano L. *Cyanidioschyzon merolae*: A new alga of thermal acidic environments. Webbia. 1978;**33**:37-44. DOI: 10.1080/00837792.1978.10670110

[7] Gross W, Schnarrenberger C. Heterotrophic growth of two strains of the acido-thermophilic red alga *Galdieria sulphuraria*. Plant and Cell Physiology. 1995;**36**:633-638. DOI: 10.1093/oxfordjournals.pcp.a078803

[8] Merola A, Castaldo R, De Luca P, Gambardella R, Musacchio A, Taddei R. Revision of *Cyanidium caldarium.* Three species of acidophilic algae. Giornale Botanica Italiana. 1981;**115**:189-195. DOI: 10.1080/11263508109428026

[9] Ciniglia C, Yoon HS, Pollio A, Pinto G, Bhattacharya D. Hidden biodiversity of the extremophilic *Cyanidiales* red algae. Molecular Ecology. 2004;**13**(7):1827-1838. DOI: 10.1093/oxfordjournals.pcp.a078803

[10] Ciniglia C, Yang EC, Pollio A, Pinto G, Iovinella M, Vitale L, et al. Cyanidiophyceae in Iceland: Plastid *rbc*L gene elucidates origin and dispersal of extremophilic *Galdieria sulphuraria* and *G. maxima* (Galdieriaceae, Rhodophyta). Phycologia. 2014;**53**(6):542-551. DOI: 10.2216/14-032.1

[11] Cozzolino A, Caputo P, De Castro O, Moretti S, Pinto G. Molecular variation in *Galdieria sulphuraria* (Galdieri) Merola and its bearing on taxonomy. Hydrobiologia. 2000;**433**:145-151. DOI: 10.1023/A:1004035224715

[12] Pinto G, Albertano P, Ciniglia C, Cozzolino S, Pollio A, Yoon HS, et al. Comparative approaches to the taxonomy of the genus *Galdieria merola* (Cyanidiales, Rhodophyta). Cryptogamie Algologie. 2003;**24**:12-32

[13] Barcytė D, Elster J, Nedbalová L. Plastid-encoded rbcL phylogeny suggests widespread distribution of *Galdieria phlegrea* (Cyanidiophyceae, Rhodophyta). Nordic Journal of Botany. 2018;**36**:e01794. DOI: 10.1111/njb.01794.1-6.

[14] Pinto G, Ciniglia C, Cascone C, Pollio P. Species composition of cyanidiales assemblages in Pisciarelli (Campi Flegrei, Italy) and

description of *Galdieria phlegrea* sp. nov. In: Seckbach J, editor. Algae and Cyanobacteria in Extreme Environments. Cellular Origin, Life in Extreme Habitats and Astrobiology. 11. Dordrecht: Springer; 2007. pp. 489-502. DOI: 10.1007/978-1-4020-6112-7_26

[15] Sentsova OJ. The study of Cyanidiophyceae in Russia. In: Seckbach J, editor. Evolutionary Pathways and Enigmatic Algae: *Cyanidium caldarium* (Rhodophyta) and Related Cells Developments in Hydrobiology. 91. Dordrecht: Kluwer Academic Publishers; 1994. pp. 167-174. DOI: 10.1007/978-94-011-0882-9_12

[16] Gross W, Heilmann I, Lenze D, Schnarrenberger C. Biogeography of the Cyanidiaceae (Rhodophyta) based on 18S ribosomal RNA sequence data. European Journal of Phycology. 2001;**36**(3):275-280. DOI: 10.1080/09670260110001735428

[17] Gross W, Küver J, Tischendorf G, Bouchaala N, Büsch W. Cryptoendolithic growth of the red alga *Galdieria sulphuraria* in volcanic areas. European Journal of Phycology. 1998;**33**(1):25-31. DOI: 10.1080/09670269810001736503

[18] Ferris MJ, Sheehan KB, Kuhl M, Cooksey K, Wigglesworth-Cooksey B, Harvey R, et al. Algal species and light microenvironment in a low-pH, geothermal microbial mat community. Applied and Environmental Microbiology. 2005;**71**(11):7164-7171. DOI: 10.1128/AEM.71.11.7164-7171.2005

[19] Skorupa DJ, Reeb V, Castenholz RW, Bhattacharya D, McDermott TR. Cyanidiales diversity in Yellowstone national park. Letters of Applied Microbiology. 2013;**57**(5):459-466. DOI: 10.1111/lam.12135

[20] Toplin JA, Norris TB, Lehr CR, McDermott TR, Castenholz RW. Biogeographic and phylogenetic diversity of thermoacidophilic cyanidiales in Yellowstone National Park, Japan, and New Zealand. Applied and Environmental Microbiology. 2008;**74**(9):2822-2833. DOI: 10.1128/AEM.02741-07

[21] Barcyte D, Nedbalová L, Culka A, Košek F, Jehlička J. Burning coal spoil heaps as a new habitat for the extremophilic red alga *Galdieria sulphuraria*. Fottea. 2018;**18**:19-29. DOI: 10.5507/fot.2017.015

[22] Gross W, Oesterhelt C, Tischendorf G, Lederer F. Characterization of a non-thermophilic strain of the red algal genus *Galdieria* isolated from Soos (Czech Republic). European Journal of Phycology. 2002;**37**(3):477-482. DOI: 10.1017/S0967026202003773

[23] Albertano P, Ciniglia C, Pinto G, Pollio A. The taxonomic position of *Cyanidium*, *Cyanidioschyzon* and *Galdieria:* An update. Hydrobiologia. 2000;**433**(1993):137-143. DOI: 10.1023/A:1004031123806

[24] Ott FD, Seckbach J. New classification of the genus *Cyanidium* Geitler 1933. In evolutionary pathways and enigmatic algae: *Cyanidium caldarium* (Rhodophyta) and related cells. In: Seckbach J, editor. Evolutionary Pathways and Enigmatic Algae: *Cyanidium caldarium* (Rhodophyta) and Related Cells Developments in Hydrobiology. 91. Dordrecht: Springer, Netherlands; 1994. pp. 113-132. DOI: 10.1007/978-94-011-0882-9_9

[25] Seckbach J. In: Seckbach J, editor. Evolutionary Pathways and Enigmatic Algae: *Cyanidium caldarium* (Rhodophyta) and Related Cells. Dordrecht: Kluwer Academic Publisher; 1994. p. 349. DOI: 10.1007/978-94-011-0882-9

[26] Reeb V, Bhattacharya D. The thermo-acidophilic Cyanidiophyceae (Cyanidiales). In: Seckbach J,

Chapman DJ, editors. Red Algae In the Genomic Age. 13. Netherlands: Springer; 2010. pp. 409-426. DOI: 10.1007/978-90-481-3795-4_22

[27] Hsieh CJ, Zhan SH, Lin Y, Tang SL, Liu SL. Analysis of rbcL sequences reveals the global biodiversity, community structure, and biogeographical pattern of thermoacidophilic red algae (Cyanidiales). Journal of Phycology. 2015;**51**:682-694. DOI: 10.1111/jpy.12310

[28] Castenholz RW, McDermott TR. The Cyanidiales: Ecology, biodiversity, and biogeography. In: Seckbach J, Chapman DJ, editors. Red Algae in the Genomic Age. Dordrecht: Springer; 2010. pp. 357-271. DOI: 10.1007/978-90-481-3795-4_19

[29] Gross W, Gross S. Hot Sulphur Springs at Sao Miguel Island (Azores): A new locality of the acidophilic red alga *Galdieria sulphuraria* (Galdieri) Merola. Algological Studies. 2002;**107**:109-116. DOI: 10.1127/algol_stud/107/2002/109

[30] Brock TD. Thermophilic Microorganisms Abd Life at High Temperatures. New York, Berlin: Springer-Verlag; 1978. DOI: 10.1002/esp.3760050112

[31] Brock TD. Lower pH limit for the existence of bluegreen algae: Evolutionary and ecological implications. Science. 1973;**179**:480-483. DOI: 10.1126/science.179.4072.480

[32] Doemel WN, Brock TD. The upper temperature limit of *Cyanidium caldarium*. Archives of Microbiology. 1970;**72**:326-332. DOI: 10.1007/BF00409031

[33] Rothschild LJ, Mancinelli RL. Life in extreme environments. Nature. 2001;**409**:1092-1101. DOI: 10.1038/35059215

[34] Donachie SP, Christenson BW, Kunkel DD, Malahoff A, Alam M. Microbial community in acidic hydrothermal waters of volcanically active White Island, New Zealand. Extremophiles. 2002;**6**:419-425. DOI: 10.1007/s00792-002-0274-7

[35] Oesterhelt C, Gross W. Different sugar kinases are involved in the sugar sensing of *Galdieria sulphuraria*. Plant Physiology. 2002;**128**(1):291-299. DOI: 10.1104/pp.010553

[36] Barbier G, Oesterhelt C, Larson MD, Halgren RG, Wilkerson C, Garavito RM, et al. Genome analysis. Plant Physiology. 2005;**137**:460-474. DOI: 10.1104/pp.104.051169.460

[37] Sentsova OY. Diversity of acido thermophilic unicellular algae of the genus *Galdieria* (Rhodophyta, Cyanidiophyceae). Botanicheskii Zhurnal. 1991;**76**:69-79

[38] Muravenko O, Selyakh I, KononenkoN,StadnichukI.Chromosome numbers and nuclear DNA contents in the red microalgae *Cyanidium caldarium* and three *Galdieria* species. European Journal of Phycology. 2001;**36**(3):227-232. DOI: 10.1080/09670260110001735378

[39] Moreira D, Lopez-Archilla A-J, Amils R, Marin J. Characterization of two new thermoacidophilic microalgae: Genome organization and comparison with *Galdieria sulphuraria*. FEMS Microbiology Letters. 1994;**122**:109-114. DOI: 10.1111/j.1574-6968.1994.tb07152.x

[40] Derelle E, Ferraz C, Lagoda P, Eychenie S, Cooke R, Regad F, et al. DNA libraries for sequencing the genome of *Ostreococcus tauri* (Chlorophyta, Prasinophyceae): The smallest free-living eukaryotic cell. Journal of Phycology. 2002;**38**(6):1150-1156. DOI: 10.1046/j.1529-8817.2002.02021.x

[41] Jain K, Krause K, Grewe F, Nelson GF, Weber AP, Christensen AC, et al. Extreme features of the *Galdieria sulphuraria* organellar genomes: A consequence of polyextremophily? Genome Biology and Evolution. 2015;7(1):367-380. DOI: 10.1093/gbe/evu290

[42] Schmidt RA, Wiebe MG, Eriksen NT. Heterotrophic high cell-density fed-batch cultures of the phycocyanin-producing red alga *Galdieria sulphuraria*. Biotechnology and Bioengineering. 2005;**90**(1):77-84. DOI: 10.1002/bit.20417

[43] Gravelholt OS, Erickson NT. Heterotrophic high-cell-density fed-batch and continuous-flow cultures of *Galdieria sulphuraria* and production of phycocyanin. Applied Microbiology and Biotechnology. 2007;**77**(1):69-75. DOI: 10.1007/s00253-007-1150-2

[44] Oesterhelt C, Schmalzlin E, Schmitt JM, Lokstein H. Regulation of photosynthesis in the unicellular acidophilic red alga *Galdieria sulphuraria*. The Plant Journal. 2007;**51**(3):500-511. DOI: 10.1111/j.1365-313X.2007.03159.x

[45] Vítová M, Goecke F, Sigler K, Řezanka T. Lipidomic analysis of the extremophilic red alga *Galdieria sulphuraria* in response to changes in pH. Algal Research. 2016;**13**:218-226. DOI: 10.1016/j.algal.2015.12.005

[46] Boyanov MI, Kelly SD, Kemner KM, Bunker BA, Fein JB, Fowle DA. Adsorption of cadmium to *Bacillus subtilis* bacterial cell walls: a pH-dependent X-ray adsorption fine structure spectroscopy study. Geochimica et Cosmochima Acta 2003;67:3299-3311. DOI: 10.1016/S0016-7037(02)01343-1

[47] Cordero B, Lodeiro P, Herrero R, Sastre de Vicente ME. Biosorption of cadmium by *Fucus spiralis*. Environment and Chemistry. 2004;**1**:180-187. DOI: 10.1071/EN04039

[48] Jiang MY, Ohnuki T, Kozai N, Tanaka K, Suzuki Y, Sakamoto F, et al. Biological nano-mineralization of Ce phosphate by *Saccharomyces cerevisiae*. Chemical Geology. 2010;**277**(1-2):61-69. DOI: 10.1016/j.chemgeo.2010.07.010

[49] Hosomomi Y, Baba Y, Kubota F, Kamiya N, Goto M. Biosorption of rare earth elements by *Escherichia coli*. Journal of Chemical Engineering of Japan. 2013;**46**:450-454. DOI: 10.1252/jcej.13we031

[50] Schönknecht G, Chen WH, Ternes CM, Barbier GG, Shrestha RP, Stanke M, et al. Gene transfer from bacteria and archaea facilitated evolution of an extremophilic eukaryote. Science. 2013;339:1207-1210. DOI: 10.1126/science.1231707

[51] Selvaratnam T, Pegallapati AK, Montelya F, Rodriguez G, Nirmalakhandan N, Van Voorhies W, et al. Evaluation of a thermo-tolerant acidophilic alga, *Galdieria sulphuraria*, for nutrient removal from urban wastewaters. Bioresource Technology. 2014;**156**:395-399. DOI: 10.1016/j.biortech.2014.01.075

[52] Minoda A, Sawada H, Suzuki S, Miyashita S, Inagaki K, Yamamoto T, et al. Recovery of rare earth elements from the sulfothermophilic red alga *Galdieria sulphuraria* using aqueous acid. Applied Microbiology and Biotechnology. 2015;**99**(3):1513-1519. DOI: 10.1007/s00253-014-6070-3

[53] Sakurai T, Aoki M, Ju X, Ueda T, Nakamura Y, Fujiwara S, et al. Profiling of lipid and glycogen accumulations under different growth conditions in the sulfothermophilic red alga *Galdieria sulphuraria*. Bioresource Technology. 2016;**200**:861-866. DOI: 10.1016/j.biortech.2015.11.014

[54] Ju X, Igarashi K, Miyashita S, Mitsuhashi H, Inagaki K, Fujii S-I, et al. Effective and selective recovery of gold and palladium ions from metal wastewater using a sulfothermophilic red alga, *Galdieria sulphuraria*. Bioresource Technolology. 2016;**211**:759-764. DOI: 10.1016/j.biortech.2016.01.061

[55] Abreu RD, Morais CA. Purification of rare earth elements from monazite sulphuric acid leach liquor and the production of high-purity ceric oxide. Minerals Engineering. 2010;**23**:536-540. DOI: 10.1016/j.mineng.2010.03.010

[56] Sethurajan M, Lens PNL, Horn HA, Figueiredo LHA, van Hullebusch ED. Leaching and recovery of metals. In: Rene ER, Lewis A, Sahynkaya E, PNL L, editors. Sustainable Heavy Metal Remediation Environmental Chemistry for a Sustainable World. 2. Cham, Switzerland: Springer; 2017. pp. 161-206. DOI: 10.1007/978-3-319-61146-4

[57] Pollmann K, Kutschke S, Matys S, Raff J, Hlawacek G, Lederer FL. Bio-recycling of metals: Recycling of technical products using biological applications. Biotechnology Advances. 2018;**36**:1048-1062. DOI: 10.1016/j.biotechadv.2018.03.006

[58] Jacinto J, Henriques B, Duarte AC, Vale C, Pereira E. Removal and recovery of critical rare elements from contaminated waters by living *Gracilaria gracilis*. Journal of Hazardous Materials. 2018;**344**:531-538. DOI: 10.1016/j.jhazmat.2017.10.054

[59] Isildar A, van Hullebusch ED, Lenz M, Du Laing G, Marra A, Cesaro A, et al. Biotechnological strategies for the recovery of valuable and critical raw materials from waste electrical and electronic equipment (WEEE): A review. Journal of Hazardous Materials. 2019;**362**:467-481. DOI: 10.1016/j.jhazmat.2018.08.050

[60] Das N, Das D. Recovery of rare earth metals through biosorption: An overview. Journal of Rare Earths. 2013;**31**(10):933-943. DOI: 10.1016/s1002-0721(13)60009-5

[61] Mehta SK, Gaur JP. Use of algae for removing heavy metal ions from wastewater: Progress and prospects. Critical Review of Biotechnology. 2005;**25**:113-152. DOI: 10.1080/07388550500248571

[62] Wang JL, Chen C. Biosorbents for heavy metals removal and their future. Biotechnology Advances. 2009;**27**(2):195-226. DOI: 10.1016/j.biotechadv.2008.11.002

[63] Čížková M, Bišová K, Zachleder V, Mezricky D, Rucki M, Vítová M. Recovery of rare earth elements from luminophores using the red alga *Galdieria*. In: 16th International Conference on Environmental Science and Technology; September 2019; Rhodes, Greece; 2019. pp. 1-3

[64] Vítová M, Čížková M, Zachleder V. Lanthanides and algae. In: Awwad NS, Mubarak AT, editors. Lanthanides. London, United Kingdom: InTech Open; 2019. pp. 87-111. DOI: 10.5772/intechopen.80260

[65] Ujaczki E, Feigl V, Molnar M, Cusack P, Curtin T, Courtney R, et al. Re-using bauxite residues: Benefits beyond (critical raw) material recovery. Journal of Chemical Technology and Biotechnology. 2018;**93**:2498-2510. DOI: 10.1002/jctb.5687

[66] Cusack PB, Courtney R, Healy MG, O'Donoghue LMT, Ujaczki E. An evaluation of the general composition and critical raw material content of bauxite residue in a storage area over a twelve-year period. Journal of Clean Products. 2019;**208**:393-401. DOI: 10.1016/j.jclepro.2018.10.083

[67] Čížková M, Mezricky D, Rucki M, Tóth TM, Náhlík V, Lanta V, et al.

Bio-mining of lanthanides from red mud by green microalgae. Molecules. 2019;**24**:1-19. DOI: 10.3390/molecules24071356

[68] Glazer AN. Phycobiliproteins—A family of valuable, widely used fluorophores. Journal of Applied Phycology. 1994;**6**:105-112. DOI: 10.1007/BF02186064

[69] Glazer AN, Stryer L. Phycofluor probes. Trends in Biochemical Sciences. 1984;**9**:423-427. DOI: 10.1016/0968-0004(84)90146-4

[70] Benemann JR, Tillett DM, Weissman JC. Microalgae biotechnology. Trends in Biotechnology. 1987;**5**(2):47-53. DOI: 10.1016/0167-7799(87)90037-0

[71] Rimbau V, Camins A, Romay C, González R, Pallàs M. Protective effects of C-phycocyanin against kainic acid-induced neuronal damage in rat hippocampus. Neuroscience Letters. 1999;**276**:75-78. DOI: 10.1016/S0304-3940(99)00792-2

[72] Romay C, Armesto J, Remirez D, González D, Ledon N, Garcìa I. Antioxidant and anti-inflammatory properties of C-phycocyanin from blue-green algae. Inflamation Research. 1998;**47**:36-41. DOI: 10.1007/s000110050256

[73] Chen F. High cell density culture of microalgae in heterotrophic growth. Trends in Biotechnology. 1996;**14**:421-426. DOI: 10.1016/0167-7799(96)10060-3

[74] Marquardt J. Effects of carotenoid-depletion on the photosynthetic apparatus of a *Galdieria sulphuraria* (Rhodophyta) strain that retains its photosynthetic apparatus in the dark. Journal of Plant Physiology. 1998;**152**(4-5):372-380. DOI: 10.1016/s0176-1617(98)80250-2

[75] Oesterhelt C, Schnarrenberger C, Gross W. Characterization of a sugar/polyol uptake system in the red alga *Galdieria sulphuraria*. European Journal of Phycology. 1999;**34**(3):271-277. DOI: 10.1080/09670269910001736322

[76] Sloth JK, Jensen HC, Pleissner D, Eriksen NT. Growth and phycocyanin synthesis in the heterotrophic microalga *Galdieria sulphuraria* on substrates made of food waste from restaurants and bakeries. Bioresource Technology. 2017;**238**:296-305. DOI: 10.1016/j.biortech.2017.04.043

[77] Sloth JK, Wiebe MG, Eriksen NT. Accumulation of phycocyanin in heterotrophic and mixotrophic cultures of the acidophilic red alga *Galdieria sulphuraria*. Enzyme and Microbial Technology. 2006;**38**(1-2):168-175. DOI: 10.1016/j.enzmictec.2005.05.010

[78] Roels JA. Applications of macroscopic principles to microbial metabolism. Biotechnology Bioengeneering. 1980;**22**:2457-2514. DOI: 10.1002/bit.22325

[79] Fernández-Rojas B, Medina-Campos ON, Hernández-Pando R, Negrette-Guzmán M, Huerta-Yepez S, Pedraza-Chaverri J. C-Phycocyanin prevents cisplatin-induced nephrotoxicity through inhibition of oxidative stress. Food & Function. 2014;**5**:480-490. DOI: 10.1039/c3fo60501a

[80] Eriksen NT. Research trends in the dominating microalgal pigments, ß carotene, astaxanthin, and phycocyanin used in feed, in foods, and in health applications. Journal of Nutrition & Food Sciences. 2016;**6**:1-6. DOI: 10.4172/2155-9600.1000507

[81] Martinez-Garcia M, Kormpa A, van der Maarel MJEC. The glycogen of *Galdieria sulphuraria* as alternative to starch for the production of slowly digestible and resistant glucose polymers. Carbohydrate Polymers. 2017;**169**:75-82. DOI: 10.1016/j.carbpol.2017.04.004

[82] Martinez-Garcia M, van der Maarel MJEC. Floridoside production by the red microalga *Galdieria sulphuraria* under different conditions of growth and osmotic stress. AMB Express. 2016;**6**:71. DOI: 10.1186/s13568-016-0244-6

[83] Martinez-Garcia M, Stuart MCA, van der Maarel MJ. Characterization of the highly branched glycogen from the thermoacidophilic red microalga *Galdieria sulphuraria* and comparison with other glycogens. International Journal of Biological Macromolecules. 2016;**89**:12-18. DOI: 10.1016/j.ijbiomac.2016.04.051

[84] Graziani G, Schiavo S, Nicolai MA, Buono S, Fogliano V, Pinto G, et al. Microalgae as human food: Chemical and nutritional characteristics of the thermo-acidophilic microalga *Galdieria sulphuraria*. Food & Function. 2013;**4**(1):144-152. DOI: 10.1039/c2fo30198a

[85] Preiss J. Bacterial glycogen synthesis and its regulation. Annual Reviews in Microbiology. 1984;**38**:419-458. DOI: 10.1146/annurev.mi.38.100184.002223

[86] Lammers PJ, Selvaratnam S, Nirmalakhandan N, Reddy H, Deng S. A route to algal biofuels via ecosystem services: the case for an extremophile, *Galdieria sulphuraria*. In: 2nd International Symposium on Energy Challenges and Mechanics; 19-21 August 2014; Aberdeen, Scotland, UK; 2014

[87] Bailey RW, Staehelin LA. The chemical composition of isolated cell wall of *Cyanidium caldarium*. Journal of General Microbiology. 1968;**54**:269-276. DOI: 10.1099/00221287-54-2-269

Chapter 7

The Colonial Microalgae *Botryococcus braunii* as Biorefinery

Edmundo Lozoya-Gloria, Xochitl Morales-de la Cruz and Takehiro A. Ozawa-Uyeda

Abstract

The growing shortage of fossil fuels caused an increase in the demand for alternative and renewable fuels. Biofuels, like bioethanol and biodiesel, have received more attention as a sustainable replacement of fossil fuels. However, these have a poor oxidative stability, little energy content by volume, and many oxygenated compounds, which may cause corrosion and damage to the engines. Therefore, they are used as a mixture with standard fuels. Some species of microalgae are candidates to produce oils as triglycerides (TGA) to produce biodiesel by transesterification; however, the problem will remain. The colonial microalgae *Botryococcus braunii* produces and accumulates a high amount of long-chain nonoxygenated hydrocarbons, similar to those obtained from the fractionated distillation of crude petroleum. This is one of the few organisms reported to have a direct contribution in the formation of the oil reserves currently in use. Additionally, *B. braunii* produces pigments and long-chain carbohydrates that have interesting properties for various industries. There are still problems to be solved in order to consider it as economically viable and profitable, but important progress is being made. Therefore, this microalga is very attractive for the synthesis of hydrocarbons and other value-added compounds, making it an interesting biorefinery organism.

Keywords: biorefinery, *Botryococcus*, exopolysaccharides, hydrocarbons, lipids, pigments

1. Introduction

Botryococcus braunii is a colonial microalga Trebouxiophyceae, distributed in brackish and sweet water [1]. It reaches densities of 1.4×10^6 colonies/L [2], and its geochemistry significance is important. Paleobotanical studies suggest that it is one of the largest sources of hydrocarbons in oil-rich deposits dating back to the Ordovician period [1, 3–5]. It is the only colonial microalga that accumulates and secrets liquid hydrocarbons (**Figure 1**), and depending on the strain and growing conditions, race B can accumulate hydrocarbons up to 85% and race A up to 61% of their dry weight.

B. braunii is related with *Characium vaculatum and Dunaliella parva* [1]. Due to the hydrocarbons and the molecular phylogeny of *B. braunii* [6], it is classified in three races (A, B, and L). Race A produces *n*-alkadienes and alkatrienes of C_{23}–C_{33}

IntechOpen

[7], although two unusual hydrocarbons have been characterized, the triene $C_{27}H_{51}$ and tetraene $C_{27}H_{48}$ [1]. Race A hydrocarbon dry weight varies from 0.4 to 61% [7, 8]. Race B produces triterpenoids hydrocarbons known as botryococcenes (C_nH_{2n-10}, n = 30-37) [9] and methylsqualenes C_{31}–C_{34} [10, 11]. The botryococcenes can be from 27 to 86% of the dry weight [12]. Race L produces a tetraterpene C_{40} known as lycopadiene and constitutes from 0.1 to 8% of the dry weight [13, 14]. This race contains 5% of lycopatriene, lycopatetraene, lycopapentaene, and lycopahexaene [15]. In addition, a race S is proposed, which synthesizes saturated *n*-alkanes C_{18} and C_{20}, and epoxy-alkanes; however, its existence is not yet fully accepted [6].

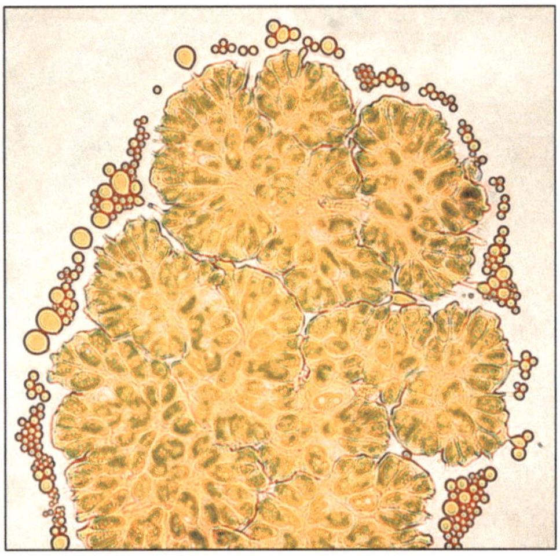

Figure 1.
B. braunii race B colony secreting liquid hydrocarbons.

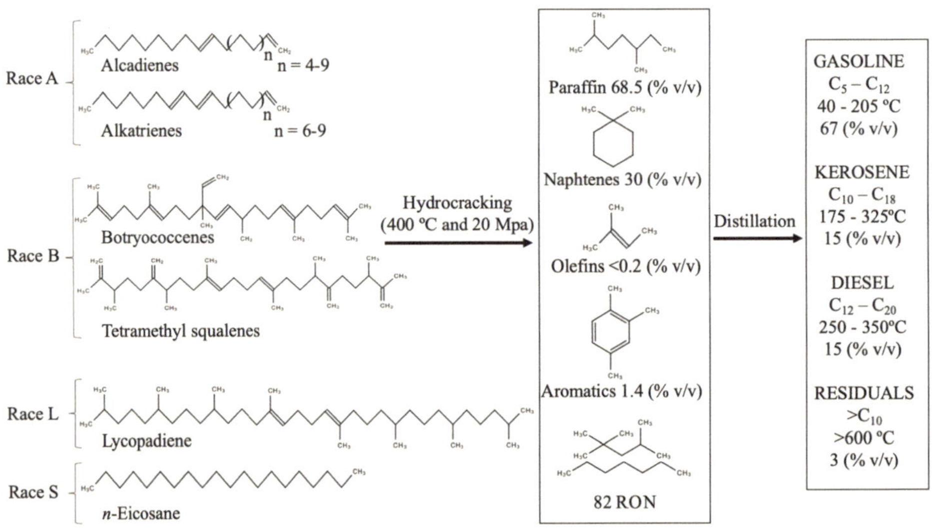

Figure 2.
Hydrocarbons produced by the B. braunii races. Biofuels derived from race B are shown. RON, research octane number = 92–98, this is a measure of autoignition resistance in a spark-ignition engine. In the USA: regular (97 RON) and premium (95 RON). Adapted from [16–18].

After the hydrocracking process and subsequent distillation, race B hydrocarbons become biofuels currently used in internal combustion engines [16] as shown in **Figure 2**.

2. Physiology and biochemistry of *Botryococcus braunii*

B. braunii races differ also by its morphological and physiological characteristics. Cells from A and B races are of 13 μm × 7–9 μm, and those of L race are 8–9 μm × 5 μm [19].

Each colony is constituted by a group of 50–100 piriform cells embedded in a hydrocarbon network and the extracellular matrix (ECM). This ECM contains three main components: (1) a fibrous cell wall surrounding each cell and having β-1,4- and/or β-1,3-glucans including cellulose; (2) the intracolonial space constituted by a network of liquid hydrocarbons; and (3) a fibrillary sheath composed mainly of arabinose and galactose polysaccharides, holding the liquid hydrocarbons [20].

B. braunii may have a hetero-, mixo-, or phototrophic grow and the morphology will depend on the C source and the amount of light [21]. The hydrocarbon production is associated with the cell division [22], likely due to the localization of the enzymes involved in the alkadienes, alkatrienes (race A), and botryococcenes (race B) biosynthesis [23].

Other difference among the races is the keto-carotenoid accumulation in the stationary phase of cultures. Races B and L change color from green-brown to orange, and race A changes from green to yellow-orange [1]. The production of carotenoids is also a stress response by environmental factors. The *DAD1* gene expression, a suppressor of programmed cell death, was reported in race B, under stress conditions at 10–60 min [24]. *B. braunii* is tolerant to desiccation and extreme temperatures, which allows its global dispersion in different environments [25]. The reproduction mechanism of *B. braunii* seems to be autosporic [26].

Symbiotic bacteria have been reported after microscopic observations, and an ectosymbiont α-proteobacteria (BOTRYCO-2) that promotes the productivity of biomass and hydrocarbons was described [2, 27].

2.1 Biosynthesis of alkadienes and alkatrienes

Characteristic alkadienes and alkatrienes of race A have double links and similar stereochemistry as oleic acid. Experiments with labeled fatty acids have shown that this one is the main precursor by the long-chain fatty acids (LCFAs) pathway, followed by a decarboxylation process [1, 17, 28, 29]. The first step is the elongation of oleic acid (18:1 *cis*-Δ9) and its isomer elaidic acid (18:1 *trans*-Δ9). The acyl-CoA reductase and decarbonylase enzymes in race A microsomes suggest an alternative mechanism where the LCFAs are reduced to aldehydes and decarbonylated to produce alkadienes and alkatrienes [17, 30]. Race A transcriptome allowed the identification of six candidate genes potentially involved in this biosynthesis [31].

2.2 Biosynthesis of botryococcenes

The analysis of race B transcriptome and other evidences suggests that the biosynthesis of isoprenoids comes from the deoxyxylulose phosphate/methylerythritol phosphate (DXP/MEP) pathway [32–34]. Expressed sequence tag (EST) markers for enzymes of the DXP/MEP pathway [34], as well as multiple isoforms of

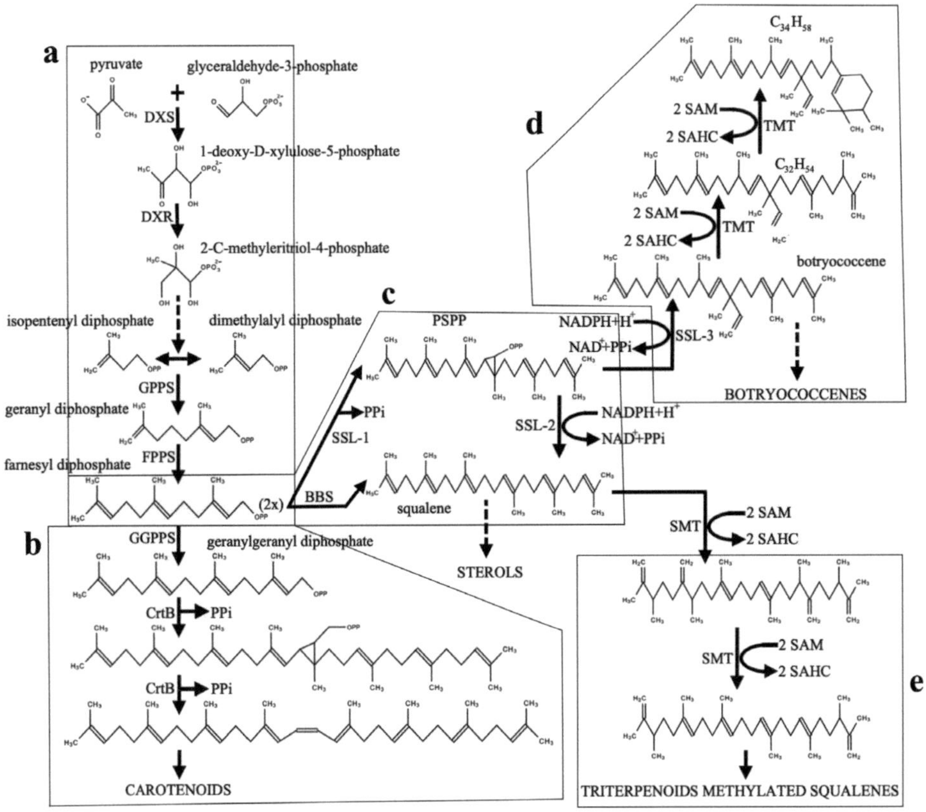

Figure 3.
Biosynthesis of tri- and tetraterpenes in B. braunii race B. (a) FPP production; (b) carotenoid production from GGPP; (c) squalene production from FPP; (d) methylated botryococcene production; (e) methylated squalene production. BSS, Botryococcus squalene synthase; CtrB, phytoene-synthase; DXR, 1-deoxy-D-xylulose-5-phosphate reductase; DXS, 1-deoxy-D-xylulose-5-phosphate synthase; FPPS, farnesyl diphosphate synthase; GPPS, geranyl diphosphate synthase; NADPH⁺ and NADP⁺, nicotinamide adenine dinucleotide phosphate (reduced and oxidized); PPi, inorganic pyrophosphate; PSPP, cyclopropyl presqualene diphosphate; SAM, S-adenosyl methionine; SAH, S-adenosyl-L-homocysteine; SSL, squalene synthase-like; SMT, squalene methyltransferase; TMT, triterpene methyltransferase. Adapted from [17, 34].

enzymes for the 3-phospho-D-glycerate biosynthesis from D-glyceraldehyde-3-phosphate and pyruvate as precursors, were identified. Some of the respective transcripts are present in high abundance (>250 reads/Kb), suggesting a high metabolic flow in *B. braunii* [31].

The first step is the formation of 1-deoxy-D-xylulose-5-phosphate (DOXP) by the DOXP synthase (DXS) (**Figure 3**).

The characterization of three DXS isoenzymes in race B shows that they are active and have similar kinetic parameters, which increases the metabolic flow for the production of terpenoids [35]. The DOXP is reduced by the DXP reductoisomerase (DXR) to 2-C-methylerythritol-4-phosphate (MEP), and converted to isopentenyl diphosphate (IPP) and dimethylallyl pyrophosphate (DMAPP). In the *B. braunii* transcriptome, only one DXR has been found [34]. The next step involves condensation of IPP and DMAPP to form geranyl diphosphate (GPP), and the addition of other IPP produces farnesyl diphosphate (FPP) [17] (**Figure 3a**). Two *B. braunii* genes code for farnesyl diphosphate synthase isoenzymes (FPPS) with an amino acid identity of 72% [34].

Addition of another IPP forms the geranylgeranyl diphosphate (GGPP), precursor of the tetraterpenoid carotenoids (**Figure 3b**). This begins with the formation of a *trans*-isoprenyl diphosphate by the phytoene synthase (CtrB) enzyme, condensing two GGPP molecules in two steps with the release of pyrophosphate. In the first step, (1R, 2R, 3R)-prephytoene diphosphate is produced from half cyclopropyl (C1′-2-3) reordered to provide 15-*cis*-phytoene, which can be converted into a wide variety of carotenoids [34, 36–38]. All are important antioxidant photoprotectors and modulators of the function of membrane proteins for photosynthetic complexes [39].

The squalene production [40] starts with the *Botryococcus* squalene synthase (BSS) enzyme, using two FPP molecules. Botryococcenes production uses also two FPP molecules but the product is the intermediary cyclopropyl presqualene diphosphate (PSPP) (**Figure 3c**). With NADPH, the PSPP has two options; one forms the botryococcene with a C3-C1 connection between the FPP molecules (**Figure 3d**). The other option forms a C1-C1′ between two FPP molecules producing squalene

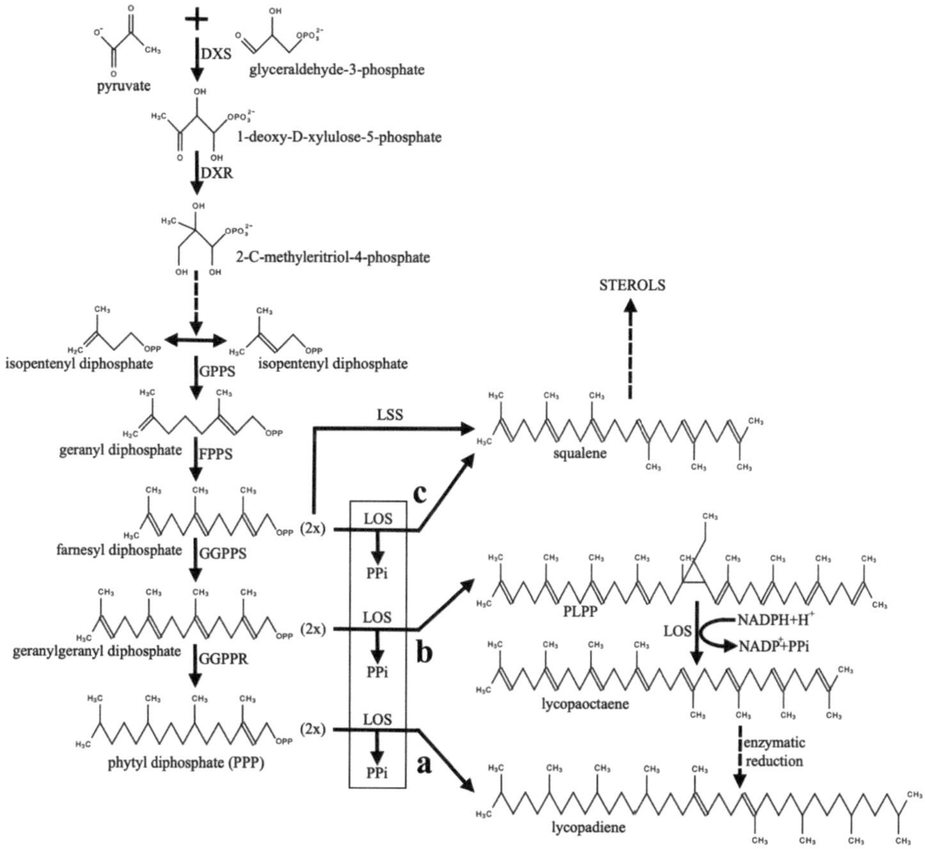

Figure 4.
Lycopadiene biosynthetic pathway. (a) Reduction of GGPP to PPP and condensation by LOS. (b) LOS condensation of GGPP to form phytyl diphosphate and reduction to lycopaoctaene. (c) FPP use by LSS or LOS for squalene production. DXR, 1-deoxy-D-xylulose-5-phosphate reductase; DXS, 1-deoxy-D-xylulose-5-phosphate synthase; FPP, farnesyl diphosphate; FPPS, farnesyl diphosphate synthase; GGPP, geranylgeranyl diphosphate; GPPS, geranyl diphosphate synthase; GPPR, geranyl diphosphate reductase; NADPH⁺ and NADP⁺, nicotinamide adenine dinucleotide phosphate (reduced and oxidized); PPi, inorganic pyrophosphate; PPP, phytyl diphosphate; PLPP, prelycopaoctaene diphosphate; LOS, lycopaoctaene synthase; LSS, B. braunii race L squalene synthase. Adapted from [15].

that will be methylated (**Figure 3e**) further on. These reactions are catalyzed by squalene synthase-like (SSL) enzymes. Three SSL genes have been identified but none is directly related with the botryococcene biosynthesis [41]. However, when the 3 SSLs enzymes were mixed *in vivo* and *in vitro*, botryococcene (SSL-1 + SSL-3) or squalene (SSL1 + SSL-2) was synthesized. SSL-1 condenses two FPP molecules to produce PSPP [42], demonstrating the versatility and potential for metabolic engineering of botryococcene biosynthesis.

Most botryococcenes are excreted to the ECM where they are methylated. The di- and tetramethyl forms are related to six genes coding for triterpene and squalene methyltransferases (TMT, SMT) [43] (**Figures 3d** and **3e**). The botryococcenes are methylated to produce C_{31}–C_{37} hydrocarbons, C_{34} being the main in race B. Three cyclic botryococcene C_{33} molecules and a trimethylsqualene isomer were recently found [44]. Also, two squalene epoxidase (*BbSQE-I* and *-II*) enzymes converting squalene into membrane sterols were identified [45]. Data of the *B. braunii* race B nuclear genome will allow the search for possible regulatory routes of this singular metabolism [46].

2.3 Biosynthesis of lycopadiene

The formation of lycopadiene of race L is similar to the squalene. In the *B. braunii* transcriptome, there are two homologous contigs to squalene synthase (SS) [31]. One encodes a squalene synthase (LSS) and the other for a lycopaoctaene synthase (LOS). LOS uses preferentially *in vivo* GGPP, and C_{15} and C_{20} prenyl diphosphates as substrates [15] (**Figure 4**).

There are two biosynthetic mechanisms for lycopadiene from C_{20} prenyl diphosphate intermediates. In one, the GGPP reduction by a GGPP-reductase produces phytyl diphosphate (PPP), and LOS condenses two PPP molecules producing lycopadiene (**Figure 4a**). In the other one, LOS condenses two GGPP molecules producing prelycopaoctaene diphosphate (PLPP), which rearranges into lycopaoctaene. Finally, lycopadiene seems to be produced by enzymatic reductions not yet identified (**Figure 4b**).

LOS may also form squalene from FPP (**Figure 4c**). These results show the plasticity of L race to synthesize squalene and lycopadiene.

2.4 Extracellular matrix (ECM) polymers

ECM contains long chains of polymerized polyacetal hydrocarbons joined to specific hydrocarbons of each race. There is a fibrillary sheath that envelops the entire colony, formed mainly by arabinose (42%) and galactose (39%). The cell wall contains β-1,4 and/or β-1,3 glucans making a cellulose-like polymer [20].

Also, there's a biopolymer resistant to nonoxidative chemical degradation as acetolysis. This biopolymer resembles sporopollenins [1] of the outer walls of pollen grains and spores of microorganisms [47]. It seems to be formed by oxidized carotenoid polymers and phenolic compounds that absorb UV-B light as *p*-coumaric and *p*-ferulic acids [48].

3. Profitability of *B. braunii* derivatives

3.1 Hydrocarbons

Both bioethanol and biodiesel have a poor oxidative stability, low energy content by volume, and high content of oxygenated compounds, which damage combustion engines and cause corrosion, erosion, and accumulation of deposits in the nozzles; because of these reasons, they are mixed with standard fuels [49, 50]. *B. braunii* accumulates hydrocarbons similar to those of the crude oil, and their direct contribution in the formation of oil reserves currently in use has been reported [3–5]. The *B. braunii* oils showed almost equal values in density and surface tension than the diesel, but with higher kinematic viscosity and distillation temperature [50]. The *B. braunii* race B oil was already converted into diesel with an 85% performance, using a simple conversion process under mild conditions of 260°C and 1 atm. The physical properties are relatively close to the specification for diesel, with 40 as estimated cetane (CN) number [51].

The limitation to use *B. braunii* as biorefinery is the slow growth rate of days in comparison with hours in other algae [49, 52]. Other factors affecting the growth

St	Culture conditions				SCGR	Dt	T_{HC}	Ref.
	°C	PAR	Php	CO_2				
Showa (B)	30	850	14:10	1	0.5	1.40	NIA	[54]
Showa (B)	25, 30	85–398	14:10	1.0–10.0	0.19–0.44	1.60–3.60	30–39	[54]
Showa (B)	23–25	250	24	0.3	0.42	1.70	24–29	[52]
Showa (B)	23	150	16:8	2	0.17	4.08^d	25	[55]
Yayoi (B)	25	240	12:12	2	0.2	3.50	40.5	[38]
AC759 (B)	23	150	16:8	2	0.07	9.90^d	21	[55]
AC761 (B)	23	150	16:8	2	0.11	6.30^d	45	[55]
IPE001 (B)	25	35	16:8	1	0.15^c	4.50^c	64.3	[61]
BOT-144 (B)	25	60^a	24	0	0.16	4.33^d	50	[62]
LB-572 (A)	26	12 Klux	24	2	0.07^c	10.60^c	28	[53]
Gottingen 807/1 (A)	25	26^b	14:10	1	0.3	2.30	40.5	[67]
AC755 (A)	23	150	16:8	2	0.05	13.86^d	16	[55]
CCALA777 (A)	23	150	16:8	2	0.06	11.55^d	10	[55]
CCALA778 (A)	23	150	16:8	2	0.17	4.08^d	0	[55]
CCAp807/2 (A)	23	150	16:8	2	0.11	6.30^d	7	[55]
765	25	150	24	20	0.13^c	5.50^c	24	[64]
765	25	120	24	ASLW	NIA	NIA	23.8	[65]
GUBIOTJTBB1	25	35	16:8	0	0.112	6.19	52.6	[66]
AP 103	23	30	16:8	0	NIA	NIA	13	[67]

ASLW, aerated swine lagoon wastewaters (not sterile); °C, temperature; CO_2, % v/v; Dt, doubling time (days); NIA, no information available; PAR, photosynthetic active radiation (μmols of photons/m^2 s); Php, photoperiod (light/dark hours); SCGR, specific cell growth rate (μ/day); μ, specific velocity of growth rate; St, strain (race); T_{HC}, total hydrocarbons (% DW, dry weight).
[a]*Blue light λ = 470 nm.*
[b]*W/m^2.*
[c]*Estimated values [54].*
[d]*Calculated values from μ, using Ln(2)/μ equation.*

Table 1.
Comparison of culture conditions and productivity of hydrocarbons between B. braunii strains at laboratory scale.

and hydrocarbon production are the strain, CO_2, light, water, nutrients, temperature, pH, and salinity [53–55, 60] (**Table 1**). A JET PASTER treatment was used to do a mechanical cell disruption and removal of the polysaccharides of the *B. braunii* colonies, increasing the hydrocarbon extraction up to 82.8%. This treatment did not affect the photosynthetic function of the cells [56]. On the other hand, a repetitive nondestructive extraction with heptane was reported as having some advantages [57]. Also, a continuous growth and extraction column of *n*-dodecane was reported recently as an efficient hydrocarbon extraction method without significant loss of the viability of the cells [58]. Considering these milking procedures and achieving a 10% rate of return, a minimum sales price (MSP) of US$3.20 per liter was calculated, and a reduction down to US$1.45 per liter was proposed, if hydrocarbon content increases and extraction procedures become more efficient [59].

There are different open and closed culture systems in photobioreactors (PBR) [63, 64], but more studies are required at pilot and industrial scale, to reduce problems by contamination and low yield of biomass and hydrocarbon production [49]. **Table 2** summarizes some data about cell growth and hydrocarbon productivity using different culture systems.

St	System	Cultures			Biomass			HCs		Ref
		°C	PAR	CO_2	SCGR	X_{max}	Px	CNT	W_{HC}	
GUBIOT JTBB1	Plain (3 L)	25	35 (16 h)	0%	0.112	NIA	13	52.6	6.8	[62]
765	Column (3 L)	25	150 (24 h)	20%	0.13[g]	NIA	92.4	24.45[g]	22.6	[64]
Showa (B)	PBR[a]	25-28	282 (15 h)	5–7%	NIA	20	1500	22.5	225-340	[68]
NIA	PBR[b]	25	270 (24 h)	Mixo-trophic	NIA	4.55	234	29.7	71.1	[69]
UTEX-LB 572 (A)	Circular (50 L)	rT	Sol r	0%	NIA	NIA	77.8	19	13.2	[70]
N-836 (B)	Rcwy (80 L)	rT	Sol r	0%	NIA	NIA	40	24	10.8	[70]
LB572 (B)	PBR[c]	20	Sol r	0%	0.04	0.3	15	NIA	2.4	[71]
AP103	Rcwy (1800 L)	29	Sol r 5 kWh/ m^2.day	0%	0.38	NIA	114	11	12.5	[67]
UTEX-LB 572 (A)	PBR[d]	25	55 (24 h)	1%	NIA	96.4	0.71[i]	NIA	NIA	[72]
FACHB 357 (B)	Attchd[e]	25	500 (24 h)	1%	NIA	62[h]	5.5–6.5[i]	19.43	1.06[i]	[73]
TN101	Rcwy sc[f]	rT	Sol r	0%	NIA	NIA	33.8[i]	22.6	8.2-13[i]	[74]

°C, temperature; CNT, content (% DW dry weight); CO_2, % v/v; HCs, hydrocarbons; PAR, photosynthetic active radiation (μmols of photons/m^2 s); PBR, photobioreactor; Php, photoperiod (light/dark hours); P_x, biomass productivity (mg/L day); NIA, no information available; Rcwy, raceway; rT, room temperature; SCGR, specific cell growth rate (μ/day); μ, specific velocity of growth rate; Sol r, solar radiation; St, strain (race); W_{HC}, weight of hydrocarbons (mg/L day); X_{max}, maximum cellular concentration (g/L).
[a]*"Tickle film" (30.5 × 16.5 in) continuous.*
[b]*"Airlift" (10 L).*
[c]*Panel (1000 L) outdoor and semicontinuous.*
[d]*"Biofilm" (0.275 m^2 or 600 mL).*
[e]*"Attached" bioreactor (0.08 m^2 or 240 mL).*
[f]*(25 m^2 or 5000 L) semicontinuous.*
[g]*Estimated values [64].*
[h]*g/m^2.*
[i]*g/m^2/day; shadow area indicates the highest reported values up to now.*

Table 2.
Comparison of culture conditions and productivity of hydrocarbons between strains of B. braunii in bioreactors.

3.2 Lipids

B. braunii also produces saturated and monounsaturated fatty acids, especially palmitic (16:0) and oleic (18:1), as well as triacylglycerols (TAGs). The percentages of total lipids as saturated, monounsaturated, and polyunsaturated fatty acids in dry biomass are around 44.97, 9.85, 79.61, and 10.54%, respectively [64, 75]. Studies *in vitro* and *in vivo* showed that these fatty acids effectively improve the absorption of lipophilic drugs like flurbiprofen, through the skin [76].

B. braunii stores TAGs and saturated fatty acids in the lag phase as an adaptation to stress conditions but most are synthesized during the stationary phase. Although highest content of these acids is intracellular, *B. braunii* secretes oily drops in small quantities observed on the surface of the cell apex [64].

The yield and lipid composition depends on the strain, the culture system used, growth conditions and cell aging, as well as nitrogen, phosphorus, and micronutrient concentrations (**Table 3**).

St	System	TRT	Biomass			Lipids			Ref.
			SCGR	X_{Max}	P_x	CNT	Yld.	Prod.	
UTEX 572 (A)	EF (125 mL)	0.04 mM NO_3	0.09	0.16	NIA	63	NIA	0.009	[77]
		0.37 mM NO_3	0.185	0.38	NIA	36	0.19	0.019	
KMITL 2 (n.d.)	EF (1 L)	86 mg/L NO_3	NIA	0.48	NIA	39.42	0.19	NIA	[78]
		222 mg/L PO_4	NIA	0.86	NIA	54.69	0.47	NIA	
		444 mg/L PO_4	NIA	1.91	NIA	23.23[a]	0.45	NIA	
		27 mg/L Fe	NIA	0.22	NIA	34.93	0.08	NIA	
KMITL 2 (n.d.)	Outdoor oval pond (150 L)	0.17 g/L NO_3	0.045	4.84	NIA	35.24	NIA	0.016	[79]
		2.5 g/L NO_3	0.049	5.62	NIA	38.60	NIA	0.0189	
LB572 (A)	FBR column (625 mL)	0083 g/L PO_4 and 0.1 g/L SO_4	NIA	NIA	0.296	64.96	NIA	0.19	[80]
		0058 g/L PO_4 and 0.09 g/L SO_4	NIA	NIA	0.304	59.56	NIA	0.18	
TRG	EF (250 mL)	Photoaut. (CO_2)	0.093	1.14	NIA	25.1	NIA	0.0241	[81]
		Heterot. (gluc 5 g/L)	0.115	1.75	NIA	29.3	NIA	0.0467	
		Mixot. (gluc 5 g/L + CO_2)	0.195	2.46	NIA	37.5	NIA	0.0645	
IBL-C117	EF (1 L)	Chu (0.75×)	0.13	0.9	0.12	47.1	NIA	NIA	[82]
		Chu (1.0×)	0.13	0.7	0.1	46	NIA	NIA	
		Chu (2.0×)	0.11	1	0.15	41.3	NIA	NIA	
LB572 (A)	EF (1 L)	Chu (0.75×)	0.15	1.3	0.18	20.2	NIA	NIA	[82]
		Chu (1.0×)	0.16	1.4	0.2	22.5	NIA	NIA	
		Chu (2.0×)	0.17	1.5	0.22	11	NIA	NIA	
2441 (A)	FBR Airlift (2 L)	(N:P = 1:1) in Chu	NIA	4.963	0.173	33.7	NIA	NIA	[83]
		(N:P = 3:3) in Chu	NIA	3.857	0.215	34.6	NIA	NIA	
		(N:P = 6:6) in Chu	NIA	3.987	0.223	32.1	NIA	NIA	

St	System	TRT	Biomass			Lipids			Ref.
			SCGR	X_{Max}	P_x	CNT	Yld.	Prod.	
BOT22 (B)	Biofilm bioreac.	Nitrocell. Memb. (diam. 25 mm and pore size 0.45 µm)	NIA	3.12[a]	0.42[b]	NIA	0.83[a]	NIA	[84]
BOT84 (L)			NIA	10.04[a]	3.8[b]	NIA	1.11[a]	NIA	
BOT7 (S)			NIA	13.6[a]	0.99[b]	NIA	0.83[a]	NIA	

CNT, content (% DW dry weight); Chu, Chu media for microalgae [8]; EF, Erlenmeyer flask; FBR, photobioreactor; N:P, proportion of nitrogen: phosphate; P_x, biomass productivity (g/L day); NIA, no information available; Prod., productivity (g/L day); Rcwy, raceway; SCGR, specific cell growth rate (µ/day); µ, specific velocity of growth rate; St, strain (race); TRT, treatment; Yld., yield (g/L); X_{max}, maximum cellular concentration (g/L).
[a]*mg/cm^2.*
[b]*$mg/cm^2/day$.*

Table 3.
Comparison of crop conditions and lipid productivity in B. braunii.

3.3 Pigments

Algae pigments have been reported to have antioxidant, anticancer, anti-inflammatory, antiobesity, and antiangiogenic properties and function as neuroprotectives [85]. So, they could replace synthetic dyes in food, cosmetic, nutraceutical, and pharmaceutical products [86].

Carotenoid pigments are unsaturated hydrocarbons, while xanthophylls have one or more functional groups containing oxygen such as lutein, canthaxanthin, and astaxanthin [85–87].

Carotenoids abound in races B and L, lutein being the main pigment (22–29%), followed by others as β-carotene, echinenone, 3-OH echinenone, canthaxanthin, violaxanthin, loroxanthin, and neoxanthin. Transition to stationary phase causes a color change in *B. braunii* from green to brown, reddish orange, and pale yellow by accumulation of carotenoids and a decrease of intracellular pigments [88]. Canthaxanthin (46%) and echinenone (20–28%) are predominant in the stationary phase in response to nitrogen deficiency [36]. The BOT-20 strain showed a dark red color during growth because of the accumulated echinenone of about 30.5% dry weight and 630 mg/L production, but with few hydrocarbons (8%) [89].

Adonixanthin was detected in race L during the stationary phase [90], and botryoxanthin A, botryoxanthin B, and braunixanthin 1 and 2 were detected in race B [37, 38, 91]. The 2-azahypoxanthine (AHX) similar to the phytohormone induced the accumulation of secondary carotenoids like botryoxanthin A and braunixanthin 1 and decreased the content of botryococcenes during the stationary phase [92], imitating a lack of nitrogen condition without inhibiting the growth.

In race A, lutein (79–84%) is the main carotenoid followed by β-carotene (1.75–2.14%), violaxanthin (6–9%), astaxanthin (3–8%), and zeaxanthin (0.32–0.78%). In salinity and high light intensity conditions, the lutein increases [53, 93]. All of these compounds shown antioxidant properties and inhibitory effect against lipid peroxidation *in vitro* and *in vivo* and activated antioxidant enzymes such as catalase [94, 95].

3.4 Polysaccharides

The aqueous extracts of *B. braunii* (strain LB 572) reduce the skin dehydration, stimulate collagen synthesis, promote the differentiation of adipocytes, and promote antioxidant and anti-inflammatory activities [96]. The extracellular polysaccharides (exopolysaccharides, EPS) constitute most of the organic material of high molecular weight released to the environment by microalgae and other microorganisms. They have antioxidant, immunomodulatory, antibacterial, antiviral, anticarcinogenic, and antihypocholesterolemic effects [97]. They are used as thickeners, emulsifiers, bioflocculants, stabilizers, and gelling agents in foods and cosmetics; are soluble in water; and modify the rheological properties of solutions increasing their viscosity to form gels [1, 98].

The ECM and the fibrillar pod are composed of mucilaginous polysaccharides [20], and other detected EPS are fucose, glucose, mannose, rhamnose, uronic acids, and unusual sugars such as 3-O-methyl fucose, 3-O-methyl rhamnose, and 6-O-methyl hexose [1]. Galactose is involved in the innate and adaptive immune system [99]. L-Arabinose is used as food additive for its sweet taste and poor absorption in humans [100] and is an antiglycemic agent by selective inhibition of invertases, reducing the glycemic response after sucrose ingestion [101]. Uronic acid is a chelating agent to remove metal ions. Fucose has high commercial value for its anticancer properties and for chemical synthesis of flavoring agents [1, 55].

Some *B. braunii* (UC 58) strains produce 4.0–4.5 g/L EPS with few hydrocarbons (5%). The EPS amount varies with the strain, race, physiological conditions, and culture. Strains of A and B races can produce up to 250 mg/L EPS, and race L up to 1 g/L plus glucose [1].

Greater EPS production correlates with minor growth by N deficiency. Urea and ammonia decrease the pH, as well as EPS production. Optimal conditions for EPS production were nitrate (8 mM) and between 25 and 30°C. Out of these temperatures, the EPS polymerization decreased significantly [1, 102]. Light/dark (16:8) photoperiod produced more hydrocarbons, but continuous light with agitation increased EPS until 1.6 and 0.7 g/L in LB 572 and SAG-30 strains, respectively [103]. EPS production increased (2–3 g/L) in low salinity levels (17–85 mM) as osmoprotectants [53]. High salinity and low N content in D medium induced EPS production (0.549 ± 0.044 g/L) in comparison to the BG11 medium (0.336 ± 0.009 g/L), but biomass was higher in BG11 (1.019 ± 0.051 g/L) than in D (0.953 ± 0.056 g/L) [104]. Modification of culture conditions could be used to increase EPS production, to facilitate the removal, and to increase hydrocarbon recovery. With *Botryococcus braunii* CCALA 778 (race A), a light:dark cycle at 26°C resulted in an increased production of EPS, and a milking procedure for these polysaccharides has been proposed [105, 106]. EPS can be used as thickening or gelling agents [107].

3.5 Other biopolymers

Algenanes are aliphatic, nonhydrolyzable, and insoluble biopolymers found in the ECM at 9 and 10% dry weight of race A and B, respectively. Due to their high resistance to degradation, they are attributed to the good preservation of colonies in sedimentary rocks [108].

Another reported biopolymer was the polyhydroxybutyrate (PHB), a biodegradable plastic with a yield of about 20% of the dry weight [109]. PHB is a polyester with thermoplastic and biodegradable properties, and it's a carbon and energy storage compound. For its similar physical properties to polypropylene and polystyrene, it is of commercial interest [110]. Under pH 7.5, 40°C, and with 60% wastewater as culture medium, a maximum yield of 247 ± 0.42 mg/L PHB was reported [111].

B. braunii (UTEX 572) was used to produce intra- and extracellular Ag nanoparticles (AgNPs) with antimicrobial properties, and analysis suggested that the exopolysaccharides were the possible reducing and capping agents [112].

4. Conclusions

Although *B. braunii* has been considered mainly as a good source of biofuels by the possibility to convert its hydrocarbons into currently used fuels, without the necessity of engine modifications, it produces many other high-value derivatives that can be exploited for their promising attractive profits. Besides, along the photosynthetic process, this alga converts 3% of solar energy into hydrocarbons [1] and can reduce CO_2 emissions up to 1.5×10^5 tons/year [113]. There are several reports about modifications of the culture conditions through vitamin addition, affecting the yield of several derivatives like biomass, hydrocarbon, and carbohydrate in *Botryococcus braunii* KMITL 5 [114]; however, those are from not clearly recognized strains and should be carefully taken. With *B. braunii* race A, B, or L, the main challenge is to accelerate the doubling rate because, depending on the race, it varies between 2 and 10 days. This results in easy contamination with faster growing microorganisms in open ponds used for industrial production, or a high cost of sterile conditions in closed bioreactors. In spite of these disadvantages, we consider that *B. braunii* is an excellent model of biorefinery. Other strategies to use *B. braunii* as biorefinery and bioreactor are being developed like the immobilization in polyester [115] or bioharvesting with *Aspergillus* sp. [116].

Acknowledgements

To the Consejo Nacional de Ciencia y Tecnología (CONACYT) Mexico by the PhD and MSc scholarships of XM-dlC and TAO-U respectively.

Conflict of interest

The authors declare no conflict of interest.

Author details

Edmundo Lozoya-Gloria*, Xochitl Morales-de la Cruz
and Takehiro A. Ozawa-Uyeda
Genetic Engineering Department, CINVESTAV-IPN Irapuato Unit, Irapuato,
Guanajuato, México

*Address all correspondence to: edmundo.lozoya@cinvestav.mx

IntechOpen

© 2019 The Author(s). Licensee IntechOpen. This chapter is distributed under the terms of the Creative Commons Attribution License (http://creativecommons.org/licenses/by/3.0), which permits unrestricted use, distribution, and reproduction in any medium, provided the original work is properly cited.

References

[1] Banerjee A, Sharma R, Chisti Y, Banerjee UC. *Botryococcus braunii*: A renewable source of hydrocarbons and other chemicals. Critical Reviews in Biotechnology. 2002;**22**:245-279. DOI: 10.1080/07388550290789513

[2] Tanabe Y, Okazaki Y, Yoshida M, Matsuura H, Kai A, Shiratori T, et al. A novel alphaproteobacterial ectosymbiont promotes the growth of the hydrocarbon-rich green alga *Botryococcus braunii*. Scientific Reports. 2015;**5**:10467. DOI: 10.1038/srep10467

[3] Moldowan JM, Seifert WK. First discovery of botryococcane in petroleum. Journal of the Chemical Society, Chemical Communications. 1980;**19**:912-914. DOI: 10.1039/C39800000912

[4] McKirdy DM, Cox RE, Volkman JK, Howell VJ. Botryococcane in a new class of Australian non-marine crude oils. Nature. 1986;**320**:57-59. DOI: 10.1038/320057a0

[5] Summons RE, Metzger P, Largeau C, Murray AP, Hope JM. Polymethylsqualanes from *Botryococcus braunii* in lacustrine sediments and crude oils. Organic Geochemistry. 2002; **33**:99-109. DOI: 10.1016/S0146-6380(01)00147-4

[6] Kawachi M, Tanoi T, Damura M, Kaya K, Watanabe MM. Relationship between hydrocarbons and molecular phylogeny of *Botryococcus braunii*. Algal Research. 2012;**1**:114-119. DOI: 10.1016/j.algal.2012.05.003

[7] Metzger P, Berkaloff C, Casadevall E, Coute A. Alkadiene and botryococcene-producing races of wild strains of *Botryococcus braunii*. Phytochemistry. 1985;**24**:2305-2312. DOI: 10.1016/S0031-9422(00)83032-0

[8] Metzger P, Largeau C. *Botryococcus braunii*: A rich source for hydrocarbons and related ether lipids. Applied Microbiology and Biotechnology. 2005; **66**:486-496. DOI: 10.1007/s00253-004-1779-z

[9] Metzger P, Casadevall E, Pouet MJ, Pouet Y. Structures of some botryococcenes: Branched hydrocarbons from the B-race of the green alga *Botryococcus braunii*. Phytochemistry. 1985;**24**:2995-3002. DOI: 10.1016/0031-9422(85)80043-1

[10] Huang Z, Poulter CD. Tetramethylsqualene, a triterpene from *Botryococcus braunii* var. Showa. Phytochemistry. 1989;**28**:1467-1470. DOI: 10.1016/S0031-9422(00)97766-5

[11] Achitouv E, Metzger P, Rager MN, Largeau C. C31-C34 methylated squalenes from a Bolivian strain of *Botryococcus braunii*. Phytochemistry. 2004;**65**:3159-3165. DOI: 10.1016/j.phytochem.2004.09.015

[12] Brown C, Knights BA, Conway E. Hydrocarbon content and its relationship to physiological state in the green alga *Botryococcus braunii*. Phytochemistry. 1969;**8**:543-547. DOI: 10.1016/S0031-9422(00)85397-2

[13] Metzger P, Allard B, Casadevall E, Berkaloff C, Couté A. Structure and chemistry of a new chemical race of *Botryococcus braunii* Chlorophyceae that produces lycopadiene, a tetraterpenoid hydrocarbon. Journal of Phycology. 1990;**26**:258-266. DOI: 10.1111/j.0022-3646.1990.00258.x

[14] Metzger P, Pouet Y, Summons R. Chemotaxonomic evidence for the similarity between *Botryococcus braunii* L race and *Botryococcus neglectus*. Phytochemistry. 1997;**44**:1071-1075. DOI: 10.1016/S0031-9422(96)00698-X

[15] Thapa HR, Naik MT, Okada S, Takada K, Molnár I, Xu Y, et al. A squalene synthase-like enzyme initiates production of tetraterpenoid hydrocarbons in *Botryococcus braunii* Race L. Nature Communications. 2016; **7**:11198. DOI: 10.1038/ncomms11198

[16] Hillen LW, Pollard G, Wake LV, White N. Hydrocracking of the oils of *Botryococcus braunii* to transport fuels. Biotechnology and Bioengineering. 1982;**24**:193-205. DOI: 10.1002/bit.260240116

[17] Cornejo-Corona I, Thapa HR, Devarenne TP, Lozoya-Gloria E. The biofuel potential of the green colonial microalga *Botryococcus braunii*. In: Torres-Bustillos LG, editor. Microalgae and Other Phototrophic Bacteria. Culture, Processing, Recovery and New Products. 1st ed. New York, USA: Nova Science Publishers, Inc.; 2015. pp. 41-58

[18] Griehl C, Kleinert C, Griehl C, Bieler S. Design of a continuous milking bioreactor for non-destructive hydrocarbon extraction from *Botryococcus braunii*. Journal of Applied Phycology. 2015;**27**:1833-1843. DOI: 10.1007/s10811-014-0472-6

[19] Metzger P, Casadevall E, Coute A. Botryococcene distribution in strains of green alga *Botryococcus braunii*. Phytochemistry. 1988;**27**:1383-1388. DOI: 10.1016/0031-9422(88)80199-7

[20] Tatli M, Ishihara M, Heiss C, Browne DR, Dangott LJ, Vitha S, et al. Polysaccharide associated protein PSAP from the green microalga *Botryococcus braunii* is a unique extracellular matrix hydroxyproline-rich glycoprotein. Algal Research. 2018;**29**:92-103. DOI: 10.1016/j.algal.2017.11.018

[21] Tanoi T, Kawachi M, Watanabe MM. Effects of carbon source on growth and morphology of *Botryococcus braunii*. Journal of Applied Phycology. 2011;**23**:25-33. DOI: 10.1007/s10811-010-9528-4

[22] Hirose M, Mukaida F, Okada S, Noguchi T. Active hydrocarbon biosynthesis and accumulation in a green alga, *Botryococcus braunii* race A. Eukaryotic Cell. 2013;**12**:1132-1141. DOI: 10.1128/EC.00088-13

[23] Suzuki R, Ito N, Uno Y, Nishii I, Kagiwada S, Okada S, et al. Transformation of lipid bodies related to hydrocarbon accumulation in a green alga, *Botryococcus braunii* Race B. PLoS One. 2013;**8**:e81626. DOI: 10.1371/journal.pone.0081626

[24] Cornejo-Corona I, Thapa HR, Browne DR, Devarenne TP, Lozoya-Gloria E. Stress responses of the oil-producing green microalga *Botryococcus braunii* Race B. Peer J. 2016;**4**:e2748. DOI: 10.7717/peerj.2748

[25] Demura M, Ioki M, Kawachi M, Nakajima N, Watanabe MM. Desiccation tolerance of *Botryococcus braunii* Trebouxiophyceae, Chlorophyta and extreme temperature tolerance of dehydrated cells. Journal of Applied Phycology. 2014;**26**:49-53. DOI: 10.1007/s10811-013-0059-7

[26] Senousy H, Beakes GW, Hack E. Phylogenetic placement of *Botryococcus braunii* Trebouxiophyceae and *Botryococcus sudeticus* isolate UTEX 2629 Chlorophyceae. Journal of Phycology. 2004;**40**:412-423. DOI: 10.1046/j.1529-8817.2004.03173.x

[27] Gouveia D, Lian J, Steinert G, Smidt H, Sipkema D, Wijffels RH, et al. Associated bacteria of *Botryococcus braunii* Chlorophyta. Peer J. 2019;**7**:e6610. DOI: 10.7717/peerj.6610

[28] Templar J, Largeau C, Casadevall E. Mechanism of non-isoprenoid hydrocarbon biosynthesis in *Botryococcus braunii*. Phytochemistry.

1984;**23**:1017-1028. DOI: 10.1016/S0031-9422(00)82602-3

[29] Templier J, Largeau C, Casadevall E. Effect of various inhibitors on biosynthesis of non-isoprenoid hydrocarbons in *Botryococcus braunii*. Phytochemistry. 1987;**26**:377-383. DOI: 10.1016/S0031-9422(00)81418-1

[30] Dennis MW, Kolattukud PE. Alkane biosynthesis by decarbonylation of aldehyde catalyzed by a microsomal preparation from *Botryococcus braunii*. Archives of Biochemistry and Biophysics. 1991;**287**:268-275. DOI: 10.1016/0003-9861(91)90478-2

[31] Unkefer CJ, Sayre RT, Magnuson JK, Anderson DB, Baxter I, Blaby IK, et al. Review of the algal biology program within the national alliance for advanced biofuels and bioproducts. Algal Research. 2017;**22**:187-215. DOI: 10.1016/j.algal.2016.06.002

[32] Sato Y, Ito Y, Okada S, Murakami M, Abe H. Biosynthesis of the triterpenoids, botryococcenes and tetramethylsqualene in the B race of *Botryococcus braunii* via the non-mevalonate pathway. Tetrahedron Letters. 2003;**44**: 7035-7037. DOI: 10.1016/S0040-4039(03)01784-2

[33] Lohr M, Schwender J, Polle JE. Isoprenoid biosynthesis in eukaryotic phototrophs: A spotlight on algae. Plant Science. 2012;**185**:9-22. DOI: 10.1016/j.plantsci.2011.07.018

[34] Molnár I, Lopez D, Wisecaver JH, Devarenne TP, Weiss TL, Pellegrini M, et al. Bio-crude transcriptomics: Gene discovery and metabolic network reconstruction for the biosynthesis of the terpenome of the hydrocarbon oil-producing green alga, *Botryococcus braunii* race B Showa. BMC Genomics. 2012;**13**:576. DOI: 10.1186/1471-2164-13-576

[35] Matsushima D, Jenke-Kodama H, Sato Y, Fukunaga Y, Sumimoto K, Kuzuyama T, et al. The single cellular green microalga *Botryococcus braunii*, race B possesses three distinct 1-deoxy-D-xylulose 5-phosphate synthases. Plant Science. 2012;**185**:309-320. DOI: 10.1016/j.plantsci.2012.01.002

[36] Grung M, Metzger P, Liaaen-Jensen S. Primary and secondary carotenoids in two races of the green alga *Botryococcus braunii*. Biochemical Systematics and Ecology. 1989;**17**:263-269. DOI: 10.1016/0305-1978(89)90001-X

[37] Okada S, Matsuda H, Murakami M, Yamaguchi K. Botryoxanthin A, a member of a new class of carotenoids from the green microalga *Botryococcus braunii* Berkeley. Tetrahedron Letters. 1996;**37**:1065-1068. DOI: 10.1016/0040-4039(95)02349-6

[38] Okada S, Tonegawa I, Matsuda H, Murakami M, Yamaguchi K. Braunixanthins 1 and 2, new carotenoids from the green microalga *Botryococcus braunii*. Tetrahedron. 1997;**53**: 11307-11316. DOI: 10.1016/S0040-4020(97)00705-9

[39] Takaichi S. Carotenoids in algae: Distributions, biosynthesis and functions. Marine Drugs. 2011;**9**: 1101-1118. DOI: 10.3390/md9061101

[40] Okada S, Devarenne TP, Chappell J. Molecular characterization of squalene synthase from the green microalga *Botryococcus braunii*, race B. Archives of Biochemistry and Biophysics. 2000; **373**:307-317. DOI: 10.1006/abbi.1999.1568

[41] Niehaus TD, Okada S, Devarenne TP, Watt DS, Sviripa V, Chappell J. Identification of unique mechanisms for triterpene biosynthesis in *Botryococcus braunii*. PNAS. 2011;**108**: 12260-12265. DOI: 10.1073/pnas.1106222108

[42] Okada S, Devarenne TP, Murakami M, Abe H, Chappell J. Characterization of botryococcene synthase enzyme activity, a squalene synthase-like activity from the green microalga *Botryococcus braunii*, Race B. Archives of Biochemistry and Biophysics. 2004;**422**:110-118. DOI: 10.1016/j.abb.2003.12.004

[43] Niehaus TD, Kinis S, Okada S, Yeo Y, Bell SA, Cui P, et al. Functional identification of triterpene methyltransferases from *Botryococcus braunii* Race B. The Journal of Biological Chemistry. 2012;**287**:8163-8173. DOI: 10.1074/jbc.M111.316059

[44] Tatli M, Naik M, Okada S, Dangott LJ, Devarenne TP. Isolation and characterization of cyclic C_{33} botryococcenes and a trimethylsqualene isomer from *Botryococcus braunii* race B. Journal of Natural Products. 2017;**80**: 953-958. DOI: 10.1021/acs.jnatprod.6b00934

[45] Uchida H, Sumimoto K, Ferriols VME, Imou K, Saga K, Furuhashi K, et al. Isolation and characterization of two squalene epoxidase genes from *Botryococcus braunii*, Race B. PLoS One. 2015;**10**: e0122649. DOI: 10.1371/journal.pone.0122649

[46] Browne DR, Jenkins J, Schmutz J, Shu S, Barry K, Grimwood J, et al. Draft nuclear genome sequence of the liquid hydrocarbon–accumulating green microalga *Botryococcus braunii* race B Showa. Genome Announcements. 2017; **5**:e00215-e00217. DOI: 10.1128/genomeA.00215-17

[47] Brooks J, Shaw G. Sporopollenin: A review of its chemistry, palaeochemistry and geochemistry. Grain. 1978;**17**: 91-97. DOI: 10.1080/00173137809428858

[48] Rozema J, Blokker P, Fuertes MM, Broekman R. UV-B absorbing compounds in present-day and fossil pollen, spores, cuticles, seed coats and wood: Evaluation of a proxy for solar UV radiation. Photochemical & Photobiological Sciences. 2009;**8**:1233-1243. DOI: 10.1039/B904515E

[49] Jin J, Dupré C, Yoneda K, Watanabe MM, Legrand J, Grizeau D. Characteristics of extracellular hydrocarbon-rich microalga *Botryococcus braunii* for biofuels production: Recent advances and opportunities. Process Biochemistry. 2016;**51**:1866-1875. DOI: 10.1016/j.procbio.2015.11.026

[50] Nagano S, Yamamoto S, Nagakubo M, Atsumi K, Watanabe MM. Physical properties of hydrocarbon oils produced by *Botryococcus braunii*: Density, kinematic viscosity, surface tension, and distillation properties. Procedia Environmental Sciences. 2012; **15**:73-79. DOI: 10.1016/j.proenv.2012.05.012

[51] Yamamoto S, Mandokoro Y, Nagano S, Nagakubo M, Atsumi K, Watanabe MM. Catalytic conversion of *Botryococcus braunii* oil to diesel fuel under mild reaction conditions. Journal of Applied Phycology. 2014;**26**:55-64. DOI: 10.1007/s10811-013-0065-9

[52] Wolf FR, Nemethy EK, Blanding JH, Bassham JA. Biosynthesis of unusual acyclic isoprenoids in the alga *Botryococcus braunii*. Phytochemistry. 1985;**24**:733-737. DOI: 10.1016/S0031-9422(00)84886-4

[53] Rao AR, Dayananda C, Sarada R, Shamala TR, Ravishankar GA. Effect of salinity on growth of green alga *Botryococcus braunii* and its constituents. Bioresource Technology. 2007;**98**: 560-564. DOI: 10.1016/j.biortech.2006.02.007

[54] Yoshimura T, Okada S, Honda M. Culture of the hydrocarbon producing

microalga *Botryococcus braunii* strain Showa: Optimal CO_2, salinity, temperature, and irradiance conditions. Bioresource Technology. 2013;**133**: 232-239. DOI: 10.1016/j.biortech.2013.01.095

[55] Gouveia JD, Ruiz J, van den Broek LA, Hesselink T, Peters S, Kleinegris DM, et al. *Botryococcus braunii* strains compared for biomass productivity, hydrocarbon and carbohydrate content. Journal of Biotechnology. 2017;**248**:77-86. DOI: 10.1016/j.jbiotec.2017.03.008

[56] Tsutsumi S, Saito Y, Matsushita Y, Aoki H. Investigation of colony disruption for hydrocarbon extraction from *Botryococcus braunii*. Fuel Processing Technology. 2018;**172**:36-48. DOI: 10.1016/j.fuproc.2017.12.004

[57] Jackson BA, Bahri PA, Moheimani NR. Response of *Botryococcus braunii* to repetitive non-destructive extraction of lipids with heptane. In: Chemeca 2018 Conference. 30 September–3 October 2018. Queenstown, NZ: Institution of Chemical Engineers; 2018. pp. 89.1-89.9. Available from: https://search.informit.com.au/documentSummary;dn=047836542036386;res=IELENG

[58] Mehta P, Jackson BA, Nwoba EG, Vadiveloo A, Bahri PA, Mathur AS, et al. Continuous non-destructive hydrocarbon extraction from *Botryococcus braunii* BOT-22. Algal Research. 2019;**41**:101537. DOI: 10.1016/j.algal.2019.101537

[59] Chaudry S, Bahr PA, Moheimani NR. Techno-economic analysis of milking of *Botryococcus braunii* for renewable hydrocarbon production. Algal Research. 2018;**31**: 194-203. DOI: 10.1016/j.algal.2018.02.011

[60] Tasić MB, Pinto LFR, Klein BC, Veljković VB, Filh RM. *Botryococcus braunii* for biodiesel production. Renewable and Sustainable Energy Reviews. 2016;**64**:260-270. DOI: 10.1016/j.rser.2016.06.009

[61] Xu L, Liu R, Wang F, Liu CZ. Development of a draft-tube airlift bioreactor for *Botryococcus braunii* with an optimized inner structure using computational fluid dynamics. Bioresource Technology. 2012;**119**: 300-305. DOI: 10.1016/j.biortech.2012.05.123

[62] Baba M, Kikuta F, Suzuki I, Watanabe MM, Shiraiwa Y. Wavelength specificity of growth, photosynthesis, and hydrocarbon production in the oil-producing green alga *Botryococcus braunii*. Bioresource Technology. 2012; **109**:266-270. DOI: 10.1016/j.biortech.2011.05.059

[63] Casadevall E, Dif D, Largeau C, Gudin C, Chaumont D, Desanti O. Studies on batch and continuous cultures of *Botryococcus braunii*: Hydrocarbon production in relation to physiological state, cell ultrastructure, and phosphate nutrition. Biotechnology and Bioengineering. 1985;**27**:286-295. DOI: 10.1002/bit.260270312

[64] Ge Y, Liu J, Tian G. Growth characteristics of *Botryococcus braunii* 765 under high CO_2 concentration in photobioreactor. Bioresource Technology. 2011;**102**:130-134. DOI: 10.1016/j.biortech.2010.06.051

[65] Liu J, Ge Y, Cheng H, Wu L, Tian G. Aerated swine lagoon wastewater: A promising alternative medium for *Botryococcus braunii* cultivation in open system. Bioresource Technology. 2013; **139**:190-194. DOI: 10.1016/j.biortech.2013.04.036

[66] Talukdar J, Kalita MC, Goswami BC. Characterization of the biofuel potential of a newly isolated strain of the microalga *Botryococcus braunii* Kützing from Assam, India. Bioresource

Technology. 2013;**149**:268-275. DOI: 10.1016/j.biortech.2013.09.057

[67] Ashokkumar V, Rengasamy R. Mass culture of *Botryococcus braunii* Kutz. under open raceway pond for biofuel production. Bioresource Technology. 2012;**104**:394-399. DOI: 10.1016/j.biortech.2011.10.093

[68] Khatri W, Hendrix R, Niehaus T, Chappell J, Curtis WR. Hydrocarbon production in high density *Botryococcus braunii* race B continuous culture. Biotechnology and Bioengineering. 2014;**111**:493-503. DOI: 10.1002/bit.25126

[69] Zhang H, Wang W, Li Y, Yang W, Shen G. Mixotrophic cultivation of *Botryococcus braunii*. Biomass and Bioenergy. 2011;**35**:1710-1715. DOI: 10.1016/j.biombioe.2011.01.002

[70] Rao AR, Ravishankar GA, Sarada R. Cultivation of green alga *Botryococcus braunii* in raceway, circular ponds under outdoor conditions and its growth, hydrocarbon production. Bioresource Technology. 2012;**123**:528-533. DOI: 10.1016/j.biortech.2012.07.009

[71] Bazaes J, Sepulveda C, Acién FG, Morales J, Gonzales L, Rivas M, et al. Outdoor pilot-scale production of *Botryococcus braunii* in panel reactors. Journal of Applied Phycology. 2012;**24**:1353-1360. DOI: 10.1007/s10811-012-9787-3

[72] Ozkan A, Kinney K, Katz L, Berberoglu H. Reduction of water and energy requirement of algae cultivation using an algae biofilm photobioreactor. Bioresource Technology. 2012;**114**:542-548. DOI: 10.1016/j.biortech.2012.03.055

[73] Cheng P, Ji B, Gao L, Zhang W, Wang J, Liu T. The growth, lipid and hydrocarbon production of *Botryococcus braunii* with attached cultivation. Bioresource Technology. 2013;**138**:95-100. DOI: 10.1016/j.biortech.2013.03.150

[74] Ashokkumar V, Agila E, Sivakumar P, Salam Z, Rengasamy R, Ani FN. Optimization and characterization of biodiesel production from microalgae *Botryococcus* grown at semi-continuous system. Energy Conversion and Management. 2014;**88**:936-946. DOI: 10.1016/j.enconman.2014.09.019

[75] Nascimento IA, Marques SSI, Cabanelas ITD, Pereira SA, Druzian J, De Souza CO, et al. Screening microalgae strains for biodiesel production: Lipid productivity and estimation of fuel quality based on fatty acids profiles as selective criteria. Bioenergy Research. 2013;**6**:1-13. DOI: 10.1007/s12155-012-9222-2

[76] Fan JY, Chiu HC, Wu JT, Chiang YR, Hsu SH. Fatty acids in *Botryococcus braunii* accelerate topical delivery of flurbiprofen into and across skin. International Journal of Pharmaceutics. 2004;**276**:163-173. DOI: 10.1016/j.ijpharm.2004.02.026

[77] Choi GG, Kim BH, Ahn CY, Oh HM. Effect of nitrogen limitation on oleic acid biosynthesis in *Botryococcus braunii*. Journal of Applied Phycology. 2011;**23**:1031-1037. DOI: 10.1007/s10811-010-9636-1

[78] Ruangsomboon S. Effect of light, nutrient, cultivation time and salinity on lipid production of newly isolated strain of the green microalga, *Botryococcus braunii* KMITL 2. Bioresource Technology. 2012;**9**:261-265. DOI: 10.1016/j.biortech.2011.07.025

[79] Ruangsomboon S. Effects of different media and nitrogen sources and levels on growth and lipid of green microalga *Botryococcus braunii* KMITL and its biodiesel properties based on fatty acid composition. Bioresource Technology. 2015;**191**:

377-384. DOI: 10.1016/j.biortech.2015.01.091

[80] Tran HL, Kwon JS, Kim ZH, Oh Y, Lee CG. Statistical optimization of culture media for growth and lipid production of *Botryococcus braunii* LB572. Biotechnology and Bioprocess Engineering. 2010;**15**:277-284. DOI: 10.1007/s12257-009-0127-7

[81] Yeesang C, Cheirsilp B. Low-cost production of green microalga *Botryococcus braunii* biomass with high lipid content through mixotrophic and photoautotrophic cultivation. Applied Biochemistry and Biotechnology. 2014; **174**:116-129. DOI: 10.1007/s12010-014-1041-9

[82] Cabanelas ITD, Marques SSI, de Souza CO, Druzian J, Nascimento IA. *Botryococcus*, what to do with it? Effect of nutrient concentration on biorefinery potential. Algal Research. 2015;**11**:43-49. DOI: 10.1016/j.algal.2015.05.009

[83] Pérez-Mora LS, Matsudo MC, Cezare-Gomes EA, Carvalho J. An investigation into producing *Botryococcus braunii* in a tubular photobioreactor. Journal of Chemical Technology and Biotechnology. 2016;**91**: 3053-3060. DOI: 10.1002/jctb.4934

[84] Wijihastuti RS, Moheimani NR, Bahri PA, Cosgrove JJ, Watanabe MM. Growth and photosynthetic activity of *Botryococcus braunii* biofilms. Journal of Applied Phycology. 2017;**29**:1123-1134. DOI: 10.1007/s10811-016-1032-z

[85] Pangestuti R, Kim SK. Biological activities and health benefit effects of natural pigments derived from marine algae. Journal of Functional Foods. 2011; **3**:255-266. DOI: 10.1016/j.jff.2011.07.001

[86] Ambati RR, Gogisetty D, Gokare RA, Ravi S, Bikkina PN, Su Y, et al. *Botryococcus* as an alternative source of carotenoids and its possible applications—An overview. Critical Reviews in Biotechnology. 2018;**38**: 541-558. DOI: 10.1080/07388551.2017.1378997

[87] Mulders KJ, Lamers PP, Martens DE, Wijffels RH. Phototrophic pigment production with microalgae: Biological constraints and opportunities. Journal of Phycology. 2014;**50**:229-242. DOI: 10.1111/jpy.12173

[88] Tonegawa I, Okada S, Murakami M, Yamaguchi K. Pigment composition of the green microalga *Botryococcus braunii* Kawaguchi-1. Fisheries Science. 1998; **64**:305-308. DOI: 10.2331/fishsci.64.305

[89] Matsuura H, Watanabe MM, Kaya K. Echinenone production of a dark red-coloured strain of *Botryococcus braunii*. Journal of Applied Phycology. 2012;**24**:973-977. DOI: 10.1007/s10811-011-9719-7

[90] Grung M, Metzger P. Algal carotenoids 53; secondary carotenoids of algae 4; secondary carotenoids in the green alga *Botryococcus braunii*, race L, new strain. Biochemical Systematics and Ecology. 1994;**22**:25-29. DOI: 10.1016/0305-1978(94)90111-2

[91] Okada S, Tonegawa I, Matsuda H, Murakami M, Yamaguchi K. Botryoxanthin B and α-botryoxanthin A from the green microalga *Botryococcus braunii* Kawaguchi-1. Phytochemistry. 1998;**47**:1111-1115. DOI: 10.1016/S0031-9422(98)80082-4

[92] Nakamura H, Matsunaga S, Kawagishi H, Okada S. Effects of 2-azahypoxanthine on extracellular terpene accumulations by the green microalga *Botryococcus braunii*, race B. Algal Research. 2016;**20**:267-275. DOI: 10.1016/j.algal.2016.10.023

[93] Ambati RR, Ravi S, Aswathanarayana RG. Enhancement of carotenoids in green alga-*Botryococcus braunii* in various autotrophic media

under stress conditions. International Journal of Biomedical and Pharmaceutical Sciences. 2010;**4**:87-92. Available from: https://www.researchgate.net/profile/Dr_Ranga_Rao_Ambati/publication/235352513

[94] Rao AR, Baskaran V, Sarada R, Ravishanka GA. *In vivo* bioavailability and antioxidant activity of carotenoids from microalgal biomass—A repeated dose study. Food Research International. 2013;**54**:711-717. DOI: 10.1016/j.foodres.2013.07.067

[95] Rao AR, Sarada R, Baskaran V, Ravishankar GA. Antioxidant activity of *Botryococcus braunii* extract elucidated *in vitro* models. Journal of Agricultural and Food Chemistry. 2006;**54**: 4593-4599. DOI: 10.1021/jf060799j

[96] Buono S, Langellotti AL, Martello A, Bimonte M, Tito A, Carola A, et al. Biological activities of dermatological interest by the water extract of the microalga *Botryococcus braunii*. Archives of Dermatological Research. 2012;**304**: 755-764. DOI: 10.1007/s00403-012-1250-4

[97] Liu L, Pohnert G, Wei D. Extracellular metabolites from industrial microalgae and their biotechnological potential. Marine Drugs. 2016;**14**:191. DOI: 10.3390/md14100191

[98] Öner ET. Microbial production of extracellular polysaccharides from biomass. In: Fang Z, editor. Pretreatment Techniques for Biofuels and Biorefineries. Berlin Heidelberg: Springer; 2013. pp. 35-56. DOI: 10.1007/978-3-642-32735-3_2

[99] Brockhausen I. The role of galactosyltransferases in cell surface functions and in the immune system. Drug News & Perspectives. 2006;**19**: 401-409. DOI: 10.1358/dnp.2006.19.7.1021491

[100] Matuso N, Kaneko S, Atsushi K, Kobayashi H, Kusakabes I. Purification and characterization and gene cloning of two α-L-arabinofuranosides from *Streptomyces chartreusis* GS 901. Biochemical Journal. 2000;**346**:9-15. DOI: 10.1042/bj3460009

[101] Shin HY, Park SY, Sung JH, Kim DH. Purification and characterization of α-L-arabinopyranosidase and α-L-arabinofuranosidase from *Bifidobacterium breve* K-110, a human intestinal anaerobic bacterium metabolizing ginsenoside Rb2 and Rc. Applied and Environmental Microbiology. 2003;**69**:7116-7123. DOI: 10.1128/AEM.69.12.7116-7123.2003

[102] Cepák V, Přibyl P. Light intensity and nitrogen effectively control exopolysaccharide production by the green microalga *Botryococcus braunii* Trebouxiophyceae. Genetics and Plant Physiology. 2018;**8**:24-37. Available from: http://www.bio21.bas.bg/ippg/bg/wp-content/uploads/2018/08/GPP_8_1-2_2018_24-37.pdf

[103] Dayananda C, Sarada R, Rani MU, Shamala TR, Ravishankar GA. Autotrophic cultivation of *Botryococcus braunii* for the production of hydrocarbons and exopolysaccharides in various media. Biomass and Bioenergy. 2007;**31**:87-93. DOI: 10.1016/j.biombioe.2006.05.001

[104] Bayona KCD, Garcés LA. Effect of different media on exopolysaccharide and biomass production by the green microalga *Botryococcus braunii*. Journal of Applied Phycology. 2014;**26**: 2087-2095. DOI: 10.1007/s10811-014-0242-5

[105] García-Cubero R, Cabanelas ITD, Sijtsma L, Kleinegris DM, Barbosa MJ. Production of exopolysaccharide by *Botryococcus braunii* CCALA 778 under laboratory simulated Mediterranean climate conditions. Algal Research.

2018;**29**:330-336. DOI: 10.1016/j.algal.2017.12.003

[106] García-Cubero R, Wang W, Martín J, Bermejo E, Sijtsma L, Togtema A, et al. Milking exopolysaccharides from *Botryococcus braunii* CCALA778 by membrane filtration. Algal Research. 2018;**34**:175-181. DOI: 10.1016/j.algal.2018.07.018

[107] Kumar D, Kastanek P, Adhikary SP. Exopolysaccharides from cyanobacteria and microalgae and their commercial application. Current Science. 2018;**115**:234. Available from: https://www.currentscience.ac.in/Volumes/115/02/0234.pdf

[108] Volkman JK. Acyclic isoprenoid biomarkers and evolution of biosynthetic pathways in green microalgae of the genus *Botryococcus*. Organic Geochemistry. 2014;**75**:36-47. DOI: 10.1016/j.orggeochem.2014.06.005

[109] Kavitha G, Kurinjimalar C, Sivakumar K, Kaarthik M, Aravind R, Palani P, et al. Optimization of polyhydroxybutyrate production utilizing waste water as nutrient source by *Botryococcus braunii* Kütz using response surface methodology. International Journal of Biological Macromolecules. 2016;**93**:534-542. DOI: 10.1016/j.ijbiomac.2016.09.019

[110] Markou G, Nerantzis E. Microalgae for high-value compounds and biofuels production: A review with focus on cultivation under stress conditions. Biotechnology Advances. 2013;**31**:1532-1542. DOI: 10.1016/j.biotechadv.2013.07.011

[111] Kavitha G, Kurinjimalar C, Sivakumar K, Palani P, Rengasamy R. Biosynthesis, purification and characterization of polyhydroxybutyrate from *Botryococcus braunii* K. International Journal of Biological Macromolecules. 2016;**89**:700-706. DOI: 10.1016/j.ijbiomac.2016.04.086

[112] Arévalo-Gallegos A, Garcia-Perez JS, Carrillo-Nieves D, Ramirez-Mendoza RA, Iqbal HM, Parra-Saldívar R. *Botryococcus braunii* as a bioreactor for the production of nanoparticles with antimicrobial potentialities. International Journal of Nanomedicine. 2018;**13**:5591. DOI: 10.2147/IJN.S174205

[113] Sawayama S, Minowa T, Yokoyama SY. Possibility of renewable energy production and CO_2 mitigation by thermochemical liquefaction of microalgae. Biomass and Bioenergy. 1999;**17**:33-39. DOI: 10.1016/S0961-9534(99)00019-7

[114] Ruangsomboon S, Sornchai P, Prachom N. Enhanced hydrocarbon production and improved biodiesel qualities of *Botryococcus braunii* KMITL 5 by vitamins thiamine, biotin and cobalamin supplementation. Algal Research. 2018;**29**:159-169. DOI: 10.1016/j.algal.2017.11.028

[115] Calderón NDG, Bayona KCD, Garcés LA. Immobilization of the green microalga *Botryococcus braunii* in polyester wadding: Effect on biomass, fatty acids, and exopolysaccharide production. Biocatalysis and Agricultural Biotechnology. 2018;**14**:80-87. DOI: 10.1016/j.bcab.2018.02.006

[116] Al-Hothaly KA. An optimized method for the bio-harvesting of microalgae, *Botryococcus braunii*, using *Aspergillus* sp. in large-scale studies. MethodsX. 2018;**5**:788-794. DOI: 10.1016/j.mex.2018.07.010

Chapter 8

Bioconcentration of Marine Algae Using Lipase Enzyme

Jithu Paul Jacob

Abstract

Marine algae rich in n-3 PUFA, being a natural and readily available resource, could be an alternative to fish oil derived n-3 PUFA; therefore, it could be of immense potentiality in nutraceutical and pharmaceutical industries. This highlights the sustainable benefits of algae and the many potential gains in creating algal bio-factories. In recent years, the use of lipase as biocatalysts had drawn considerable attention. Lipase is an enzyme that hydrolyzes lipids, the ester bonds in triglycerides, to form fatty acids and glycerol. Among the lipases assayed, the enzyme from the yeast *Candida cylindracea* is of special interest, as these are proved to be a nonspecific catalyst for many (commercially) interesting reactions such as the modification of oils and fats, reactions in organic solvents, and resolution of racemic mixtures. Hence, the enrichment of microalgae using biolipase from the source *Candida cylindracea* is of particular attention. Lipase action of *Candida cylindracea* is investigated as a function of time. It is observed that the lipases display a significant preference to saturated fatty acids; however, the resistance to release EPA and DHA was less as the hydrolysis reaction progresses.

Keywords: marine algae, lipase, PUFA, phospholipids, biofuels

1. Introduction

Poly unsaturated fatty acids (PUFAs) are unit fatty acids with a protracted chain contains 20 carbons or more, and primary covalent linkage situated on the third position carbon atom at the methyl end. PUFAs, together with EPA and DHA, proposed long before projected to bequeath health edges by rising blood pressure [1], appeasing symptoms of rheumatoid arthritis and depression, as well as attenuating the progression of Alzheimer's disease [2]. Although plant-derived α-linolenic acid (ALA) is obtained from dairy products and margarines [3] and with the help of desaturase and elongase enzyme convert it to EPA and DHA in humans, where the process is inefficient (0.04–2.84%), and the conversion is restricted by high dietary intake of EPA, DHA and linoleic acid. Also, low delta-6 desaturase activity in humans may be the reason for poor conversion of ALA to EPA and DHA [4]. However, the intake of PUFA-enriched foods or marine oil supplements containing these fatty acids through diet can increase the levels of EPA and DHA. Whereas, the diet of developed countries major sources of PUFAs are fish, red meat and poultry [5] where combinations of these foods contribute high levels (>75% of the total intake derived from 29 different food groups) of DHA (fish and poultry), EPA (fish and red meat), and docosapentaenoic acid (DPA; red meat, poultry and fish).

Marine algae, like other algae, have chlorophyll a photosynthetic system and thus considering as a diverse group of photosynthetic organisms. However, they possess simple structural moiety; their reproductive structures lack sterile cells so also do not form any embryos. For broader classification generally algae are divided into eight major groups or divisions based on their difference in their photosynthetic pigments, carbohydrate reserves, and cell structures. These algal groups contain unicellular members (collectively called microalgae) and multicellular members (macroalgae or seaweeds). The application of marine algae varied greatly from human food to animal fodder, source of phycocolloids and bioactive products to even recently use for biofiltration. The economic utilization of both marine macroalgae and microalgae has been explored for some time. Since 1940, it has been used as a source of liquid fuels and single cell proteins. During 1960s, with the invention of the extraordinarily halophilic algae Dunaliella could be considered as the most effective natural supply of carotene, started the business utilization of microalgae gained impetus.

Nowadays, microalgae provide a wide range of use as fine chemicals, oils, and polysaccharides, as soil conditioners, waste treatment and aquaculture. As a result of their usable products, the natural resources of algae cannot meet the demand and they are overexploited in their natural habitats [6]. The cultivation of microalgae is presently one of the most productive and environmentally friendly forms of livelihood among the coastal populations. Algal culture is being investigated to be used in house vehicles as a way of air revitalisation, food production, and waste treatment.

2. Recovery of PUFA from microalgae

The process has three main steps: (1) combined extraction-transesterification of fatty esters from the algal biomass; (2) a silver ion column chromatography step; and (3) a chlorophyll removal step [7]. Optimal processing conditions, the scale up of recovery, and the relative economics of producing microalgal EPA are important. The quality and stability characteristics of EPA from microalgae area unit has established. Previously, many advance process schemes introduced to purify polyunsaturated fatty acids from complex mixtures. To obtain good purity, these schemes perpetually employed as many processing operations which reduce the overall recovery and magnified costs. In several other cases, these methods had other problems which omitted it from uses. There in another study which involves complicated procedure of a two-step winterization, saponification, and urea fractionation of sardine oil successfully recovered 90% fraction of EPA and DHA, but failed to resolve those two compounds. Selective extraction of PUFA can be achieved by using aqueous solutions of silver nitrate through a water immiscible organic phase. However, this approach is questionable such that suitably it does not allow purification of a single compound such as EPA from complex mixture such as esterified oils. Similarly, PUFAs could also by selection obtained by surface assimilation of the esterified oil on aminopropyl warranted silicon oxide columns and selective extraction of saturated and oleic acid esters with solvent. The polyunsaturated fatty acid esters are then eluted with dichloromethane.

This method again had the drawback that it does not resolve highly pure EPA from the other polyunsaturated esters. For the fractionation of PUFA there is another variant of column chromatography envisions which uses aluminum oxide stationary phase and supercritical or liquid carbon dioxide as the mobile phase, but only few clear details have been published so far. Fractionation of fish oil and whole triglyceride extracts of other marine organisms can be done directly on silica gel and Ag-impregnated silica gel, whereas, initial fractionation of whole triglycerides is not efficient for at the same time getting a high purity associate degreed smart recovery of most of the Eicosapentenoic acids in an oil.

A newer novel approach used kinetic resolution to separate EPA from fish oil. Kinetic resolution is based on differences in selectivity and rates of lipase catalyzed esterification of different fatty acids in a mixture. Whereas this approach has allowed high recovery of Eicosapentenoic acids (up to 75%) from the oil, the purity of the product did not exceed 18%. When free fatty acids were used as the starting material rather than the triglycerides, EPA recovery by kinetic resolution improved to 93% but purity declined to less than 8%. Obviously, kinetic resolution as a method of purification has limited capabilities [8]. In addition, kinetic resolution using lipases under anhydrous conditions is difficult to economically implement in practice and the process is comparatively slow. Other PUFA recovery schemes, mostly useful only in the laboratory, have been reviewed elsewhere. Because of their survival in a variety of environmental conditions and widespread availability, studies on the mass culture of algae have been largely confined to freshwater species of Chlorella and Scenedesmus. According to the use of intend it is needed to screen a large number of unicellular algae from the point of view of nutritional composition, toxicity, resistance to contaminants, growth rates in mass culture, suitability etc. It is possible to mass culture in the laboratory with good results, in that cases the conditions must be good not like in primitive area where glass jars or tubes, artificial illumination, and equipment for sterilizing large volumes of seawater is not available, and in locations where the large scale culture of marine species is not possible [9].

It is important therefore to find unsophisticated methods for the production of large quantities of marine species. Other investigations into the culture of marine protoctist have trusted natural brine, each in open tanks and in closed controlled systems. Although impregnated natural saltwater created an honest crop of plant life, the inoculated culture was sometimes replaced rapidly by other organisms introduced with the seawater, such as motile and non-motile chlorophytes, colorless flagellates, ciliates or other zooplankters. Even underneath controlled laboratory conditions, little protoctist species in saltwater often experience filters and become established in culture carboys. It would be advantageous to use an artificial saltwater medium which would eliminate the immediate introduction of undesirable organisms.

Marine algae rich in n-3 PUFA being natural and readily available resource could be an alternative to fish oil derived n-3 PUFA; therefore, it could be of immense potentiality in nutraceutical and pharmaceutical industry. Lipids and protein produce during algal growth may be used as biodiesel, biomass for oil sources, and also as animal feed. This uses highlights the suitable benefits of algae and many potential gains while developing algal bio-factories. Limit factors for the potential use of marine algal oils on large scale are cost, extraction and purification methods.

In addition, extra experimentation is needed to confirm best growth conditions for enhancing macromolecule biogenesis. More over algae-derived oils are vegetarian-friendly and easy to grow on a large scale due to their small size. n-3 PUFA are typically associated with marine organisms, and algae, as the basis of the marine tropic chain that poses a very promising source of PUFA. In recent years, the use of lipase as biocatalysts had drawn considerable attention.

Lipase is associate degree catalyst that hydrolyzes lipids, the organic compound bonds in triglycerides, to create fatty acids and glycerin. Currently they account for 25% of total enzymes used in biotechnology, and this is because of the great versatility of the enzyme in catalyzing reactions of hydrolysis and synthesis, interesterification and transesterfication. Among the lipases, the enzyme from the yeast *Candida cylindracea* is of particular interest, since these are proved to be a nonspecific catalyst for many (commercially) appealing reactions such as the modification of oils and fats, resolution of racemic mixtures and reactions in organic solvents [10]. Hence, the enrichment of microalgae using biolipase from the source *Candida cylindracea* is of particular attention. The partial hydrolysis of the sardine oil by *Candida cylindracea* lipase indicates a strong discrimination by the lipase against DHA, so the DHA present in triglycerides does not get hydrolyzed, in effect get concentrated [11].

On the other hand, this lipase has only moderate discrimination against EPA, so the concentration percentage of EPA is comparatively lower than that of DHA with a moderate enrichment. Thus, the partial hydrolysis values of the sardine oil by *Candida cylindracea* lipase indicate higher specificity of lipase towards DHA than towards EPA. Lipase action of *Candida cylindracea* is investigated as a function of time. It is observed that the lipases display a significant preference to saturated fatty acids, however, the resistance to release EPA and DHA was less as the hydrolysis reaction progresses. It has been reported that because n 3 PUFA is located in the second position of triglyceride, hydrolysis of sardine oil with 1,3 specific lipase should produce PUFA rich 2-monoglycerides and 1,2 diacyl glycerides [12, 13]. The presence of cis carbon–carbon double-bonds in the fatty acids result in bending of the chains. Therefore, the terminal methyl group of the fatty acids lies so close with the ester bond that can cause a stearic hindrance effect on lipases. Due to the presence of five and six double bonds there is a high bending effect of EPA and DHA, enhancing the stearic hindrance effect; consequently lipases cannot reach the ester-linkage between these fatty acids and glycerol. However, saturated and monounsaturated fatty acids of triglycerides do not possess any barriers to lipases so they can be easily hydrolyzed. Therefore, fatty acid selectivity of the lipases for EPA and DHA allows their separation and concentration from other components left behind portion of marine oils. In addition to it, the lipases have been frequently used for the discrimination between EPA and DHA in concentrates containing both of these fatty acids, thus providing the possibility of producing omega3-PUFA concentrates with dominance of either EPA or DHA.

The enzyme lipase was first commercially successfully introduced by Novo Nordisk in 1988 under the trade name "Lipolase". It was actually originated from the fungus *Humicola lanuginosa*. Again, in 1995 two bacterial lipases were introduced—"Lumafast" and "Lipomax" from *Pseudomonas mendocina* and *Pseudomonas alcaligens*, respectively both produced by Genencor International. Currently, industrial enzymes are manufactured by three major suppliers, they are Novozymes, Denmark, Genencor International Inc., US and DSM NV, Netherlands. Lipases are marketed by various brand names like Lipopan, Lipozyme, Novozyme, Patalase, Greasex, Lipolase and Lipoprime. Lipases of microbial origin have gained considerable attention in the field of biotechnology and a large number of microbial strains have been used for the enzyme production.

The production of extracellular lipase by *Candida cylindracea* in a batch bioreactor is influenced by aeration, substrate type and concentration. Both olive oil and oleic acid when used as the carbon sources gave almost identical activity while the production of extracellular lipase was growth associated. For optimum lipase production it required the enrichment of air flow by pure oxygen by maintaining the oxygen concentration at the recommended value. The optimal growth conditions for lipase production by *Candida cylindracea* is influenced by agitation speeds and aeration in a fermentor [14]. Maximum lipolytic activity was observed when the microorganisms were at the beginning of the stationary growth phase. For the production of lipase submerged cultivations using yeast has been found to be the most suitable process. Meanwhile, these processes are influenced by a variety of parameters and also their interactions. Due to the complex morphology of the cells, industrial processes involving submerged fermentation require greater attention.

The fermentation runs with 200 rate yielded higher protein activity compared with the fermentations runs at 400 rates. Higher bad hat speed light-emitting diode to the formation of cells to aggregates with vacuolation. The vacuoles formed clumps which lowered the enzyme production. When the agitator speed was 200 rpm the vacuoles will be separated, resulting in increased enzyme production [15]. Aeration and dissolved oxygen also affected the morphology of the cell. Yeasts are considered as important sources for lipase production. There are problems such as changes in morphology of cells during agitation and while enzyme production the combined effect of operational parameters have negative effect on enzyme preparation in fermentation steps.

For the enrichment or bio concentration of DHA and EPA in marine oils lipases are used. These can be applied in free fatty acids (FFA) or in simple esters of marine oil fatty acids also. One of the major advantages is that the lipases can be operated under mild conditions, so that products such as EPA and DHA are preferable since they are prone to oxidation. The choice of lipase and raw material used were depended upon the structure of the desired lipid and ratio of EPA to DHA in the final product. Specificity of the lipase towards the fatty acid should also be considered. The location of the fatty acids is in triacylglycerols (TG), and then the regiospecificity and TG specificity also have an effect on the enrichment. Thus, the positional distribution of the fatty acids on the acylglycerol molecule structure may have an effect on the ability of the lipase to enrich DHA and/or EPA in either the substrate or product. Different strategies are employed using lipases for the concentration of EPA and/or DHA of marine origin. Lipases from *Candida rugosa* (formerly *Candida cylindracea*) and *Rhizomucor miehei* discriminate against DHA than EPA. But those from porcine pancreas, *Chromobacterium viscosum*, *Pseudomonas* sp., *Pseudomonas cepacia* and *Pseudomonas fluorescens* do vice versa, i.e.: discriminate against EPA than DHA. Lipase from *Rhizomucor miehei* can be applied for the enrichment of DHA in FFA from fish oil by alcoholysis of the oil with butanol or glycerol. By ethanolysis of fish oil this lipase succeeded in separating DHA into the acylglycerol fraction and EPA into the ethyl ester fraction [16]. Lipase from *Candida rugosa* also got similar application in catalyzing the enrichment of DHA in the acylglycerol fraction by the hydrolysis of fish oil. Both DHA and EPA enrichment has been successful in the acylglycerol fraction obtained by ethanolysis or hydrolysis of fish oil catalyzed by the lipase from *Pseudomonas* sp., *Pseudomonas fluorescens* and *Geotrichum candidum* [12, 13]. The lipase from *Rhizopus delemar* has been used to catalyze the esterification between FFA and lauryl alcohol to concentrate DHA in the FFA from fish oil. To concentrate both DHA and EPA in FFA another approach is to esterify FFA from marine origin with glycerol catalyzed by the lipase from Pseudomonas sp., *Pseudomonas fluorescens*, *Thermomyces lanuginosus* (formerly *Humicola lanuginosa*) or *Rhizopus oryzae*.

3. Analysis of enriched fatty acid in marine algae

The reagents commonly used for acid-catalyzed transesterification are methanolic, hydrochloric and sulfuric acid, and boron trifluoride in methanol. All of them are suitable for lipid transesterification and also free-fatty-acid methylation. However, at ambient temperature neither acid-catalyzed nor boron-fluoride-catalyzed reactions proceed; in both cases the reaction requires heating. Among the mentioned reagents, boron trifluoride-methanol reagent (12–14% w/v) is the most often used for transesterification of all types of lipids and it is being the best and very useful reagent for lipid esterification. Under the conditions recommended (heating at 100°C), transesterification is complete within 2 min for free fatty acids, within 10 min for phosphoglycerides, within 30 min for triglycerides and within 90 min for sphingomyelin. For biological samples boron fluoride-methanol reagent is used for transesterification of lipids following the procedure given below: into a screw-capped tube (Teflon cap liner) a small aliquot of the lipid extract (dissolved in chloroform) is added; to it add 0.5–1 ml boron fluoride-methanol reagent (140 g/l, containing an required amount of BHT as antioxidant); the tube is then closed and heated at 90°C for 2 h. This methodology proved to be administered complete transesterification of all lipids; the addition of BHT can prevent rancidity of PUFA.

After the transesterification reaction completed, FAMEs are extracted by adding n-hexane and water twice to the proportion of sample. This is the most simple and effective procedure. Even though, this method is most popular, boron tri-fluoride-methanol has few disadvantages. Unless refrigerated, the reagent has only a limited shelf life [17]. The use of recent or too focused solutions might lead to the assembly of artifacts or loss of PUFAs.

If the sample contains plasmalogens, aldehydes area unit liberated by the chemical agent and area unit regenerate into dimethyl acetals (DMAs), that area unit nearly not possible to break free the major carboxylic acid methyl esters including methyl palmitate. There is wide usage of anhydrous methanolic hydrochloric acid and methanolic sulfuric acid for lipid esterification varying under different conditions, in particular, different reaction temperatures, different acid concentrations, and different reaction times. The methanolic acid boron fluoride-methanol works as methylating free fatty acids very rapidly so can be used to transesterify all the lipids which are typically present in the biological samples. With a concentration of 5% methanolic hydrochloric acid complete transesterification can be carried out by heating the sample in the reagent for about 2 h under refluxion. This reaction can also be carried out at 50°C overnight. In the same way a solution of 1–2% (v/v) concentrated sulfuric acid in methanol can be used for the transesterification of lipid samples. This method using methanolic, hydrochloric and sulfuric acids also have the disadvantage like boron trifluoride-methanol that DMAs are formed during transesterification from plasmalogens. Acetyl chloride and aluminum chloride are the other reagents used for transesterification. Both these reagents shown complete transesterification in the samples without prior extraction of the lipid, whereas, aluminum chloride has the disadvantage that it does not esterify free fatty acids.

4. PUFA as phospholipids fractions

Phospholipids are major constituents of cell membranes and play essential roles in biochemistry and physiology of the cell functions. Phospholipids in fish and marine species are highly enriched with the long-chain n-3 type polyunsaturated fatty acids. About 40–50% content of EPA and DHA is not uncommon in some phospholipid classes in fish. The role of n-3 polyunsaturated fatty acids in phospholipid moiety is in adjusting the membrane integrity and functions presumably at lower temperature, and also to the membrane fluidity and mobility as a result of their higher unsaturation. In the case of fish among the phospholipids, phosphatidylcholine (PC) and phosphatidylethanolamine (PE) are by far the most abundant in the flesh, especially PC make up to 50–60% of the total phospholipid content [18]. The composition of individual phospholipid classes is remarkably similar among fish species as is the characteristic fatty acid composition of each class. Lecithins present in plant and vegetable origin are popularly using as health supplements. The vegetable oil is highly enriched with n-6 fatty acids, which so as in the case of the n-3 fatty acid fish oils. On the other hand, purified fish lecithins, which are highly enriched with n-3 polyunsaturated fatty acids phospholipids, are not available on the market at all. This is because tedious extraction procedure is required for obtaining lecithin from fish oil unlike plant or vegetable lecithins. Here, certain attempts for the preparation of such phospholipids, highly enriched with EPA and DHA, from the more readily available plant or animal lecithins is explained. Pure phosphatidylcholine can be obtained from egg yolk after purification by preparative HPLC and was treated under the acidolysis reaction using the *Mucor miehei* lipase. There is observed reaction as anticipated in which the rate of the reaction involving the phospholipids that possess the zwitterionic head group. This is much lower as compared to the natural triacylglycerol substrates. Therefore, large quantities of lipase were required, which will resulted in high extent of hydrolysis side reaction. The optimal reaction conditions is offered in a mixture of phospholipids of approximately 40% desired for phosphatidylcholine and lysophosphatidylcholine (LPC) whereas, 20% of glycerol phosphatidylcholine (GPC). In LPC only one of the acyl moiety will be hydrolyzed and in glycerol phosphatidylcholine both acyl groups will get hydrolyzed. When pure EPA was used, both the PC and the LPC fractions were highly enriched with EPA, particularly the LPC fraction, 58 and 69%, respectively.

5. Application of PUFA

PUFA had many helpful effects for human health so it considered as unit important elements in human nutrition. The intake of PUFA in diet, together with n-6 fatty acids, is understood to modulate the inflammatory processes among different cell functions. Although many of the species exhibited high amounts of SFA, some *Phaeophyta* and *Rhodophyta* species show higher concentrations of PUFA, and PUFA/SFA ratios higher than 1 (*H. scoparia*, 1.46; *T. atomaria*, 1.33; *C. spongiosus*, 1.77; *Peyssonnelia* sp., 1.33). Whereas, the lowest ratios were discovered in algae from the phylum Chlorophyta (0.27–0.68) [19]. It seems that this phylum incorporates a lower potential, examination to the opposite two phyla studied, as a nutritional source of PUFA for human consumption. However, not all PUFA are associated with the promotion of health benefits. For example, in the inflammation process, eicosanoids derived from n-6 PUFA are generally considered as pro-inflammatory or as promoters of other cell harmful effects, whereas n-3 PUFA derivatives are considered less inflammatory or even anti-inflammatory [20].

Since the synthesis pathway of those fatty acids depends on identical enzymes for n-3 and n-6 PUFA, the health promoting effects area unit keen about the n-6/n-3 magnitude relation of PUFA obtained through diet.

The World Health Organization (WHO) recommends a $\sum$n-6/$\sum$n-3 magnitude relation not up to 10. Almost all algae can be considered as a good source of dietary PUFA, since they showed ratios ranging between 0.29 and 6.73 [21]. The exception was *Chaetomorpha* sp., during which the $\sum$n-6/$\sum$n-3 magnitude relation was the best from all the studied species (31.25) and in *D. spiralis* during which no n-3 fatty acids were detected. Besides associate degree applicable nutritionary profile, these macroalgae can also be exploited for pharmaceutical purposes.

Many of the PUFA thought-about powerful molecules against many diseases and area unit already employed in totally different medical specialty applications. For example, several reports suggest that n-3 fatty acids, mainly EPA and DHA, may have a significant potential in the treatment of autoimmune and inflammatory diseases. Rhodophyta was the phylum with the highest percentage of n-3 fatty acids (16–27% of total FAME), followed by Phaeophyta (0–15%), in which significant amounts of n-3 were also present. Aside from *Ulva* sp. that had 18 of n-3 FAME, Chlorophyta macroalgae conferred very cheap values of n-3 fatty acids (1–9%). Conversely, the detected n-6 fatty acids were lower in Rhodophytes (8–15%), thanks to the low concentration of linoleic acid, except for *Peyssonnelia* sp., where n-6 concentration was approximately 28% of total FAME. Phaeophytes showed the highest contents of n-6 fatty acids (23–44%), whereas chlorophytes presented mid-range values (6–27%).

Considering absolutely the concentrations of PUFA within the varied species tested, *Ulva* sp., *T. atomaria*, *C. spongiosus*, *Peyssonnelia* sp. and *B. secundiflora* possess the best contents of n-3 PUFA, 1.07, 1.38, 1.19, 1.06 and 1.42 mg/g, severally. Apart from genus *Ulva* sp., during which ALA dominated, the n-3 profile of the remaining strains was basically composed of independent agency. Nevertheless, Peyssonnelia sp. exhibited a relatively high content of DHA, 0.22 mg/g of dry biomass, coupled with an EPA concentration of 0.84 mg/g. A variety of potential applications area unit delineated for independent agency and DHA, which hold significant potential for pharmaceutical purposes, namely cancer treatment, asthma, psoriasis, rheumatoid arthritis, antibiotic, inflammatory bowel disease, depression, allergies, cardiovascular diseases, among others [22].

More recently, PUFA verified to own a robust potential in drug delivery; additionally to the delineated toxicity of a number of PUFA, PUFA enable a more efficient penetration of specific molecules through the cell membranes of tumor cells, due to their unique lipophilic characteristics. In fact, several studies show that tumor cells display faster PUFA intake than normal cells, as demonstrated for the conjugated taxoid DHA-paclitaxel. The nutritionary and pharmaceutical edges of PUFA, however, contrast with the increasing difficulty in finding sustainable sources of n-3 VLCPUFA, which traditionally were obtained from fish and fish oil [23].

Declining fish stocks caused by decades of overfishing makes ever more urgent to find non-traditional alternatives for the western world. As VLCPUFA are usually absent from terrestrial higher plants, traditional crops can also be excluded as viable sources of these FA. Though this deficiency can be overcome by applying genetic engineering, transgenic foods are not always well accepted by the general public. Therefore, n-3 VLCPUFA are typically associated with marine organisms, and algae, as the basis of the marine trophic chain, come out as a very promising source of VLCPUFA. In fact, large scale farming of marine algae has been accomplished successfully for hundreds of years. Approximately, 220 protoctist species area unit presently cultivated and harvested everywhere the globe for various functions. Though principally used as food for human consumption, particularly in Asia, macroalgae are also the primary source of hydrocolloids such as agar, carrageenan and alginate, which have numerous industrial applications, such as gelling, stabilizing or binding agents. The next step may somewhat be the property exploitation of marine macroalgae as different sources of VLCPUFA, not solely in Asia, however conjointly within the western world.

Author details

Jithu Paul Jacob
St. Albert's College (Autonomous), Ernakulam, Kerala, India

*Address all correspondence to: jithupaul007@gmail.com

IntechOpen

© 2019 The Author(s). Licensee IntechOpen. This chapter is distributed under the terms of the Creative Commons Attribution License (http://creativecommons.org/licenses/by/3.0), which permits unrestricted use, distribution, and reproduction in any medium, provided the original work is properly cited.

References

[1] Fortin PR, Lew RA, Liang MH, Wright EA, Beckett LA, Chalmers TC, et al. Validation of a meta-analysis: The effects of fish oil in rheumatoid arthritis. Journal of Clinical Epidemiology. 1995;**48**:1379-1390

[2] Martins JG. EPA but not DHA appears to be responsible for the efficacy of omega-3 long chain polyunsaturated fatty acid supplementation in depression: Evidence from a meta-analysis of randomized controlled trials. Journal of the American College of Nutrition. 2009;**28**:525-542

[3] Astorg P, Arnault N, Czernichow S, Noisette N, Galan P, Hercberg S. Dietary intakes and food sources of n- and n-3 PUFA in French adult men and women. Lipids. 2004;**39**:527-535

[4] Ballesteros A, Bemabe M, Cruzado C, Martin-Lomas M, Otero C. Regioselective deacylation of 1,6-anhydro-beta-D-galacto-pyranose derivatives catalyzed by soluble and immobilized lipases. Tetrahedron. 1989;**45**:7077-7082

[5] Howe P, Meyer B, Record S, Baghurst K. Dietary intake of long-chain ω-3 polyunsaturated fatty acids: Contribution of meat sources. Nutrition. 2006;**22**:47-53

[6] Li CY, Chen SJ, Cheng CY. Production of *Acinetobacter radioresistens* lipase with repeated fed-batch culture. Biochemical Engineering Journal. 2005;**25**:195-199

[7] Deleuze H, Langrand G, Millet H, Baratti I, Buono G, Triantaphylides C. Lipase catalyzed reactions in organic media: Competition and applications. Biochimica et Biophysica Acta. 1987;**911**:117-120

[8] Monot F, Borzeis F, Bardin M, Vandecasteele JP. Enzymatic esterification in organic media: Role of water and organic solvent in kinetics and yield of butyl butyrate synthesis. Applied Microbiology and Biotechnology. 1991;**35**:759-765

[9] Noor IM, Hasan M, Ramachandran KB. Effect of carbon and nitrogen sources on the production of lipase by *Candida cylindracea* 2031 in batch fermentation. In: Proceedings of the 1st International Conference on Natural Resources Engineering and Technology; 2006. pp. 158-166

[10] Felony G, Armas JC, Mendoza JCD. Production of extracellular lipase from *Aspergillus niger* by solid-state fermentation. Food Technology and Biotechnology. 2006;**44**:235-240

[11] Fadiloglu S, Erkmen E. Effects of carbon and nitrogen sources on lipase production by *Candida rugosa*. Turkish Journal of Engineering and Environmental Sciences. 2001;**26**:249-254

[12] Otero C, Pastor E, Ballesteros A. Synthesis of monobutyryl-glycerol by transesterification with soluble and immobilized lipases. Applied Biochemistry and Biotechnology. 1990;**26**:35-44

[13] Ciafardini G, Zullo BA, Iride A. Lipase production by yeasts from extra virgin olive oil. Journal of Food Microbiology. 2006;**23**:60-67

[14] Carta G, Gainer JL, Benton AH. Enzymatic synthesis of esters using an immobilized lipase. Biotechnology and Bioengineering. 1991;**37**:1004-1009

[15] Benzonana G, Esposito S. On the positional and chain specifities of *Candida cylindracea* lipase. Biochimica et Biophysica Acta. 1971;**231**:15-22

[16] Otero C, Pastor E, Fernandez VM, Ballesteros A. Influence of the support

on the reaction course of tributyrine hydrolysis catalyzed by soluble and immobilized lipases. Applied Biochemistry and Biotechnology. 1990;**23**:237-247

[17] Sokolovska I, Albasi C, Bales V. Production of extracellular lipase by *Candida cylindracea* CBS 6330. Bioprocess Engineering. 1998;**19**:179-186

[18] Burdge GC, Finnegan YE, Minihane AM, Williams CM, Wootton SA. Effect of altered dietary n-3 fatty acid intake upon plasma lipid fatty acid composition, conversion of [13C]α-linolenic acid to longer-chain fatty acids and partitioning towards β-oxidation in older men. The British Journal of Nutrition. 2003;**90**:311-321

[19] Morris MC, Evans DA, Bienias JL, Tangney CC, Bennett DA, Wilson RS, et al. Consumption of fish and n-3 fatty acids and risk of incident Alzheimer disease. Archives of Neurology. 2003;**60**:940-946

[20] Block RC, Harris WS, Pottala JV. Determinants of blood cell omega-3 fatty acid content. The Open Biomarkers Journal. 2008;**1**:1-6

[21] O'Sullivan TA, Ambrosini G, Beilin LJ, Mori TA, Oddy WH. Dietary intake and food sources of fatty acids in Australian adolescents. Nutrition. 2011;**27**:153-159

[22] Geleijnse JM, Giltay EJ, Grobbee DE, Donders ART, Kok FJ. Blood pressure response to fish oil supplementation: Metaregression analysis of randomized trials. Journal of Hypertension. 2002;**20**:1493-1499

[23] Singer P, Berger I, Wirth M, Gödicke W, Jaeger W, Voigt S. Slow desaturation and elongation of linoleic and α-linolenic acids as a rationale of eicosapentaenoic acid-rich diet to lower blood pressure and serum lipids in normal, hypertensive and hyperlipidemic subjects. Prostaglandins, Leukotrienes, and Medicine. 1986;**24**:173-193

Chapter 9

Concepts and Trends for Extraction and Application of Microalgae Carbohydrates

Maiara Priscilla de Souza, Andrea Sanchez-Barrios, Tiele Medianeira Rizzetti, Lisianne Brittes Benitez, Michele Hoeltz, Rosana de Cassia de Souza Schneider and Fábio de Farias Neves

Abstract

The proposed chapter aims to provide a more in-depth explanation of the composition of carbohydrates in microalgae biomass, focusing on separation methods, chemistry, molecular characterization, as well as their application in several areas. The purpose of this review chapter is to show that biological products from microalgae have potential in health, food, and industry applications (materials and biofuel production). Steps for extraction and purification will be discussed, as well as the relationship between the type of microalgae and its composition, as a way of optimizing protocol selection and product making, without breaking down the cell to begin with (total carbohydrate extraction present in the cell). An overall overview of the current and prospective trends and methodologies for the use of microalgae carbohydrate will be included as starting points to shed light on some of the possible issues that currently do not allow the development and feasibility of microalgae biorefineries.

Keywords: microalgae, carbohydrates, analytical chemistry, biorefinery, trends

1. Introduction

The constant necessity of looking for new alternatives to produce sustainable versions of products has led to the discovery and development of new technique and biological models. Organisms with the potential of producing molecules that can be used for the development of bioproducts in different areas (food, beauty, health, and biodiesel, among others) have led to the discovery, study, and use of new organisms. Algae and microalgae have become potential and promising model organisms to be used for carbohydrate production and use, but besides that, it is also deeply studied due to its lipids, proteins, and photosynthetic activity, making them a prospective source of bioenergy production [1, 2].

The problems associated to the first- and second-generation feedstock for biofuels have seen to become more and more complicated to solve considering the food crisis and complex conversion of lignocellulosic materials [3] researchers, has led to the exploration of a third-generation feedstock, mostly represented by photosynthetic organisms, primarily algae/microalgae [4].

Microalgae are considered to have the potential to produce third-generation biodiesel (due to its capability of fixing carbon dioxide (CO_2), which is eventually converted to biomass and other products), which can also be referred as third-generation feedstock, providing mainly lipids, proteins, and carbohydrates. The utilization of these molecules in a sequential way allows the treatment of biomass in biorefineries, including its use in fermentative production of a range of platform biochemical [5]. Through this process, carbohydrates are used as a fundamental piece for the production of certain products.

Besides the cell wall and reserves of photosynthetic reaction, the carbohydrates can be excreted by the cell. The exopolysaccharides (EPS) are complex carbohydrates produced from some microalgae, which are long chains composed of sugar derivative structures, mucilaginous and with reactive functional groups, such as sulfate, hydroxyl, or carboxylic [6]. The major components of EPS include mainly the polysaccharides and others as proteins, nucleic acid, and lipids [7]. Addition of these molecules is considered to be of extreme importance for enriching the nutritional value of food items [8]. Although these characteristics are beneficial, the extraction of these compounds from microalgae becomes a real challenge. In this context, several treatments can be performed for the disruption of microalgal cells, including chemical modifications and mechanical, thermal, or ultra-sonication processes [9].

Although promising, it is still hard to manage the cost and work that developing new technologies have for investors (in the industry for applied approaches and academia for basic development and standardization), which presents some limitations for the advance of research in this area. On the long run, an implementation of the use of microalgae as a substitute for many of the other crop options still used will have an important impact on the economics, environment, and more sustainable practices.

Due to the vast diversity of species of microalgae, we present an analysis of the current trends and importance and potential of the use of carbohydrates present in

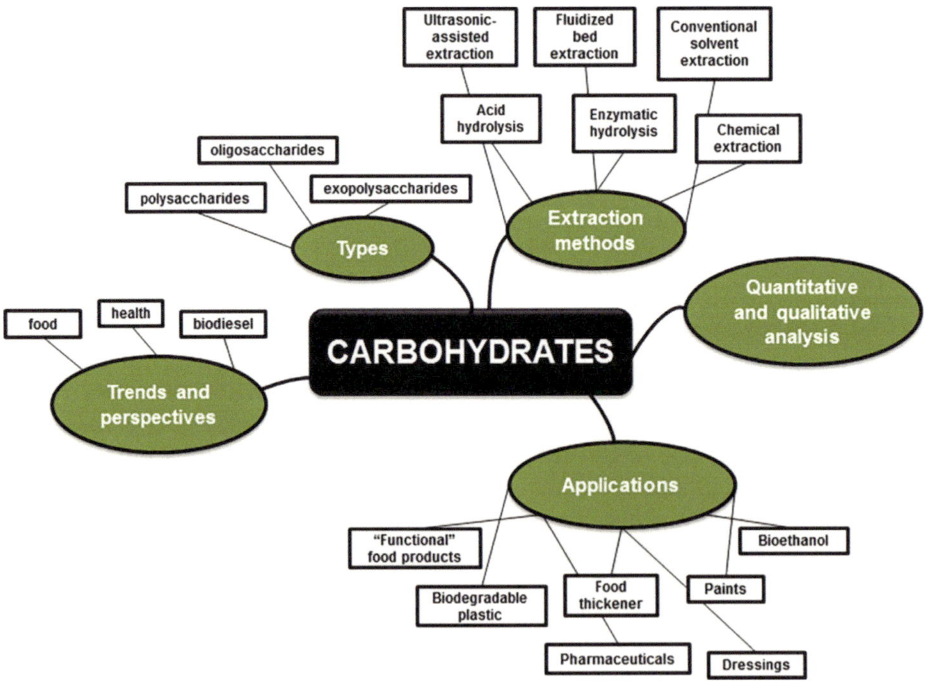

Figure 1.
Diagram of the main topics involved in the harnessing of microalgae carbohydrates.

microalgae. The aim of highlighting these points is to bring awareness and present new alternatives of methods that will allow the use of carbohydrates in microalgae without the breakdown of these carbohydrates, as a way of improving and lowering cost and making the use of these organisms a more feasible way of developing new products and technologies [10]. The main topics that will be discussed in this chapter can be seen in **Figure 1**.

2. Carbohydrate importance and composition

Carbohydrates from microalgae are considered to have a great application in industry, which has led to the development of new techniques and studies but that has exposed new challenges to industry [11, 12]. The diversity of microalgae, the composition, and cell organization are some of the few trials that many scientists are currently facing.

Carbohydrates are poly- or oligosaccharides that can be present in vacuoles and cell walls or that could also be excreted as exopolysaccharides (EPS) [13]. Microalgae come to be an interesting key organism to study due to the high content

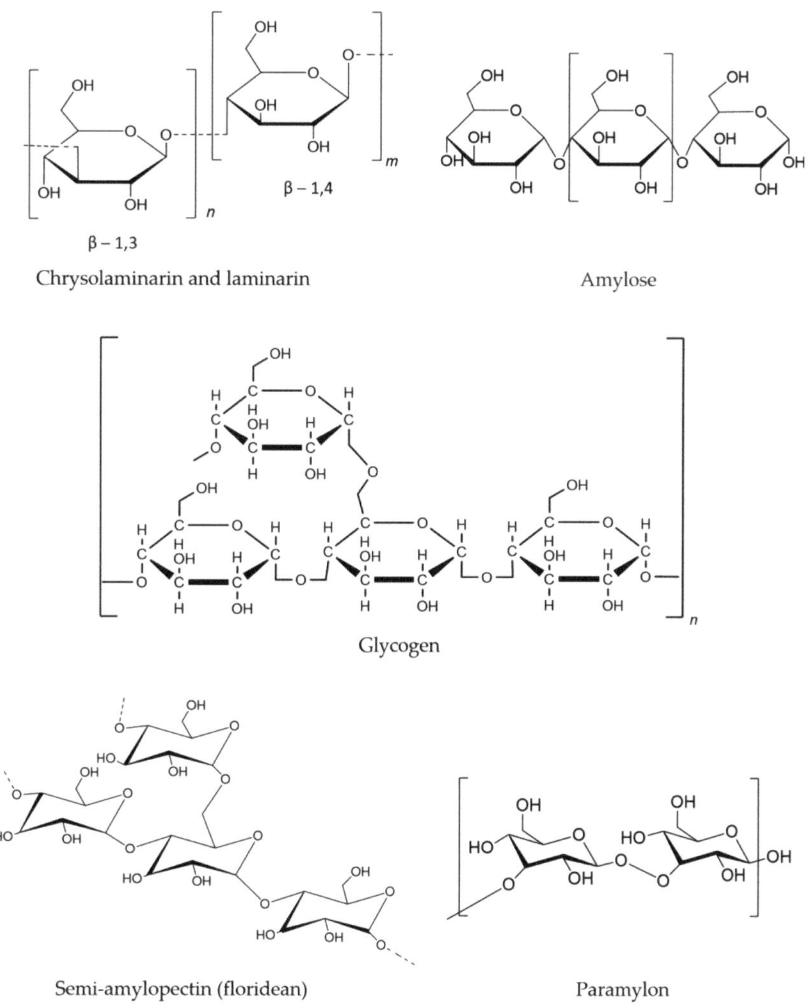

Figure 2.
Glucose polymers found in microalgae.

of carbohydrates that some of them have. Some examples are *Desmodesmus spp.* (41%) from wastewater and landfill leachate treatment [14], *Chlorella stigmatophora* (~55%) [5], or *Chlorella vulgaris* (60% [15] or >52% [16]) cultivated under conditions of nitrogen depletion.

Similarly, indigenous microalgae species have been bioprospected in nitrogen-depleted environments, such as *Desmodesmus sp.* (57%), an unidentified one with more carbohydrate content (70%) [17], and *Arthrospira platensis* that accumulated up to 74% of carbohydrates [18].

Despite their vast potential, using different species imposes a few challenges when trying to establish a consistent methodology for extraction and use. Based on the differences in metabolism that each species presents, a possible and feasible approach is to consider that they all have reserves made out of polymers of glucose (glucans), such as chrysolaminarin (1:11), laminarin (3:1) (β-1,3 and β-1,6 branches), paramylon(β-1,3), glycogen-type, cyanophycean, floridean (semi-amylopectin: α-1,4 and α-1,6 branches), and amylose-type starch (α-1, 4) or both (**Figure 2**). The external covering of cell with polysaccharides could be peptidoglycan matrices, cellulosic wall, and galactose polymer matrix, and others [18–20].

The potential of using microalgae for the development of various products, using carbohydrates as the main source, exposes some of the challenges that using these organisms can have. The main of it is the process of extraction of the diverse saccharides present in each species, but in comparison with plant-derived products, it is actually a much easier path, since they do not present lignification of the cell wall [11, 21]. In the same way, having various species together as a main feedstock source may require adaptations to a proper and more efficient method of extraction.

3. Trends for microalgae use and production, using carbohydrates as the main molecule of interest

Microalgae use for development of new products that could be beneficial to multiple industries has raised many questions in the quest of finding what will be the next technology that will be developed, what products will change our industries, and how this could benefit populations. As we know, the use of microalgae carbohydrates could be a solution to many current limitations that agriculture, pharmacy, nutrition, and other areas are facing while trying to develop better solutions to fulfill the needs of people [22].

One of the main reasons for the development of new technologies and the arise of trends is to be able to satisfy the demands of the public and produce the income that the market will like to have in return for investing in these areas. Unfortunately, it is still a difficult task since there is still a long way to go for developments of products outside of the food industry.

Carbohydrates/starch is considered to be a positive/beneficial step for the chemical development of products, and when produced in high quantities, its use for fermentation is considered a better option. To date, the most common way to induce the production of specific molecules (in this case, carbohydrates and starch) is by creating stress conditions in the environment where the alga is growing, mainly by altering nutrient concentrations or by changing light, temperature, and other parameters [23, 24].

Photosynthetic electron transport approach seems to be the most researched area [25] since genetic modification of algae becomes challenging due to the great variety of species. Cyanobacteria are being used more for genetic engineering, but no complete success has come out it yet. Besides the interest cyanobacteria, some algae species have been explored as potential species for biofuel production through modification of metabolic pathways that could increase the production of carbohydrates, lipids, and other compounds, but the progress in this area has seen to be slow, which although discouraging is still considered one of the main focus for the future of biotechnologies in microalgae [26].

Microbiome studies of microalgae populations seem to focus more on the ecology aspects and the importance of their use in multiple biological processes. Toxicity of waters seems to attract the attention of most research groups. Its importance lies in not only the understanding of structure communities and safety of fresh water resources (which is the main source of drinking water for the world) but also the discovery of new species that may be hard to culture through regular isolation practices.

Similarly, human microbiome studies seem to be interested in the potential use of carbohydrates (including exopolysaccharides) derived from microalgae, due to its prebiotic potential, improving the health of adults and infants that cannot consume breast milk in their early stages, strengthening their immune responses [12, 27].

4. Carbohydrate extraction and analysis

Some microalgae species are capable of accumulating a large amount of intracellular starch and have structural polysaccharides in their cell walls [28]. However, the development of standard analytical procedures for the characterization of microalgae biomass has been difficult because of the existence of several microalgae species with different cellular structures and chemical compositions [28]. Even a specific microalgae species may present variability in their chemical composition because they can strongly be affected by factors of cultivation conditions such as temperature, salinity, and nutrient availability [9]. Thereby, sample preparation for the determination of carbohydrates in microalgae is very complex.

The analytical method applied depends on the intended usage of the algal carbohydrates. Some require the qualitative or quantitative composition of the monosaccharides and others the total carbohydrate analysis [29].

In algal biomass, the carbohydrate profile analysis requires preliminary acid hydrolysis. Acid hydrolysis allows depolymerizing the intracellular starch and structural polysaccharides into their monomers, which are then further quantified [28]. The optimal hydrolysis conditions (acid, temperature, and time) should ensure complete hydrolysis of the polysaccharides and at the same time avoid excessive degradation of monosaccharides [29].

In this context, the selective cleaving of algal polysaccharides by enzymatic hydrolysis is another promising approach once the formation of degradation products like furfural or hydroxymethylfurfural is avoided [29]. However, for some microalgae species, the composition of the cell wall is complicated and unknown; also, some enzymes are very expensive [30]. The cell wall present in the microalgae limits extraction yields of high-value products or results in a low bioavailability of intracellular components [9, 30].

The methodology widely employed in acid hydrolysis is the procedure published by the National Renewable Energy Laboratory (NREL) based on two-step hydrolysis using H_2SO_4 [31, 32]. Despite being a reliable method, it is a multistep procedure, increasing the chance of experimental errors, and also presents a high time of sample preparation. Northcote et al. [33] proposed another method based on one-step acid hydrolysis with dilute H_2SO_4 and use smaller biomass samples.

Other methods use chemical extraction method like alkaline pretreatment [34] and physical method, such as hot-water treatment, microwave-assisted extraction, and ultrasonic-assisted extraction [30, 35]. The choice of extraction method is that the pretreatment is effective qualitatively and quantitatively, and the technology is simple to operate and economical for scale-up [30].

Zhao's research team [30] investigated three methods of conventional solvent extraction (CSE), fluidized bed extraction (FBE), and ultrasonic-assisted extraction (UAE) to obtain an effective extraction method of carbohydrates/glucose. The CSE employed lyophilized microalgae extracted with distilled water and agitation in a vortex. For FBE, the *Chlorella* sp. culture was harvested and washed with distilled water and then diluted using distilled water and added into a fluidized reactor with air aeration. To UAE, the algal cells were harvested and washed with distilled water, diluted, and taken to the ultrasonic processor. The ultrasonic-assisted extraction was more effective than the other methods.

Information in the literature related to the amount of cell wall microalgae polysaccharides is scarce. Usually, the quantification of polysaccharides in microalgal is made by analyzing the total carbohydrate, thus including storage polysaccharides (SPS) and cell wall-related polysaccharides, which exhibit different functions in the microalgal cell [9].

According to Bernaerts et al. [9], the insight into the composition of cell wall-related polysaccharides, such as the monosaccharide profile or the degree of sulfation, is not only desired in terms of process optimization but also as a potential for several biotechnological. Thereby, the authors investigated to apply a universal procedure for extraction of the total cell wall-related polysaccharides, including cell wall polysaccharides (CWPS) and extracellular polymeric substances (EPMS), of 10 commercially available microalgae species followed by a characterization of the monosaccharide profile, uronic acid content, and sulfate [9]. Initially, the procedure consisted of dry biomass suspended in saline solution incubated for 16 h at 25°C, followed by a two-step centrifugation. Afterward, the supernatant was submitted for extraction of EPMS and the residual biomass (pellet) for extraction of CWPS. Ethanol was added to the supernatant precipitating EPMS; the solution was vacuum filtered, and the insoluble residue was dialyzed against demineralized water for 48 h, and finally, the extracts were lyophilized.

In the extraction of CWPS, the pellets were suspended in MOPS buffer, and the cells were disrupted using UHPH after cold ethanol was added to the suspensions and for pellet recovery. Lipids were removed by addition of hexane/isopropanol to the pellet, mixed and centrifuged to remove the upper solvent layer. Afterward, SPS and protein were enzymatically removed using endo-β-1,3-glucanase or a combination of α-amylase/amyloglucosidase and *Subtilisin* A protease, respectively. The mixtures were incubated and after addition of cold ethanol they were centrifuged. The pellet was finally washed in acetone, vacuum filtered, and dried overnight at 40°C, and this residue was considered as CWPS. Monosaccharide and uronic acid composition of CWPS and EPMS were hydrolyzed according to De Ruiter et al. [36] using methanolysis combined with trifluoroacetic acid (TFA) hydrolysis.

After sample preparation, carbohydrate analysis is a very complex field. Usually, after microalgal acid hydrolysis, the total carbohydrate content of the hydrolysate can be determined using colorimetric procedures like the phenol-sulfuric acid [37–39] or anthrone-based [40–42]. These methods are available, giving excellent and robust results with low effort in a very short time. Nevertheless, detailed information about the monosaccharide composition cannot be generated [29]. Qualitative investigations can be performed using TLC methods with silica-based separation materials making the separation of most monosaccharides possible. However, quantification with the TLC methods is not possible [29], and for quantification of monosaccharides, analytical methods such as high performance liquid chromatography (HPLC) are often used. The HPLC equipped with a refractive index detector (RID) [32] and HPLC combined with pulsed amperometric detection (HPAEC-PAD) [9, 28] and liquid chromatography mass spectrometer (LC–MS) [29] were related.

The trends for carbohydrate analysis are the exploration of methods that study both cell wall polysaccharides and extracellular polymeric substances. The developed methods always aim to use small amounts of sample, reagents, and shorter analysis. Liquid chromatography has been highlighted in the carbohydrate determination since it presented a good separation and quantification of these compounds. Although it is a more expensive analytical technique, it provides data on the composition of individual monomeric sugars that make it of interest for this type of analysis.

5. Applications

Microalgae have several types of polysaccharides in their composition, such as phycocolloids, agar, alginate, carrageenan, fucoidan, ulvana, and cellulose, among others. phycocolloids can be formed by different monomers such as glucose, galactose, mannuronic acid, guluronic acid, mannitol, and laminarin. These carbohydrates can be inserted into functional beverages and food products such as functional bread, ready to serve soups, functional snack foods and a variety of sauces, creams, bakery products, and additional food products [43, 44].

Due to the high carbohydrate content, poultry and aquaculture feed is one of the main study targets for the use of microalgae biomass. In 2007, around 30% of the world's current algae production was sold for animal feed application [8]. Microalgae are also a suitable alternative for growing fish, larvae, and zooplankton. *Chlorella* is one of the main examples of microalgae that can play a key role in food and feed due to the properties of its biomass, which can simultaneously provide high concentrations of carbohydrates, vitamins, and proteins [45]. Besides *Chlorella*, other species used in aquaculture can be highlighted: *Tetraselmis*, *Isochrysis*, *Pavlova*, *Phaeodactylum*, *Chaetoceros*, *Nannochloropsis*, *Skeletonema*, and *Thalassiosira*. *Spirulina* and *Chlorella* microalgae can be applied in the feeding of cats, dogs, aquarium fish, ornamental birds, horses, birds, cows, and breeding bulls. The most common genera of larval microalgae include *Chaetoceros*, *Thalassiosira*, *Tetraselmis*, *Isochrysis*, and *Nannochloropsis* [46].

1,3-β-glucan is an important carbohydrate present in microalgae composition due to its applications in the food industry as a thickener, and health applications, especially in the protection against infections and also to inhibit cancer cell growth in vivo [44, 47]. According to [48], the global β-glucan market was valued at USD 307.8 million in 2016, and it is predicted that in 2022, the global carbohydrate market could reach up to USD 476.5 million, which indicates the huge potential for development in many different types of applications.

According to Koller et al. [49], sulfated polysaccharides produced by microalgae can be applied in therapies against bacterial infections. Carrageenan polysaccharide, also known as food additive E407, can be used in pharmaceutical applications. Marine carbohydrates have been widely used in the cosmetics industries due to their chemical and physical properties. Brown algal fucoidans/alginates, green algal ulcers, and red algal carrageenans/agar are used as gelling, thickening, and stabilizing agents. In addition, marine carbohydrates have potential skin benefits, and biological activities are linked to their structure as determined by molecular weights or the presence of sulfate groups and other sugars [50].

Red algae, such as *Chondrus* sp., *Gigartina* sp. *Eucheuma* sp., *Hypnea* sp., and *Furcellaran* sp., are widely used for the production of carrageenan. This compound can be used in food and pharmaceutical industries for applications in fruit gel, fruit juices, sweets, and jellies, among others. Another carbohydrate group molecule is fucoidan, which is associated with brown algal cell wall components (*Phaeophyceae*). Among the bioactivities derived from this molecule, the anticoagulant, antitumor, antivirus, and antioxidant properties stand out, making it attractive for pharmaceutical applications [51].

Besides these applications, the remaining biomass of microalgae presents carbohydrate-rich molecules, which have been widely used in the production of bioplastics, agar, sugars, and other high-added value chemicals. However, despite being a growing area, the biorefinery stage must be studied in order to extend its applicability on an industrial scale [51]. According to Mihranyan [52], the rheological behavior of cellulose found in *Cladophora* algae is similar to micro fibrillated cellulose. Because this cellulose is very robust and not susceptible to chemical reactions, the properties of cellulose found in these algae provide excellent rheological properties making this material interesting in food, pharmaceuticals, paints, dressings, and biodegradable plastic applications.

The high carbohydrate content and low-ash values make microalgae more suitable for conversion to biofuels [43]. The production of bioethanol from microalgae gained importance due to their high biomass productivity, diversity, variable chemical composition, and high photosynthetic rates of these organisms [53]. Due to the large amount of carbohydrates/polysaccharides and cellulose walls, these microorganisms become favorable for the production of this biofuel [54, 55]. In many countries, ethanol is produced on a large scale from crops containing sugars and starches in its composition through fermentation. The biomass is ground, and the starch is converted into sugars by different methods. Polysaccharide starch is also accessible as a storage material for various algal species and can be anaerobically converted into bioethanol [49].

6. Conclusions

Microalgae biomass conversion technologies involve carbohydrates as the main source in the production of biofuels and other compounds of high commercial value. Changes in metabolic pathways aiming at increased carbohydrate production are seen as a potential for enhancing microalgae biotechnology. Extraction methods and trends in analytical methodologies focus on microalgae cell wall polysaccharides and the polymers excreted by these microorganisms. The high carbohydrate content makes microalgae excellent candidates for the production of numerous biocomposites, especially beta-glucan, which is on the international market, indicating its strong potential for its use in different biotechnological applications.

Author details

Maiara Priscilla de Souza[1], Andrea Sanchez-Barrios[1], Tiele Medianeira Rizzetti[1], Lisianne Brittes Benitez[1], Michele Hoeltz[1], Rosana de Cassia de Souza Schneider[1] and Fábio de Farias Neves[2]*

1 University of Santa Cruz do Sul, Santa Cruz do Sul, Brazil

2 Santa Catarina State University, Laguna, Brazil

*Address all correspondence to: fabio.neves@udesc.br

IntechOpen

© 2019 The Author(s). Licensee IntechOpen. This chapter is distributed under the terms of the Creative Commons Attribution License (http://creativecommons.org/licenses/by/3.0), which permits unrestricted use, distribution, and reproduction in any medium, provided the original work is properly cited. (cc) BY

References

[1] Das P, Aziz SS, Obbard JP. Two phase microalgae growth in the open system for enhanced lipid productivity. Renewable Energy. 2011;**36**(9):2524-2528

[2] Markou G, Nerantzis E. Microalgae for high-value compounds and biofuels production: A review with focus on cultivation under stress conditions. Biotechnology Advances. 2013;**31**(8):1532-1542

[3] Lavoie J-M, Marie-Rose S, Lynch D. Non-homogeneous residual feedstocks to biofuels and chemicals via the methanol route. Biomass Conversion and Biorefinery. 2013;**3**(1):39-44

[4] Huang G et al. Biodiesel production by microalgal biotechnology. Applied Energy. 2010;**87**(1):38-46

[5] Mayers JJ et al. Identifying a marine microalgae with high carbohydrate productivities under stress and potential for efficient flocculation. Algal Research. 2018;**31**:430-442

[6] Cunha C et al. Marine vs freshwater microalgae exopolymers as biosolutions to microplastics pollution. Environmental Pollution. 2019;**249**:372-380

[7] Xiao R, Zheng Y. Overview of microalgal extracellular polymeric substances (EPS) and their applications. Biotechnology Advances. 2016;**34**(7):1225-1244

[8] Becker EW. Micro-algae as a source of protein. Biotechnology Advances. 2007;**25**(2):207-210

[9] Bernaerts TMM et al. Comparison of microalgal biomasses as functional food ingredients: Focus on the composition of cell wall related polysaccharides. Algal Research. 2018;**32**:150-161

[10] Fasaei F et al. Techno-economic evaluation of microalgae harvesting and dewatering systems. Algal Research. 2018;**31**:347-362

[11] Wells ML et al. Algae as nutritional and functional food sources: Revisiting our understanding. Journal of Applied Phycology. 2017;**29**(2):949-982

[12] de Jesus Raposo MF, de Morais AMMB, de Morais RMSC. Emergent sources of prebiotics: Seaweeds and microalgae. Marine Drugs. 2016;**14**(2):27

[13] Larronde-Larretche M, Jin X. Microalgal biomass dewatering using forward osmosis membrane: Influence of microalgae species and carbohydrates composition. Algal Research. 2017;**23**:12-19

[14] Hernández-García A et al. Wastewater-leachate treatment by microalgae: Biomass, carbohydrate and lipid production. Ecotoxicology and Environmental Safety. 2019;**174**:435-444

[15] Uyaguari-Diaz MI et al. A comprehensive method for amplicon-based and metagenomic characterization of viruses, bacteria, and eukaryotes in freshwater samples. Microbiome. 2016;**4**(1):20

[16] Chen C-Y, Chang H-Y, Chang J-S. Producing carbohydrate-rich microalgal biomass grown under mixotrophic conditions as feedstock for biohydrogen production. International Journal of Hydrogen Energy. 2016;**41**(7):4413-4420

[17] Sanchez Rizza L et al. Bioprospecting for native microalgae as an alternative source of sugars for the production of bioethanol. Algal Research. 2017;**22**:140-147

[18] Depraetere O et al. Harvesting carbohydrate-rich *Arthrospira platensis*

by spontaneous settling. Bioresource Technology. 2015;**180**:16-21

[19] Gaignard C et al. New horizons in culture and valorization of red microalgae. Biotechnology Advances. 2019;**37**(1):193-222

[20] Sheath RG, Wehr JD. Introduction to the freshwater algae. In: Wehr JD, Sheath RG, Kociolek JP, editors. Freshwater Algae of North America. 2nd ed. Boston: Academic Press; 2015. pp. 1-11

[21] Mišurcová L et al. Amino acid composition of algal products and its contribution to RDI. Food Chemistry. 2014;**151**:120-125

[22] Jutur PP, Nesamma AA, Shaikh KM. Algae-derived marine oligosaccharides and their biological applications. Frontiers in Marine Science. 2016;**3**(83):1-5

[23] Jerez CG et al. Effect of nutrient starvation under high irradiance on lipid and starch accumulation in *Chlorella fusca* (Chlorophyta). Marine Biotechnology. 2016;**18**(1):24-36

[24] Sun X et al. Effect of nitrogen-starvation, light intensity and iron on triacylglyceride/carbohydrate production and fatty acid profile of Neochloris oleoabundans HK-129 by a two-stage process. Bioresource Technology. 2014;**155**:204-212

[25] Rockwell NC, Lagarias JC. A brief history of phytochromes. ChemPhysChem. 2010;**11**(6):1172-1180

[26] Rabinovitch-Deere CA et al. Synthetic biology and metabolic engineering approaches to produce biofuels. Chemical Reviews. 2013;**113**(7):4611-4632

[27] Castro-Bravo N et al. Interactions of surface exopolysaccharides from bifidobacterium and lactobacillus within the intestinal environment. Frontiers in Microbiology. 2018;**9**:1-15

[28] Souza MFD et al. Neutral sugars determination in chlorella: Use of a one-step dilute sulfuric acid hydrolysis with reduced sample size followed by HPAEC analysis. Algal Research. 2017;**24**:130-137

[29] Schulze C et al. Carbohydrates in microalgae: Comparative determination by TLC, LC-MS without derivatization, and the photometric thymol-sulfuric acid method. Algal Research. 2017;**25**:372-380

[30] Zhao G et al. Ultrasound assisted extraction of carbohydrates from microalgae as feedstock for yeast fermentation. Bioresource Technology. 2013;**128**:337-344

[31] Van Wychen S, Laurens LML. Determination of Total Carbohydrates in Algal Biomass: Laboratory Analytical Procedure (LAP). United States: NREL/TP-5100-60957, National Renewable Energy Laboratory (NREL); 2016

[32] Hanifzadeh M, Garcia EC, Viamajala S. Production of lipid and carbohydrate from microalgae without compromising biomass productivities: Role of Ca and Mg. Renewable Energy. 2018;**127**:989-997

[33] Northcote DH, Goulding KJ, Horne RW. The chemical composition and structure of the cell wall of *Chlorella pyrenoidosa*. The Biochemical Journal. 1958;**70**(3):391-397

[34] Correia PR, Beirão-da-Costa ML. Starch isolation from chestnut and acorn flours through alkaline and enzymatic methods. Food and Bioproducts Processing. 2012;**90**(2):309-316

[35] Velmurugan R, Muthukumar K. Ultrasound-assisted alkaline pretreatment of sugarcane bagasse for fermentable sugar production: Optimization through response surface

methodology. Bioresource Technology. 2012;**112**:293-299

[36] de Ruiter GA et al. Carbohydrate analysis of water-soluble uronic acid-containing polysaccharides with high-performance anion-exchange chromatography using methanolysis combined with TFA hydrolysis is superior to four other methods. Analytical Biochemistry. 1992;**207**(1):176-185

[37] DuBois M et al. Colorimetric method for determination of sugars and related substances. Analytical Chemistry. 1956;**28**(3):350-356

[38] Braga VDS et al. Cultivation strategy to stimulate high carbohydrate content in Spirulina biomass. Bioresource Technology. 2018;**269**:221-226

[39] Cea-Barcia G et al. A cost-effective strategy for the bio-prospecting of mixed microalgae with high carbohydrate content: Diversity fluctuations in different growth media. Bioresource Technology. 2014;**163**:370-373

[40] Lupatini AL et al. Protein and carbohydrate extraction from *S. platensis* biomass by ultrasound and mechanical agitation. Food Research International. 2017;**99**:1028-1035

[41] Batista AP et al. Comparison of microalgal biomass profiles as novel functional ingredient for food products. Algal Research. 2013;**2**(2):164-173

[42] Osborne DRVP. In: Acribia E, editor. Analisis de nutrientes de los alimentos. Zaragoza: Editorial Acribia; 1986

[43] Sudhakar MP et al. A review on bioenergy and bioactive compounds from microalgae and macroalgae-sustainable energy perspective. Journal of Cleaner Production. 2019;**228**:1320-1333

[44] Yaakob Z et al. An overview: Biomolecules from microalgae for animal feed and aquaculture. Journal of Biological Research (Thessaloniki). 2014;**21**(1):6

[45] Guccione A et al. Chlorella for protein and biofuels: From strain selection to outdoor cultivation in a Green Wall panel photobioreactor. Biotechnology for Biofuels. 2014;**7**(1):84

[46] Priyadarshani I, Rath B. Commercial and industrial applications of micro algae–A review. Journal of Algal Biomass Utilization. 2012;**3**(4):89-100

[47] Carballo C et al. Yeast β-glucans and microalgal extracts modulate the immune response and gut microbiome in Senegalese sole (*Solea senegalensis*). Fish and Shellfish Immunology. 2019;**92**:31-39

[48] Bozbulut R, Sanlier N. Promising effects of β-glucans on glyceamic control in diabetes. Trends in Food Science and Technology. 2019;**83**:159-166

[49] Koller M, Muhr A, Braunegg G. Microalgae as versatile cellular factories for valued products. Algal Research. 2014;**6**:52-63

[50] Kim J et al. Beneficial effects of marine algae-derived carbohydrates for skin health. Marine Drugs. 2018;**16**(11):459

[51] Khanra S et al. Downstream processing of microalgae for pigments, protein and carbohydrate in industrial application: A review. Food and Bioproducts Processing. 2018;**110**:60-84

[52] Mihranyan A. Cellulose from cladophorales green algae: From environmental problem to high-tech composite materials. Journal of Applied Polymer Science. 2011;**119**(4):2449-2460

[53] Sharma P, Sharma N. Industrial and biotechnological applications of algae: A review. Journal of Advances in Plant Biology. 2017;**1**(1):01

[54] Abdulla R et al. Microalgae chlorella as a sustainable feedstock for bioethanol production. In: Green Engineering for Campus Sustainability. Singapore: Springer; 2020. pp. 81-103

[55] Smachetti MES et al. Microalgal biomass as an alternative source of sugars for the production of bioethanol. In: Principles and Applications of Fermentation Technology. United States: Scrivener Publishing LLC; 2018. pp. 351-386

Chapter 10

Changes in Photochemical Efficiency and Differential Induction of Superoxide Dismutase in Response to Combined Stresses of Chilling Temperature and Relatively High Irradiation in Two *Chlorella* Strains

Lee-Feng Chien and Wen-Hao Lin

Abstract

The green algae *Chlorella* sp. DT (DT) and *Chlorella pyrenoidosa* 211-8b (8b) had similar cell growth rates and photochemical efficiency (F_v/F_m) when they were cultivated under a moderate irradiance of 120 µmol photons m^{-2} s^{-1} in combination with a series of temperatures that decreased from 32 to 7°C. Upon shifting the cultures to the relatively high irradiance of 240 µmol photons m^{-2} s^{-1}, DT exhibited higher cell growth rates than 8b under the chilling temperatures of 20°C and 15°C and differences in the F_v/F_m and Chl *a*/*b* ratios from 8b. In particular, DT possessed more new differentially induced SOD isoforms than 8b.

Keywords: *chlorella*, chilling temperature, relatively high irradiance, chlorophyll content, F_v/F_m, Chl *a*/*b*, superoxide dismutase

1. Introduction

The unicellular green alga *Chlorella* is a popular nutraceutical that is produced industrially in Taiwan. *Chlorella* requires a moderate climate, including ample sunshine and high temperatures of about 25–38°C for optimal large-scale outdoor growth. However, in winter, the temperature can range from 4 to 15°C, which is unsuitable for algal growth. In order to maintain productivity, it would be helpful to understand how green algae overcome chilling temperatures and a mimicking high irradiance resulted from chilling temperature [1–3].

Photosynthesis is the energy source for the growth and development of photosynthetic organisms. Photosynthetic efficiency is reliant on environmental conditions such as light and temperature. At low temperatures, algae experience reduced photosynthetic efficiency, whereas in high-light environments, they absorb more energy than they can consume in the photosynthetic processes [4]. The absorption of too much energy can lead to an increase in the production of reactive oxygen species (ROS), which can damage the photosynthetic apparatus and further decrease photosynthetic efficiency [5]. Therefore, in response to wide daily and seasonal fluctuations in temperature and light, algae must possess some protective and regulatory systems to avoid this "energy excess" [6–9].

Upon initial exposure to low temperature or high irradiation, excessive excitation pressure may be induced between the rate of energy absorbed via the photosynthetic antenna and energy utilization [4, 5, 10–12]. One of the protection mechanisms that algae and higher plants employ to avoid receiving too much light energy is to adjust their chlorophyll (Chl) a/b ratios and the structure of the photosystem I and II (PSI and PSII) antenna complexes in response to different combinations of light intensity and temperature [2, 13–15]. Light-harvesting complexes (LHCs) with modified Chl composition have the ability to absorb different levels of light energy depending on the environmental conditions [16–18]. Another protective mechanism of algae and plants after receiving too much light energy is to adjust the antioxidant response of the scavenging system such that any excess excitation pressure is transferred to the superoxide radical ($O_2 \cdot^-$) pathway and other derived reactive oxygen species (ROS) [19]. Superoxide dismutase (SOD, EC 1.15.1.1) is known as the first line of cellular defense against oxidative stress, and it catalyzes the dismutation of $O_2 \cdot^-$ to H_2O_2 and O_2. There are three distinct types of SOD classified on the basis of their metal cofactors: the copper/zinc (CuZnSOD), iron (FeSOD), and manganese (MnSOD) isoenzymes [20]. SOD activity increases in cells in response to diverse environmental stresses including high light and chilling temperatures [21–23].

The primary objective of this present work was to explore combinations of light and temperature in algal cultures that may inform optimization of production system in manufacturing [24, 25]. The two warm-climate green algae, *Chlorella* sp. DT (DT) and *Chlorella pyrenoidosa* 211-8b (8b), were compared in their photosynthetic activity and antioxidant enzymatic responses under relatively high irradiance and various chilling temperatures [26, 27]. To determine the capacity of these algae to absorb light, their Chl contents and Chl a/b ratios were measured. Photochemical efficiency and the extent of photodamage were assessed by quantifying the chlorophyll fluorescence emission of PSII [28, 29]. The responses of SOD antioxidant enzymes to chilling and high-light acclimation were also examined because they enabled correlation with cell growth and photosynthetic activity [30].

2. Materials and methods

2.1 Chlorella culture and growing conditions

Chlorella sp. DT (DT) was discovered on the surface of power transmission cables near a mountain in central Taiwan, and *Chlorella pyrenoidosa* 211-8b (8b) was acquired from the Algal Collection Center at the University of Gottingen, Germany [26, 27]. Stock cultures were maintained at an initial concentration of 4 μg Chl mL^{-1} in 200 mL *Chlorella* medium in a 6 × 50 cm column at 32 ± 1°C in a water bath with continuous irradiance of 120 μmol photons m^{-2} s^{-1} and bubbling of 4% CO_2. In this study, six different temperatures and two different irradiance levels were used, but the growth conditions were similar.

2.2 Chlorophyll (Chl) determination

The growth of algal cultures was monitored by measuring the total Chl (Chl *a* + Chl *b*) content according to the method of Hoffman and Werner [28].

An algal culture of 5 mL was centrifuged for 5 min at 2000 × g (Sigma, MK-201, Germany). After the supernatant was removed, the algal cell pellet was collected, 5 mL of 100% methanol was added, and the mixture was heated at 60°C for 3 min. After centrifugation at 2000 × g for 10 min to remove any cell debris, the Chl extract was obtained. To determine the total Chl and Chl *a*/*b* ratios, the concentrations of Chl *a* and Chl *b* were measured spectrophotometrically according to Hoffman and Werner's equations:

$$\text{Chl } a \left(\mu\text{g mL}^{-1}\right) = 16.5 \times A_{665\text{nm}} - 8.3 \times A_{650\text{nm}}; \tag{1}$$

$$\text{Chl } b \left(\mu\text{g mL}^{-1}\right) = 33.8 \times A_{650\text{nm}} - 12.5 \times A_{665\text{nm}}; \tag{2}$$

$$\text{Total Chl} \left(\mu\text{g mL}^{-1}\right) = 4 \times A_{650\text{nm}} + 25.5 \times A_{665\text{nm}}. \tag{3}$$

The cell specific growth rate (μ, day^{-1}) was derived from the difference in cellular Chl content over time as follows:

$$\mu = \frac{\ln[\text{Chl}]_{t_2} - \ln[\text{Chl}]_{t_1}}{t_2 - t_1} \tag{4}$$

where $[\text{Chl}]t_1$ and $[\text{Chl}]t_2$ are the initial and final concentrations, at time t_1 and t_2, respectively.

2.3 Measurement of chlorophyll fluorescence parameters

Chlorophyll fluorescence was measured using a modulated chlorophyll fluorometer (Hansatech Instruments Ltd., Norfolk, UK). Algal samples were collected at the indicated times, adjusted to a concentration of 4 μg Chl mL^{-1}, and dark adapted for a period of 10 min at room temperature before measurement. The fiber-optic probe of the fluorometer was then placed into the chamber with 2 mL of the algal samples. The minimum fluorescence (F_0) with all open PSII reaction centers was determined by a weak non-actinic modulated light (<0.1 μmol m^{-2} s^{-1}). The maximum fluorescence (F_m) with all closed PSII reaction centers was induced by a saturating pulse of white light (1 s, 13,000 μmol m^{-2} s^{-1}) and measured after applying the actinic light (300 μmol m^{-2} s^{-1}). The variable fluorescence (F_v) was calculated as F_m-F_o, and the ratio of the variable to maximum fluorescence (F_v/F_m) represented the maximal photochemical efficiency [29, 30].

2.4 Spectrophotometrical assay of SOD activity

Algal cells from the log phase were collected by centrifugation. The cell pellet was washed three times and resuspended with extraction buffer (100 mM K_2PO_4 and 5 mM EDTA, pH 7.0). Algal cells were then broken down by sonication (V500, SONIC, USA) and centrifuged at 10,000 × g (MK-201, Sigma, Germany) for 15 min. The supernatant obtained was the algal crude extract. Protein concentrations of algal crude extracts were determined by the method of Lowry et al. [31] using BSA as the standard. The algal crude extracts were then subjected to a SOD assay.

The activity of the SODs was determined by measuring the inhibited reduction of cytochrome c (cyt c) because the SOD competed with cyt c for superoxide radicals, thus inhibiting cyt c reduction [32]. The reduction of cyt c was measured by monitoring the change in absorbance at 549 nm (cyt c absorption) when xanthine oxidase was added to a reaction mixture of K_2PO_4 (pH 7.8), xanthine, and cyt c (oxidized) at room temperature. After a few seconds, the algal crude extract was added to the reaction mixture. SOD activity was then calculated as 50% of the inhibited reduction rate of cyt c. An extinction coefficient of ε_{549nm} = 21 mM^{-1} cm^{-1} for cyt c was used.

2.5 Native PAGE analysis of SOD

About 5–20 μg of protein from algal cell crude extracts suspended in sample buffer comprising 12.5 mM Tris–HCl (pH 6.8), 0.02% (w/v) bromophenol blue, and 4% (v/v) glycerol was loaded into each well. SODs in the algal extract were separated by native polyacrylamide gel electrophoresis (native PAGE) (10%). After electrophoresis the gels were washed with 100 mM K_2PO_4 buffer (pH 7.8) for 10 min and incubated in staining buffer composed of 20 mM K_2PO_4 buffer (pH 7.0), 0.05 mM riboflavin, 0.1 mM nitroblue tetrazolium (NBT), and 0.2% (w/v) TEMED in the dark at room temperature for 30 min [33]. Then gels were washed twice with 100 mM K_2PO_4 butter (pH 7.8) and exposed to light until the development of colorless bands. The colorless bands on the purple-stained gel indicated the existence of SOD because the free radicals produced by riboflavin are removed by the SOD, and as a consequence the colorless oxidized NBT in the SOD band is not converted into its purple reduced form. The reaction was then stopped by immersing the gels in deionized water. The SOD isoenzymes were identified on the basis of their sensitivity to KCN (5 mM) or H_2O_2 (10 mM), which were added into the staining buffer when required. MnSODs are resistant to both inhibitors, Cu/ZnSODs are sensitive to both inhibitors, and FeSODs are resistant to KCN but sensitive to H_2O_2.

3. Results

3.1 Enhanced inhibition of cell growth by a combined stress of doubled irradiance and chilling

The Chl content was monitored during acclimation because it represented not only the level of cell growth but also the capacity for light absorption [34]. Under a moderate irradiance of 120 μmol photons m^{-2} s^{-1} at 32°C (**Figure 1A**), the Chl content of DT algal culture increased with acclimation time. The DT culture exhibited a maximum specific growth rate of 2.2 μg Chl mL^{-1} day^{-1} and reached a stationary phase on Day 3 at a content of 92 μg Chl mL^{-1}. At 20°C, DT growth was inhibited on Day 1, but growth resumed at a slower rate than the control on Day 2. When the temperature was lowered to 15, 10, or 7°C, DT stopped growing and even showed negative growth rates; 15°C seemed to be a critical temperature at which no net growth was observed. Under a doubled irradiance of 240 μmol photons m^{-2} s^{-1} at 32°C (**Figure 1B**), DT cells exhibited a maximum specific growth rate of 2.7 μg Chl mL^{-1} day^{-1} and reached the stationary phase on Day 1 at a content of 70 μg Chl mL^{-1}. At 20°C, DT cells initially ceased growth under this doubled irradiance but resumed growth on Day 3. Transferring cultures to temperatures below 20°C promoted cell death, while the critical temperature for avoiding the negative growth rate now rose to 17°C, two degrees higher than for moderate irradiation.

A similar response was observed in 8b cells during acclimation at the various temperatures. The 8b exhibited maximum specific growth rates of 2.1 and 2.7 μg Chl mL^{-1} day^{-1} at 32°C on Day 1 under irradiance of 120 and 240 μmol photons m^{-2} s^{-1}, respectively, and reached the stationary phase at about 98 and 69 μg Chl mL^{-1} on Day 3 (**Figure 2A, B**). However, once moved to temperatures below 20°C

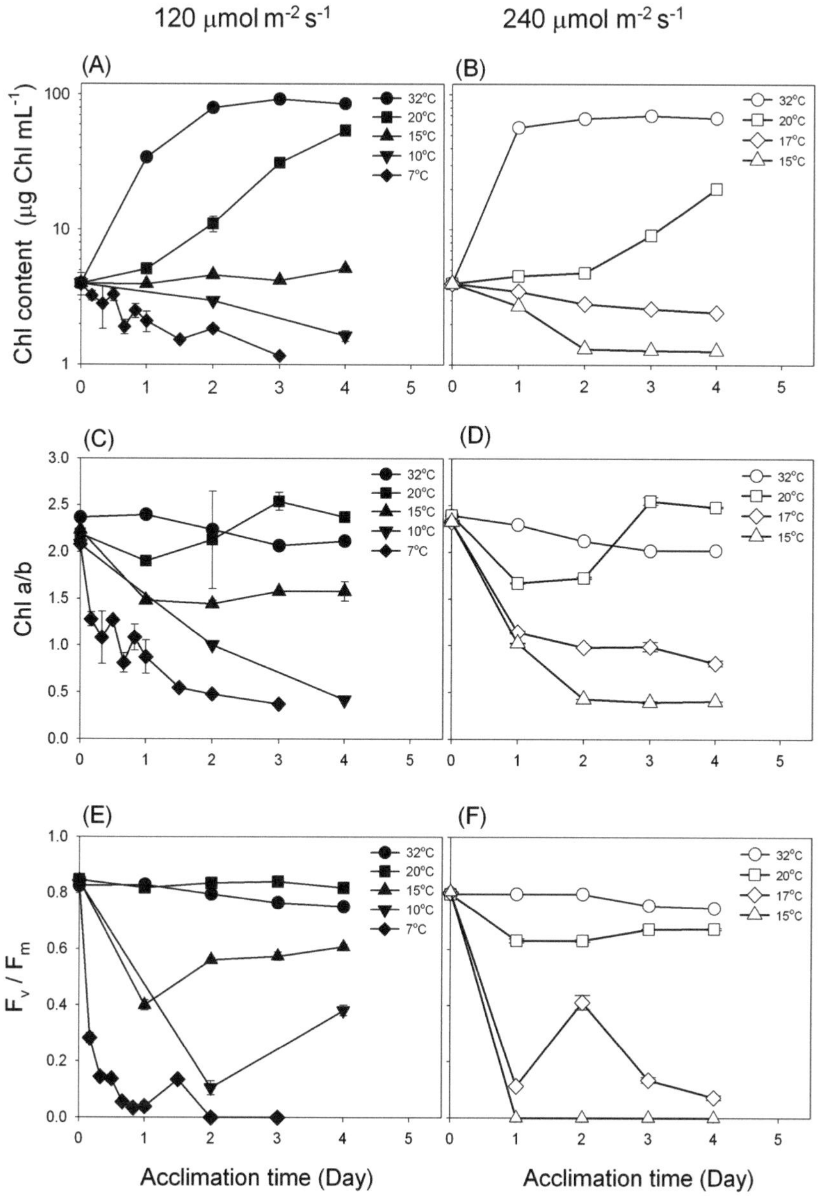

Figure 1.
Changes in the Chl content and photosynthetic activity of DT under irradiance of 120 or 240 μmol photons m^{-2} s^{-1} during cultivation between 32 and 7°C. the total Chl content (A, B), the Chl a/b ratio (C, D), and the F_v/F_m ratio (E, F) of DT were measured each day. The initial cultivation concentration was 4 μg Chl mL^{-1}. Each point represents the mean ± SD (n = 4) from duplicate cultures (where not visible, error bars are smaller than the symbol).

under 240 μmol photons m^{-2} s^{-1}, the 8b culture after Day 3 produced slightly lower Chl content than DT after Day 3. For 8b, the critical temperatures below which no net cell growth occurred and cell death was observed under both 120 and 240 μmol photon m^{-2} s^{-1} irradiance were 15°C and above 17°C, respectively.

Therefore, although a slightly higher maximum specific growth rate was observed initially at 32°C, lower temperatures induced enhanced inhibition of cell growth in both DT and 8b, and this was further inhibited under doubled irradiance.

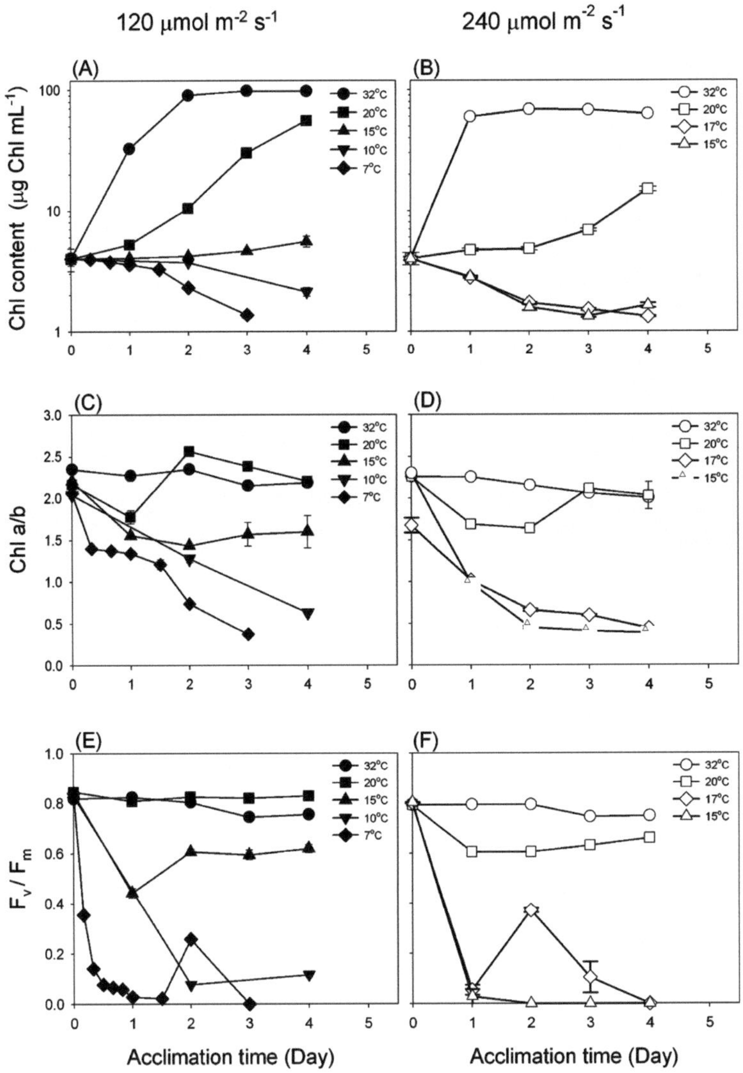

Figure 2.
Changes in the Chl content and photosynthetic activity of 8b under irradiance of 120 or 240 µmol photons m^{-2} s^{-1} during cultivation between 32 and 7°C. the total Chl content (A, B), the Chl a/b *ratio (C, D), and the* F_v/F_m *ratio (E, F) of 8b were measured each day. The initial cultivation concentration was 4 µg Chl* mL^{-1}*. Each point represents the mean ± SD (n = 4) from duplicate cultures (where not visible, error bars are smaller than the symbol).*

3.2 The readjustment of Chl a/b ratios

In order to understand the influence of Chl composition on excitation energy transfer, the Chl *a*/*b* ratio was analyzed. Under 120 µmol photon m^{-2} s^{-1} irradiance, the variation in the Chl *a*/*b* ratios of DT (**Figure 1C**) and 8b (**Figure 2C**) had similar trends at various temperatures. At 32°C, during acclimation, the Chl *a*/*b* ratios of both strains decreased slightly with time but remained between 2.4 and 2.1. When the cultures were moved to lower temperatures, the Chl *a*/*b* ratios changed dramatically. At 20°C, the Chl *a*/*b* ratios of both strains decreased to 1.7 by Day-1 but climbed back to about 2.1 by Day 2. At 15°C, the Chl *a*/*b* ratios of both strains were reduced to 1.0 on Day 1 and then remained at this value until the end of the experimental period. Under the lower temperatures of 10 and 7°C, the Chl *a*/*b* ratios of both strains rapidly declined to 0.4.

Under 240 μmol photon m^{-2} s^{-1} irradiance, the Chl *a*/*b* ratios of DT (**Figure 1D**) and 8b (**Figure 2D**) acclimated to 32°C were similar to the ratios observed under 120 μmol photons m^{-2} s^{-1}. At lower temperatures, the Chl *a*/*b* ratios decreased with time, while it returned to a value of 2.6 on Day-3 and was higher than the value of 2.4 recorded for 8b.

3.3 Change in PSII photochemical efficiency

To assess the photochemical efficiency of PSII, the ratio of the variable to maximum fluorescence (F_v/F_m) was measured [28, 29]. In both algal cultures, the F_v/F_m ratios of the DT and 8b controls stayed initially in the 0.83–0.85 range at 32°C with an irradiance of 120 μmol photons m^{-2} s^{-1}, but decreased slightly to 0.74–0.75 by the end of the acclimation period (**Figures 1E, 2E**). The F_v/F_m ratios of both *Chlorella* strains were higher than those of most green algae but close to those of healthy green leaves of higher plants [28, 35–37]. This may be due to the antenna sizes of *Chlorella* PSII being different from those of other algae but similar to higher plants because the measured Chl fluorescence is assumed to originate from PSII [29, 38]. Algal cells grown at 20°C exhibited almost constant F_v/F_m ratios, which were similar to those at 32°C, although the cell growth rates were slower than those at 32°C. Once the cultures were transferred to 15°C, a significant decrease in the F_v/F_m ratios was observed, first falling to 0.40 for DT and 0.42 for 8b on Day 1 but by Day 2 recovering to 0.57 and 0.60 and staying at this value throughout the rest of the cultivation period. When the cultures were transferred to lower temperatures, the F_v/F_m ratios of DT and 8b fell rapidly to 0.08 and 0.10 at 10°C and 0.04 and 0.03 at 7°C, respectively, on Day 1 and continued to decrease to nearly zero by the end of the acclimation period.

Under 240 μmol photon m^{-2} s^{-1} irradiance, at 32°C the F_v/F_m ratios of the DT and 8b strains also remained in 0.79–0.80 range (**Figures 1F, 2F**). However, the F_v/F_m ratios changed dramatically with lower temperatures. At 20°C, the F_v/F_m ratio of DT decreased to 0.20 on Day 2 but returned to 0.70 on Day 3, while in 8b it decreased to 0.40 but returned to 0.65 on Day 3. At 10 or 7°C, the F_v/F_m ratios of both strains declined to zero on Day 1, indicating that photosynthetic activity was immediately and completely inhibited. The Day 2 F_v/F_m ratios of both 8b and DT at 17°C and 7°C (**Figures 1E, F, 2E, F**) showed peaks that were probably due to experimental variations.

3.4 Shielding effects of high algal cell concentrations on light absorption

In order to understand whether the initial concentration of algal cells affected light absorption during chilling acclimation, cell growth was measured at different initial concentrations of 2, 4, and 6 μg Chl mL^{-1} under the doubled irradiance of 240 μmol photons m^{-2} s^{-1}.

By Day 3 following the initial cessation of growth at 20°C (**Figure 3A, B**), the DT and 8b cultures that started at concentrations of 4 and 6 μg Chl mL^{-1} were quicker to resume growth than those at 2 μg Chl mL^{-1} (**Figure 3A, B**). The Chl *a*/*b* ratios of DT decreased to 1.2, 1.7, and 2.0 with respect to initial concentrations of 2, 4, and 6 μg Chl mL^{-1} by Day 1, but then they increased close to control values by Day 3 (**Figure 3C**). The Chl *a*/*b* ratios of 8b showed similar variations with concentration to DT, with the exception of 2 μg Chl mL^{-1} (**Figure 3D**). The F_v/F_m ratios of DT and 8b initially decreased to 0.58 and 0.60 on Day-1; however, the ratios soon recovered and by Day-3 were 0.67 for DT and 0.63 for 8b (**Figure 3E, F**).

At 15°C, the cell growth of DT and 8b gradually declined with time regardless of the initial Chl concentrations. Nevertheless, at initial concentrations of 4 and 6 μg

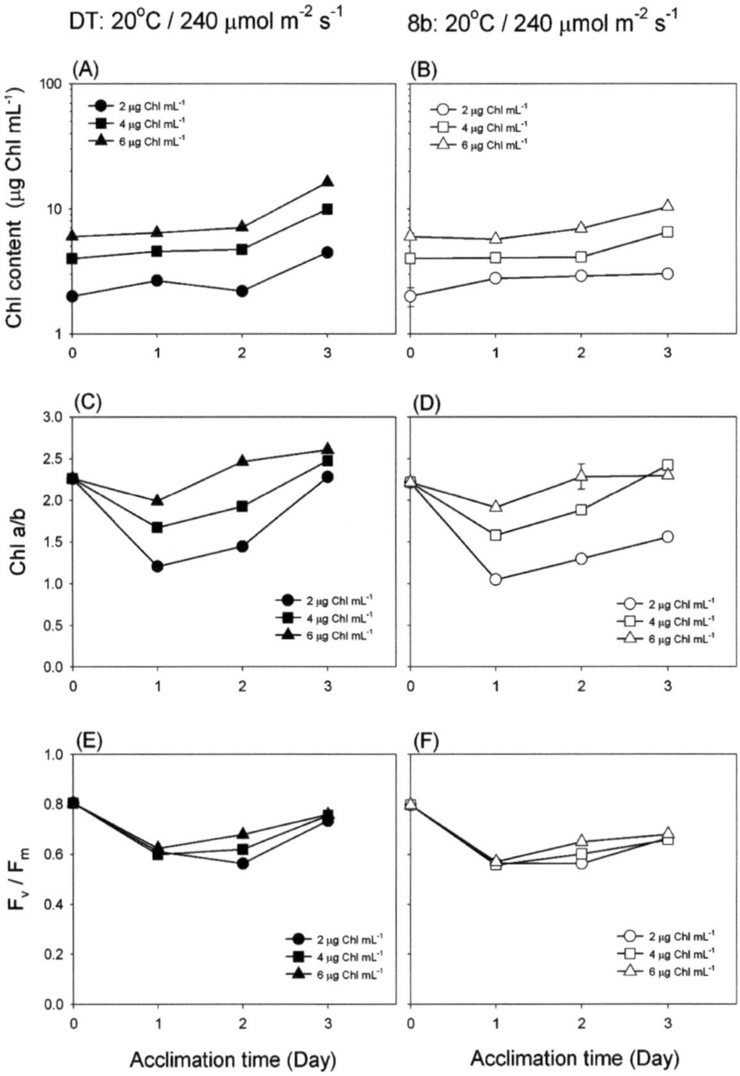

Figure 3.
Effect of initial cultivation concentration on photosynthesis under 20°C and 240 μmol photon m^{-2} s^{-1} irradiance. The total Chl content (A, B), the Chl a/b ratio (C, D), and the F_v/F_m ratio (E, F) of Chlorella *DT and 8b were measured in cultures with initial concentrations of 2, 4, and 6 μg Chl mL^{-1} at 20°C with irradiation of 240 μmol photons m^{-2} s^{-1}. Each point represents the average of two measurements from duplicate cultures (where not visible, error bars are smaller than the symbol).*

Chl mL^{-1}, both strains were slower to die than cultures starting out at 2 μg Chl mL^{-1} (**Figure 4A, B**). Neither DT nor 8b at 2 μg Chl mL^{-1} resumed growth at 15°C, and no significant difference was recorded between the two strains. The Chl *a*/*b* ratios of DT and 8b (**Figure 4C, D**) rapidly decreased on Day 1 from 2.32 to 0.50, 1.03, and 1.65 with respect to the initial concentrations of 2, 4, and 6 μg Chl mL^{-1}, and no increases were observed for the duration of the acclimation period. The F_v/F_m ratios of DT and 8b fell dramatically to near zero on Day 1 regardless of the initial concentrations, and no significant recovery was seen (**Figure 4E, F**).

The results suggested that the initial concentration (2, 4, and 6 μg Chl mL^{-1}) of algal cells did affect light absorption, but temperature was the major factor determining cell growth (**Figures 3, 4**). DT had a slightly greater tolerance at 20°C than 8b because its Chl a/b ratios attained levels higher than the control (Day 0), even though the Chl a/b ratios of 8b also returned to slightly above the control level.

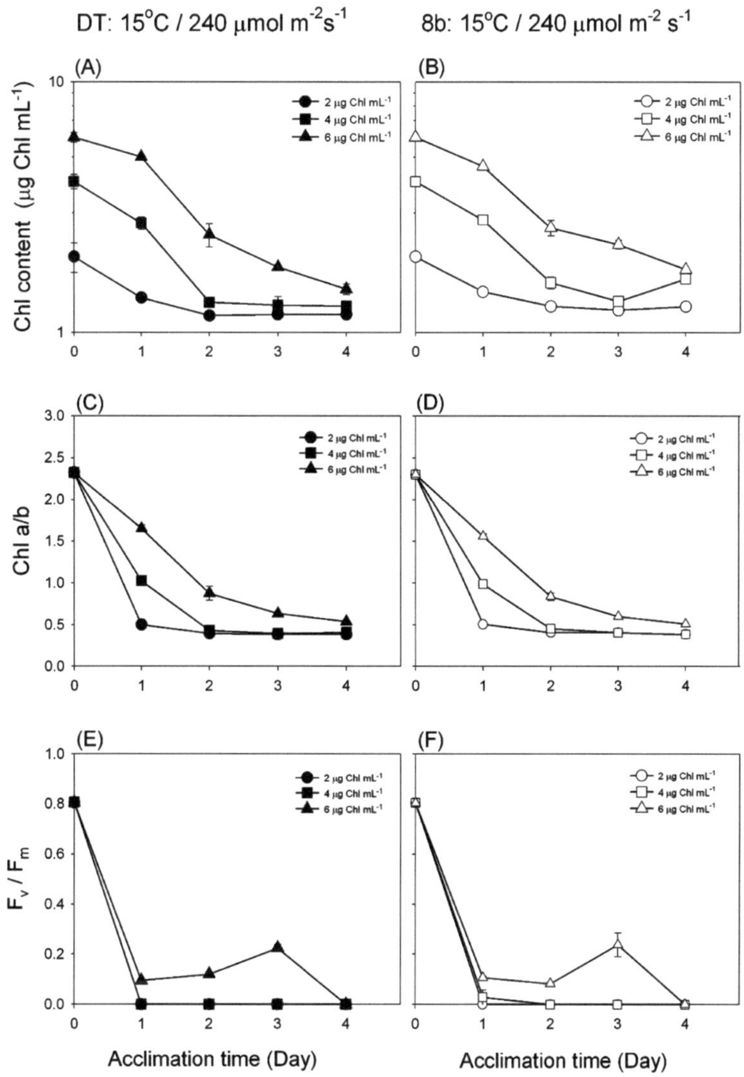

Figure 4.
Effect of initial cultivation concentration on photosynthesis under 15°C and 240 μmol photon m^{-2} s^{-1} irradiance. The total Chl content (A, B), the Chl a/b ratio (C, D), and the F_v/F_m ratio (E, F) of Chlorella *DT and 8b were measured in the cultures with initial concentrations of 2, 4, and 6 μg Chl mL^{-1}. Each point represents the average of two measurements from duplicate cultures (where not visible, error bars are smaller than the symbol).*

However, neither DT nor 8b could overcome the stress of low temperatures of 15°C and below combined with the doubled irradiance of 240 μmol photons m^{-2} s^{-1}.

3.5 Differential induction of multiple SOD isoforms

For the duration of the 15°C acclimation with 120 μmol photon m^{-2} s^{-1} irradiation, the specific growth rate of algal cells remained zero, implying that the energy input and output seemed to reach a critical point. To understand the contribution of antioxidants in scavenging ROS produced during chilling acclimation, the SOD activities were assayed with a spectrophotometrical method. It was found that DT had an approximately twofold higher rate of SOD activity (0.46 μmol mg^{-1} protein sec^{-1}) than 8b (0.21 μmol mg^{-1} protein sec^{-1}). Moreover, when the expression of SOD isoforms was examined after activity staining on native PAGE, three distinct colorless bands were observed in DT (**Figure 5**) while only two bands were observed in 8b (**Figure 6**). The SOD activities of both strains were generally amplified with time and decreasing temperature. At the same time, some new SODs were induced, and some were diminished.

As shown in **Figure 5**, the DT control contained two DTMnSODs and three DTFeSODs, which were verified with inhibitors of H_2O_2 and KCN. Once the culture was moved to 15°C, the SOD activities of the DT increased greatly on Day-1 and reached a maximum on Day 2. By Day 4, at least 10 SOD isoforms were observed in DT including two new DTMnSODs and three new DTFeSODs. However, by Day 8, SOD activities declined, and some isoforms disappeared, leaving only three DTFeSODs and two DTMnSODs present. Similarly, as shown in **Figure 6**, the 8b control contained two 8bMnSODs and two 8bFeSODs. The SOD activities of 8b were amplified on Day 2 and reached a maximum on Day 4, while one new 8bMnSOD and two new 8bFeSODs were induced. At the end of acclimation, the SOD activity declined, and only two 8bFeSODs and two 8bMnSODs were present.

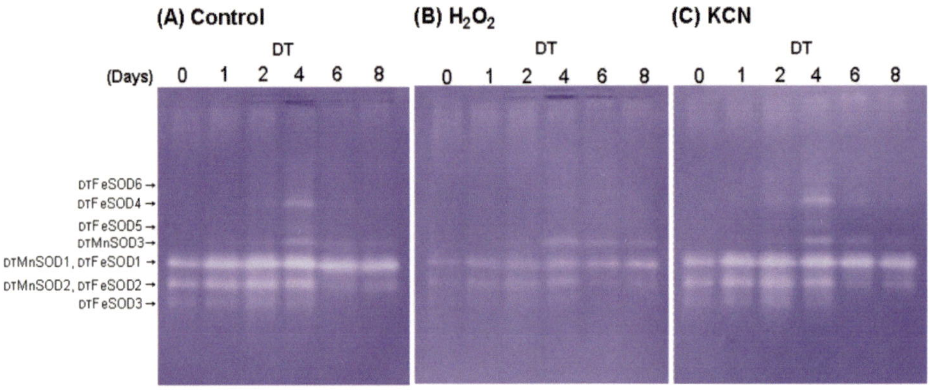

Figure 5.
Native PAGE analysis of SOD from a crude extract of DT grown at 15°C under 120 μmol photon m^{-2} s^{-1} irradiance. In each well 5 μg of crude extract proteins was loaded. In comparison to the control (A), SOD isoforms were recognized by adding the inhibitors H_2O_2 (5 mM) (B) and 2 mM KCN (2 mM) (C). In total, nine SODs were induced differentially in DT with regard to six FeSODs and three MnSODs. The numbers represent the order of discovery.

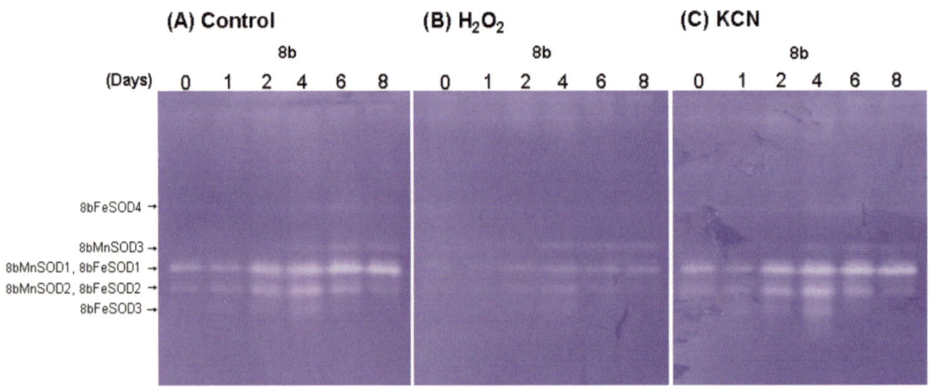

Figure 6.
Native PAGE analysis of SODs from crude extracts of 8b grown at 15°C under 120 μmol photon m^{-2} s^{-1} irradiance. In each well 5 μg of crude extract proteins was loaded. In comparison to the control (A), SOD isoforms were recognized by adding the inhibitors H_2O_2 (5 mM) (B) and 2 mM KCN (2 mM) (C). Seven SODs were induced differentially in DT with regard to original four FeSODs and three MnSODs in 8b. The numbers represent the order of discovery.

These results suggest that DT and 8b utilize different strategies for scavenging $O_2{\cdot}^-$. We found that MnSOD1 and FeSOD1 were the most abundant isoforms in both *Chlorella* strains, accounting for about 60–70% of the estimated total SOD activity. The other 30% is made up of other isoforms. The main FeSOD in both strains was particularly responsive to temperature [39]. Although there are three distinct types of SOD isoenzymes, only FeSOD and MnSOD were found in both *Chlorella* stains. Our observation of no CuZnSOD in either strain agrees with Asada et al. [40].

For further identification of which SOD isoforms responded to light stress and to temperature stress, the SODs were analyzed under lower temperature or doubled irradiance. In the DT culture, the original SOD isoforms of DTFeSOD1, DTFeSOD2, DTMnSOD1, and DTMnSOD2 were amplified in response to both the lower temperature of 10°C (**Figure 7A**) and to a doubled irradiance of 240 μmol photons m^{-2} s^{-1} (**Figure 7B**). DTFeSOD3 disappeared on Day 1, probably because it was sensitive to both higher light and lower temperature. A newly induced DTFeSOD4 appeared on Day 1 in response to doubled irradiance, but it was not detected until

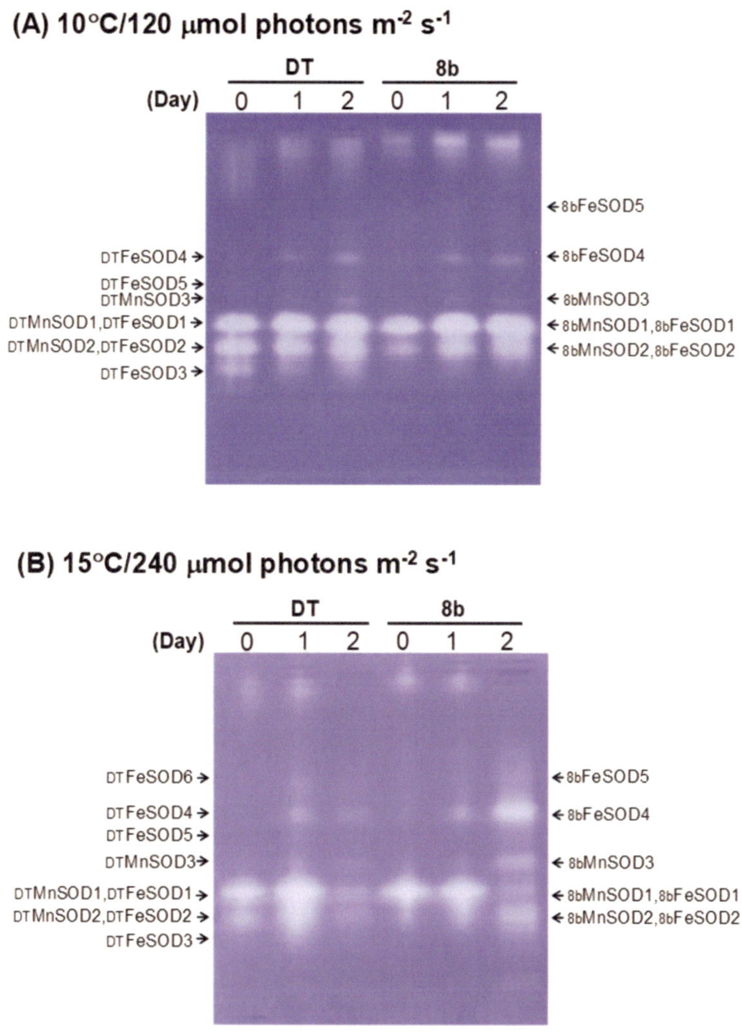

Figure 7.
Native PAGE analysis of SOD from crude extract of DT and 8b grown at 10°C under 120 μmol photon m^{-2} s^{-1} irradiance (A) or 15°C under 240 μmol photon m^{-2} s^{-1} irradiance (B). In (A), 15 μg of crude extract proteins was loaded in each well; in (B), 10 μg of crude extract proteins (except 1 μg proteins of DT on Day-2) was loaded. The numbers represent the order of discovery.

Day 1 under moderate irradiance, implying that DTFeSOD4 was probably more sensitive to light than to low temperatures. In 8b culture, in addition to the original SOD isoforms of 8bFeSOD1, 8bFeSOD2, 8bMnSOD1, and 8bMnSOD2, some new isoforms were induced. They were amplified in response to the lower temperature of 10°C (**Figure 7A**) and the doubled irradiance of 240 μmol photons $m^{-2} s^{-1}$ (**Figure 7B**) on Day 1. However, 8bMnSOD1 declined on Day 2. In spite of new SOD isoforms being amplified and in spite of the expectation that the SODs would prevent cell death, under the two combined stresses, the algal cells were still dying.

4. Discussion

4.1 Imbalance in excitation pressure

The specific growth rates on Day 1 from DT and 8b were plotted as a function of the cultivation temperatures (**Figure 8**). This showed that the specific growth rates decreased exponentially with decreasing temperatures from 32 to 10°C. Our results did not follow the previous observation of Sandnes et al. [41] where the specific growth rate of the green alga *Nannochloropsis oceanica* increased linearly with increasing low irradiance in the 17–26°C range. The curves fitted for the 120 μmol photon $m^{-2} s^{-1}$ irradiance data are dispersed from the 240 μmol photon $m^{-2} s^{-1}$ doubled irradiance data. Obviously, doubling the irradiance did not simply double the effect of the temperature reduction on the specific growth rate.

Furthermore, the relationship of specific growth rates versus cultivation temperatures was theoretically simulated in accordance with the excessive excitation pressure. The temperature coefficient (Q_{10}) represents the factor by which the speed of a biochemical reaction approximately doubles for every 10°C rise. Although some evidence indicated that Q_{10} in plants is temperature dependent [42], a Q_{10} of 2 was used here to theoretically estimate excessive excitation pressure. Therefore, the excessive excitation pressure due to the reduction in biochemical processes was calculated as $2^{(32°C-T_i)/10}$ (T, temperatures below 32°C) so that the theoretical excessive excitation pressure of

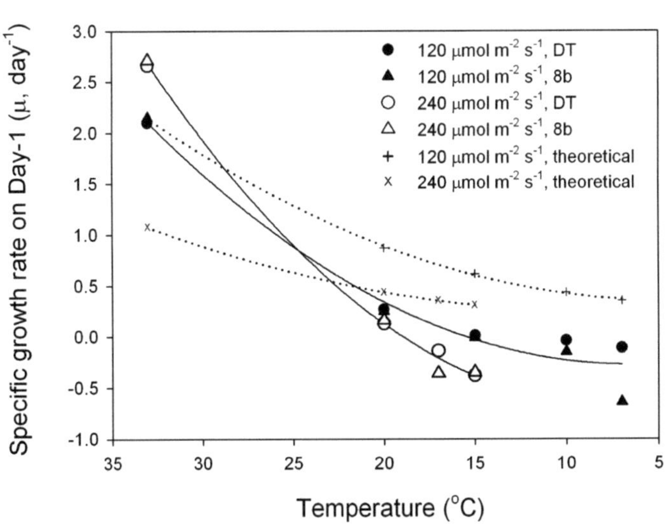

Figure 8.
Plots of measured and theoretical specific growth rates versus temperatures in DT and 8b. The solid line curves represent the measured specific growth rates at 120 (●, DT; ▲, 8b) and 240 (○, DT; Δ, 8b) μmol photon $m^{-2} s^{-1}$ irradiance. The dotted line curves represent theoretical specific growth rates at 120 (+) and 240 (×) μmol photon $m^{-2} s^{-1}$ irradiance.

Irradiation at 120 μmol photons $m^{-2} s^{-1}$				Irradiation at 240 μmol photons $m^{-2} s^{-1}$			
Temperature	Theoretical excessive excitation pressure (fold)	DT cell specific growth rate (μ) on Day-1 (μg Chl day^{-1})	8b cell specific growth rate (μ) on Day-1 (μg Chl day^{-1})	Temperature	Theoretical excessive excitation pressure (fold)	DT cell specific growth rate (μ) on Day-1 (μg Chl day^{-1})	8b cell specific growth rate (μ) on Day-1 (μg Chl day^{-1})
33°C	1-fold	2.15	2.07	33°C	2-fold	2.66	2.71
20°C	2.5-fold	0.25	0.27	20°C	4.9-fold	0.13	0.16
15°C	3.9-fold	−0.01	0.01	17°C	6.1-fold	−0.14	−0.36
10°C	4.9-fold	−0.13	−0.04	15°C	7.0-fold	−0.36	−0.35
7°C	6.1-fold	−0.69	−0.11				

Excessive excitation pressure was calculated upon the assumptions of temperature factor Q_{10} equaling to 2 for biochemical processes and light pressure factor equaling to 2 for double irradiance.

Table 1.
Theoretical excessive excitation pressure.

2.3-fold at 20°C, of 3.3-fold at 15°C, and so on were calculated relative to the control (onefold at 32°C) (**Table 1**). The diminished activities caused by theoretical excessive excitation pressure were plotted as a function of acclimation temperature (**Figure 8**). Subsequently, another plot was obtained for the doubled irradiance of 240 μmol photons m^{-2} s^{-1}, assuming that the excessive excitation pressure was twice the value under 120 μmol photon m^{-2} s^{-1} irradiance. However, the experimental curves did not follow the theoretical ones, implying that regulation of the response to the combined light and temperature stresses was more complicated than expected.

In our experiments, under a moderate irradiance of 120 μmol photons m^{-2} s^{-1}, DT and 8b showed no significant differences in growth rates and photochemical efficiency when subjected to various low temperatures. However, under a doubled irradiance of 240 μmol photons m^{-2} s^{-1}, DT had a slightly higher growth rate than 8b at temperatures below 20°C. This suggests that DT might possess a more efficient energy dissipation system against the combined stress of low temperatures and high irradiation than 8b. These results are in agreement with reports that the impact from photoinhibition due to low temperature and high light varies greatly across different green algal species [41, 43–45]. Although a greater specific growth rate was obtained under 240 μmol photon m^{-2} s^{-1} irradiance compared to 120 μmol photon m^{-2} s^{-1} irradiance, neither DT nor 8b favored high irradiance because a smaller Chl content was found during the stationary phase, that is, less biomass was generated.

In order to control light energy absorption and transfer, the LHC must modify the pigment composition of the Chl *a*/*b* ratio, and this is related to alterations in the photosynthetic apparatus under various conditions [16–18]. In the present study, decreases in both the Chl content and the Chl *a*/*b* ratio under low temperatures and high lights occurred simultaneously, suggesting a degradation of Chl molecules or the rearrangement of the LHCII complex [12]. A Chl *a*/*b* ratio of about 2.5 was obtained in both DT and 8b, which was similar to the green alga *Dunaliella salina* (2.3) [16], smaller than in *Chlorella vulgaris* (7.2) [2], and larger than in *Bryopsis maxima* (1.5) [38]. The lowering of Chl *a*/*b* ratios in DT and 8b is likely a mechanism to avoid absorbing too much light during acclimation [17]. The restoration of the Chl *a*/*b* ratio to 2.6 during 20°C acclimation might derive from the bleaching of Chl *b*, which is expected to absorb higher light excitation energy.

Despite the apparent decrease in the F_v/F_m ratios in our 10 and 7°C acclimation experiments, an initial increase and then a quenching of F_o was observed (data not shown). This phenomenon has been found in *C. vulgaris* and is suggested as being due to a rise in the xanthophyll cycle for dissipating excessive energy [43]. The reduction in both F_m and F_o implied changes in antenna size, thereby minimizing the absorbance of incident light [43]. Because F_o originates from the Chl *a* of the PSII-associated antenna, an increase in F_o is indicative of decreased energy transfer from LHCII to PSII. A large reduction in F_o has generally been regarded as a symptom of serious damage to the photosynthetic apparatus.

4.2 Differential SOD response

Since SOD is the first line of cellular defense against oxidative stress to remove $O_2\cdot^-$, monitoring how SOD responds to photoinhibition during acclimation may provide more information about photoprotection [20]. It is known that SOD activity increases in cells in response to diverse environmental stresses including high light intensities and low temperatures and that SOD isoforms are expressed differently to protect against a subset of oxidative stresses under various environmental conditions [46, 47]. In particular, each of the SOD isoforms is independently regulated according to the degree of oxidative stress experienced in the respective subcellular compartments [48].

At 15°C acclimation and 120 μmol photon m^{-2} s^{-1} irradiation, which was the point where the specific growth rate of the algal cells was zero, DT possessed higher SOD activities and more isoforms than 8b. To clarify further which SOD isoform responded to light or temperature, SOD activities were measured under the lower temperature conditions of 10°C and 120 μmol photon m^{-2} s^{-1} irradiance (**Figure 7A**) and at 15°C under the doubled irradiance of 240 μmol photon m^{-2} s^{-1} (**Figure 7B**). The results showed that the original SOD isoforms, which are likely sensitive to low temperature, were amplified by at 10°C and the newly induced SOD isoforms, which are likely sensitive to light, appeared under the doubled irradiance treatment.

Our data also suggested that the regulation of the antioxidant response to chilling was different from the response to irradiation. This raises the interesting question of why the regulation of antioxidant defenses is so highly complex and varied under a range of oxidative stresses even though they are targeting the same $O_2\cdot^-$ substrate [20–23].

5. Conclusion

The green algae *Chlorella* species DT (DT) and *Chlorella pyrenoidosa* 211-8b (8b) were very alike in their cell growth rate (total Chl), light energy absorption regulation (Chl *a*/*b* ratio), and photochemical efficiency (F_v/F_m) under optimal conditions of 120 μmol photons m^{-2} s^{-1} and as temperatures decreased from 32 to 7°C. Upon exposure of the cultures to a doubled irradiance of 240 μmol photons m^{-2} s^{-1}, DT exhibited higher cell growth rates than 8b at chilling temperatures of 20°C and 15°C. It was also found that under the combined stresses of chilling temperature and relatively high irradiance, DT possessed higher SOD activity and more new SOD isoforms for removing free radicals than 8b.

Acknowledgements

The authors acknowledge that this work was partly supported by grants to Lee-Feng Chien from the National Science Council (now Ministry of Science and Technology) of Taiwan (NSC89-2312-B-005-007 and NSC93-2311-B-005-017). This article is dedicated in memory of Professor Pao-Chung Chen.

Abbreviations

PS	photosystem
Chl	chlorophyll
F_m	maximum fluorescence
F_v	variable fluorescence
LHC	light-harvesting complex
SOD	superoxide dismutase
ROS	reactive oxygen species

Author details

Lee-Feng Chien* and Wen-Hao Lin
Department of Life Sciences, National Chung Hsing University, Taichung, Taiwan

*Address all correspondence to: lfchien@dragon.nchu.edu.tw

IntechOpen

© 2019 The Author(s). Licensee IntechOpen. This chapter is distributed under the terms of the Creative Commons Attribution License (http://creativecommons.org/licenses/by/3.0), which permits unrestricted use, distribution, and reproduction in any medium, provided the original work is properly cited.

References

[1] Allen DJ, Ort DR. Impact of chilling temperatures on photosynthesis in warm-climate plants. Trends in Plant Science. 2002;**6**:36-42

[2] Maxwell DP, Falk S, Trick CG, Hüner NP. Growth at low temperature mimics high-light acclimation in *Chlorella vulgaris*. Plant Physiology. 1994;**105**:535-543

[3] Singh SP, Singh P. Effect of temperature and light on the growth of algae species: A review. Renewable and Sustainable Energy Reviews. 2015;**50**:431-444

[4] Hüner NPA, Öquist G, Sarhan F. Energy balance and acclimation to light and cold. Trends in Plant Science. 1998;**3**:224-230

[5] Ort DR. When there is too much light. Plant Physiology. 2001;**125**:29-32

[6] Niyogi KK. Photoprotection revisited: Genetic and molecular approaches. Annual Review of Plant Physiology and Plant Molecular Biology. 1999;**50**:333-359

[7] Thomashow MF. Plant cold acclimation: Freezing tolerance genes and regulatory mechanisms. Annual Review of Plant Physiology and Plant Molecular Biology. 1999;**50**:571-599

[8] Mittler R. Oxidative stress, antioxidants and stress tolerance. Trends in Plant Science. 2002;**7**:405-410

[9] Suzuki N, Mittler R. Reactive oxygen species and temperature stresses: A delicate balance between signaling and destruction. Physiologia Plantarum. 2006;**126**:45-51

[10] Wilson KE, Ivanov AG, Öquist F, Grodzinski B, Sarhan F, Hüner NPA. Energy balance, organellar redox status, and acclimation to environmental stress. Canadian Journal of Biotany. 2006;**84**:1355-1370

[11] Guy C, Kaplan F, Kopka J, Selbig J, Hincha DH. Metabolomics of temperature stress. Physiologia Plantarum. 2008;**132**:220-235

[12] Takahashi S, Murata N. How do environmental stresses accelerate photoinhibition? Trends in Plant Science. 2008;**4**:178-182

[13] Wilson KE, Hüner NP. The role of growth rate, redox-state of the plastoquinone pool and the trans-thylakoid delta pH in photoacclimation of *Chlorella vulgaris* to growth irradiance and temperature. Planta. 2000;**212**:93-102

[14] Finazzi G, Minagawa J. High light acclimation in green microalgae. In: Demmig-Adams B, Garab G, Adams W III, Govindjee, editors. Non-Photochemical Quenching and Energy Dissipation in Plants, Algae and Cyanobacteria. Advances in Photosynthesis and Respiration (Including Bioenergy and Related Processes), Vol. 40. Dordrecht, Netherland: Springer; 2014. pp. 445-469

[15] Karsten U, Herburger K, Holzinger A. Living in biological soil crust communities of African deserts-physiological traits of green algal *Klebsormidium* species (Streptophyta) to cope with desiccation, light and temperature gradients. Journal of Plant Physiology. 2016;**194**:2-12

[16] Nishigaki A, Ohshima S, Nakayama K. Characterization of three forms of light-harvesting chlorophyll a/b-protein complexes of photosystem II isolated from the green alga, *Dunaliella salina*. Plant Cell Physiology. 2000;**41**:591-599

[17] Goss R, Wilhelm C, Garab G. Organization of the pigment molecules in the chlorophyll a/b/c containing alga *Mantoniella squamata* (Prasinophyceae)

studied by means of absorption, circular and linear dichroism spectroscopy. Biochimica et Biophysica Acta. 2000;**1457**:190-199

[18] Andreeva A, Stoitchkova K, Busheva M, Apostolova E. Changes in the energy distribution between chlorophyll-protein complexes of thylakoid membranes from pea mutants with modified pigment content: I. changes due to the modified pigment content. Journal of Photochemistry and Photobiology, B: Biology. 2003;**70**:153-162

[19] Foyer CH, Noctor G. Oxygen processing in photosynthesis: Regulation and signaling. New Phytology. 2000;**146**:359-388

[20] Pospíšil P. Production of reactive oxygen species by photosystem II as a response to light and temperature stress. Frontier in Plant Sciences. 2016;7:1950

[21] Bowler C, van Montagu M, Inzé D. Superoxide dismutase and stress tolerance. Annual Review of Plant Physiology and Plant Molecular Biology. 1992;**43**:83-116

[22] Clare DA, Rabinowitch HD, Fridovich I. Superoxide dismutase and chilling injury in *Chlorella ellipsoidea*. Archives of Biochemistry and Biophysics. 1984;**231**:158-163

[23] Wolfe-Simon F, Grzebyk D, Schofield O, Falkowski PG. The role and evolution of superoxide dismutases in algae. Journal of Phycology. 2005;**41**:453-465

[24] Ota M, Takenaka M, Sato Y, Lee Smith R, Inomata H. Effects of light intensity and temperature on photoautotrophic growth of a green microalga, *Chlorococcum littorale*. Biotechnology Reports. 2015;7:24-29

[25] Terada R, Matsumoto K, Borlongan IA, Watanabe Y, Nishihara GN, Endo H, et al. The combined effects of PAR and temperature including the chilling-light stress on the photosynthesis of a temperate brown alga, *Sargassum patens* (Fucales), based on field and laboratory measurements. Journal of Applied Phycology. 2018;**30**:1893-1904

[26] Lai CL, Chen TE, Chen JC, Chen PC. Physiological adaptation to dehydration of green alga chlorella sp. strain DT. In: Mathis P, editor. Photosynthesis: From Light to Biosphere. Vol. 4. Dordrecht, Netherland: Kluwer Academic Publishers; 1995. pp. 605-608

[27] Chen PC, Lorenzen H. Changes in productivity and nuclear divisions in synchronous *Chlorella* and circadian rhythm. Plant Cell Physiology. 1987;**27**:1423-1427

[28] Hoffman P, Werner D. Spectrophotometric chlorophyll determination having special regard to various types of equipment. Jena Review. 1966;**7**:300-303

[29] Maxwell K, Johnson GN. Chlorophyll fluorescence — A practical guide. Journal of Experimental Botany. 2000;**51**:659-668

[30] Baker NR. Chlorophyll fluorescence: A probe of photosynthesis in vivo. Annual Review of Plant Biology. 2008;**59**:89-113

[31] Lowry OH, Rosebrough NJ, Farr AL, Randall RJ. Protein measurement with the Folin phenol reagent. Journal of Biological Chemistry. 1951;**193**:265-275

[32] McCord JM, Fridovich I. The utility of superoxide dismutase in studying free radical reactions. I. Radicals generated by the interaction of sulfite, dimethyl sulfoxide, and oxygen. Journal of Biological Chemistry. 1969;**244**:6056-5663

[33] Huang CC, Chen MW, Hsieh JL, Lin WH, Chen PC, Chien LF. Expression of mercuric reductase from *Bacillus megaterium* strain MB1 in eukaryotic microalga *Chlorella sp*. DT: An approach for mercury phytoremediation. Applied Microbiology and Biotechnology. *2006*;**72**:197-205

[34] Fischer BB, Wiesendanger M, Eggen RIL. Growth condition-dependent sensitivity, photodamage and stress response of *Chlamydomonas reinhardtii* exposed to high light conditions. Plant Cell Physiology. 2006;**47**:1135-1145

[35] Gordillo FJL, Jiménez C, Chavarría J, Niell FX. Photosynthetic acclimation to photon irradiance and its relation to chlorophyll fluorescence and carbon assimilation in the halotolerant green alga *Dunaliella viridis*. Photosynthesis Research. 2001;**68**:225-235

[36] Ho TY, Quigg A, Finkel ZV, Milligan AJ, Wyman K, Falkowski PQ, et al. The elemental composition of some marine phytoplankton. Journal of Phycology. 2003;**39**:1145-1159

[37] Darehshouri A, Affenzeller M, Lütz-Meindl U. Cell death upon H_2O_2 induction in the unicellular green alga *Micrasterias*. Plant Biology. 2008;**10**:732-745

[38] Yamazaki JY, Suxuki T, Maruta E, Kamimura Y. The stoichiometry and antenna size of the two photosystems in marine green algae, *Bryopsis maxima* and *Ulva pertusa*, in relation to the light environment of their natural habitat. Journal of Experimental Botany. 2005;**56**:1517-1523

[39] Li T, Huang X, Zhou R, Liu Y, Li B, Nomura C, et al. Differential expression and localization of Mn and Fe superoxide dismutases in the heterocystous cyanobacterim *Anabaena* sp. strain PCC 7120. Journal of Bacteriology. 2002;**184**:5096-5103

[40] Asada K, Kanematsu S, Uchida K. Superoxide dismutases in photosynthetic organisms: Absence of the cuprozinc enzyme in eukaryotic algae. Archives of Biochemistry and Biophysics. 1977;**179**:243-256

[41] Sandnes JM, Kallqvist T, Wenner D, Gislerod HR. Combined influence of light and temperature on growth rates of *Nannochloropsis oceanica*: Linking cellular responses to large-scale biomass production. Journal of Applied Phycology. 2005;**17**:515-525

[42] Atkin OK, Tjoelker MG. Thermal acclimation and the dynamic response of plant respiration to temperature. Trends in Plant Science. 2003;**8**:343-351

[43] Maxwell DP, Falk S, Huner NP. Photosystem II excitation pressure and development of resistance to photoinhibition (I. light-harvesting complex II abundance and zeaxanthin content in *Chlorella vulgaris*). Plant Physiology. 1995;**107**:687-694

[44] Teoh ML, Chu WL, Phang SM. Effect of temperature change on physiology and biochemistry of algae: A review. Malaysia Journal of Science. 2010;**29**:82-97

[45] Hollis L, Ivanov AG, Hüner NPA. *Chlorella vulgaris* integrates photoperiod and chloroplast redox signals in response to growth at high light. Planta. 2019;**249**:1189-1205

[46] Haghjou MM, Shariati M, Smirnoff N. The effect of acute high light and low temperature stresses on the ascorbate-glutathione cycle and superoxide dismutase activity in two *Dunaliella salina* strains. Physiologia Plantarum. 2009;**135**:272-280

[47] Ndong C, Danyluk J, Huner NPA, Sarhan F. Survey of gene expression in winter rye during changes in growth temperature, irradiance or excitation

pressure. Plant Molecular Biology. 2001;**45**:691-703

[48] Janknegt PJ, Rijstenbil JW, van de Poll WH, Gechev TS, Buma AG. A comparison of quantitative and qualitative superoxide dismutase assays for application to low temperature microalgae. Journal of Photochemistry and Photobiology, B: Biology. 2007;**87**:218-226

Section 4

Microalgae Secondary Metabolites

Chapter 11

Microalgae Cultivation for Secondary Metabolite Production

Facundo J. Márquez-Rocha, Diana Palma-Ramírez, Pedro García-Alamilla, Jenny F. López-Hernández, Ivonne S. Santiago-Morales and Abelardo I. Flores-Vela

Abstract

Microalgae including cyanobacteria have been recognized as an excellent source of fine chemicals, renewable fuels, vitamins, and proteins and usually are found in health food stores around the world. However, the accumulation of these compounds generally occurs at end of the exponential growth phase; furthermore, biomass density in cultivation commonly is low. Open cultures have been used for pigment, biofuels, and biomass production, but these types of culture system are not a good choice for the production of fine chemicals, due to contamination problems and the expensive production costs. Closed photobioreactors can be operated in a continuous cultivation providing an increase on biomass density and contamination-free condition and generally working at a maximum growth rate under specific conditions; besides, these systems can recycle the consumed culture medium at least three times before a new enriched medium is supplied, generating a more cost-effective production system. In addition, microalgae metabolism can be manipulated to provoke a specific secondary metabolite accumulation by the addition of organic carbon source or changing light intensity or both. In other words, photobioreactors can operate in continuous mode, with efficient light supply and the supplementation of organic carbon source to produce fine biochemicals such as anticancer, antibacterial, antioxidant, lectins, antiviral compounds, and biofuels.

Keywords: microalgae metabolism, mixotrophy, continuous cultivation, secondary metabolites, fine chemicals, biofuels, pigments

1. Introduction

Industrial reactors for microalgae cultivation have been generally constructed using channels with movement and adapted for a better gas exchange. One of the biggest problems in this culture system is the low density of microalgae cells; they are constructed between 15 and 30 cm deep along the canal, limiting therefore the available light in addition to increasing the potential for contamination. A system proposed to solve the problem of low density and pollution has been found in closed polyethylene pipe systems, having the geometric design of the reactor as its main objective. Some strategies addressing three aspects have been developed to improve cultivation of microalgae in photobioreactors and produce fine chemicals:

(1) the culture medium design—it is necessary to fix the nutrient composition to provide the right source of carbon and energy depending on the microalgae strain and secondary metabolite to be produced; (2) reducing adverse conditions for culture, such as oxygen accumulation, CO_2 efficient supply, and sufficient light distribution. For this purpose, studies on the photobioreactor prototype should be performed; (3) once that photobioreactor prototype works well, critical factor criteria for scale-up bioengineering process should be fixed [1–3].

2. Microalgae metabolism and mixotrophic growth kinetics

Biomass and product productivity are significantly affected by the culture condition; energy and carbon supply impacts directly biomass and product concentration. In effect, different metabolic growth modes for microalgae have been recognized: (a) autotrophy, in which light is the sole source of energy and inorganic carbon is the sole source of carbon; (b) heterotrophy, in which energy and carbon are both obtained exclusively from an organic carbon source, such as glucose, glycerol, and acetate, and growth can proceed without light supply; (c) mixotrophy, in which the photosynthetic microorganisms obtain energy from light and organic carbon sources and carbon is obtained from organic and inorganic carbon sources [4, 5]; and (d) photoheterotrophy, in which carbon can be obtained from organic compound but strictly with a light supply [5]. Chojnacka and Noworyta designed an empirical mathematical model to describe mixotrophic growth; in this model heterotrophic and autotrophic cultures are fractions of mixotrophic growth, but the metabolic interaction of photosynthesis and heterotrophy is important to improve biomass density and consequently secondary metabolite productivity [6].

Light as a source of energy for photosynthetic organisms is the main limiting factor during cultivation process of these organisms. In light intensities above the light saturation point, photosynthesis rate is directly proportional to the incident light supplied. The photosynthetic system of many microalgae becomes saturated to a radiation close to 30% of the total solar irradiance, i.e., between 1700 and 2000 $\mu Em^{-2} s^{-1}$. Some species of phytoplankton grow to optimal intensities of 50 $\mu Em^{-2} s^{-1}$ and are photoinhibited at around 130 $\mu Em^{-2} s^{-1}$. The culture limitation by photoinhibition is the most important problem for commercial cultivation of microalgae. A possible solution is to assume that the heterotrophic metabolism in photosynthetic cells occurs, replacing or supplementing energy and carbon requirements from organic sources. Some studies suggest that mixotrophic, autotrophic, and heterotrophic metabolic activities occur simultaneously during cell growth [7]. The relative contribution of autotrophy to biomass production increases by increasing the light supply coefficient (kJ kg $m^2 s^{-1}$) or with an increase in the supply of CO_2 and a decrease of organic carbon source supply. For example, at a light supply coefficient of 0.5 at 0.03 and 10% of CO_2 concentration, the ratio of contribution of autotrophy (heterotrophy/autotrophy) to the biomass production was of 98:2 and 70:30, respectively [8]. A respirometric procedure has been proposed to obtain half saturation constant values for several nutrients; it is useful for modeling bioprocess for photosynthetic microorganisms [9]. These methods can be useful to evaluate organic substrates to be used in cultures, in a practical way.

2.1 Contribution of autotrophic and heterotrophic metabolisms during the mixotrophic cultivation performance

Prior works [8, 10] conducted a detailed analysis of the heterotrophic and autotrophic modes simultaneously in *Euglena gracilis* and *Spirulina platensis*, respectively, under strict control of culture conditions; mixotrophic biomass concentration and growth rates resulted in the sum of the biomass or growth rates of heterotrophic and autotrophic cells in growing culture in parallel. The yields produced approximately the same amount of biomass produced in mixotrophic conditions; a mathematical approach can be summarized in Eqs. (1)–(8) [8, 10]. Assuming that mixotrophic growth is derived from cells growing in autotrophic and heterotrophic conditions, it can be described mathematically by defining the contribution of both metabolic growth modes. This can be simplified in Eq. (1):

$$X_M = X_A + X_H \tag{1}$$

Supposing that α is the heterotrophic fraction in mixotrophic growth, X_H (biomass from heterotrophy) may be expressed as follows:

$$X_H = \alpha X_M \text{ with the condition } 0 < \alpha < 1 \tag{2}$$

Combining Eqs. (1) and (2), biomass in mixotrophic (X_M) conditions can be expressed as follows:

$$X_M = \frac{X_A}{1 - \alpha} \tag{3}$$

in which X_A is the biomass from autotrophy; the growth rate ratio can be expressed as follows:

$$1 \approx \left[\frac{dX_M/dt}{dX_A/dt + dX_H/dt}\right] \left(g\ d^{-1}\right) \tag{4}$$

The α coefficient is low when the incident light reaches all cells in the photobioreactor, where cell density should be low enough to avoid hiding among cells. Once that cell density increases, heterotrophic fraction increases as well, in the presence of an organic substrate as carbon source. To estimate α at any time, the value of α can be represented by an α_i which can be constant for a period of time, Δt_i.

Then, Eq. (4) can be expressed as follows:

$$\frac{\Delta X_M}{\Delta t_i} = \frac{1}{1 - \alpha_i} \frac{\Delta X_A}{\Delta t_i} \tag{5}$$

By replacing in Eq. (5) of autotrophic growth, the equation included incident light:

$$\frac{\Delta X_M}{\Delta t_i} = \frac{1}{1 - \alpha_i} K l_0 \tag{6}$$

where $K = \frac{Y_{kJ} A}{V}$.

Integrating Eq. (6) results in

$$\int_{X_{Mi}}^{X_{Mi+1}} \Delta X_M = \frac{K l_o}{1 - \alpha_i} \int_{t_i}^{t_{i+1}} \Delta t_i \tag{7}$$

And then, α_i at any time can be calculated:

$$\alpha_i = 1 - \frac{K l_o}{X_{M_{i+1}} - X_{M_i}} \left(t_{i+1} - t_i\right) \tag{8}$$

With Eq. (8), α value was 0.02 at the beginning of culture, independent of incident light, to 0.61 after several hours of cultivation. Carbon balance showed inorganic carbon and carbon from glucose were consumed simultaneously; CO_2 produced from respiration was used as carbon source during autotrophic growth [11]. Ogbonna and McHenry observed similar behavior in *Euglena gracilis*. Heterotrophic and autotrophic growth occurred simultaneously and independently [8]; this work defined two fractions instead of one as follows:

$$\frac{dX_M}{dt} = \frac{dX_A}{dt} + \frac{dX_H}{dt} \quad (9)$$

The mixotrophic growth rate dX_M/dt was equal to the sum of the heterotrophic growth rate dX_H/dt and the rate of autotrophic growth dX_A/dt. However, when two metabolic activities interact and when the presence of an organic carbon source affects the autotrophic metabolism or when the light affects the heterotrophic metabolic activity, then the heterotrophic rate in the mathematical description can be modified in accordance with Eq. (10), where dX_M/dt is the total mixotrophic growth rate and β and α are coefficients of the autotrophic and heterotrophic fractions of the total mixotrophic growth rate, respectively:

$$\frac{dX_M}{dt} = \beta\frac{dX_A}{dt} + \alpha\frac{dX_H}{dt} \quad (10)$$

The values of α and β can be calculated on the basis of dX_A/dt autotrophic growth rate and dX_H/dt heterotrophic growth rate, both during mixotrophic culture.

$$\beta = \frac{dX_A/dt}{dX_M/dt} \quad (11)$$

$$\alpha = \frac{dX_H/dt}{dX_M/dt} \quad (12)$$

The sum of the values of β and α is 1.0 when the growth proceeds independently and simultaneously; when the sum is more than 1.0, there is an effect of promotion; and when it is <1.0, there is an inhibitory effect. It is clear that when mixotrophic growth occurs, both growth modes, namely, autotrophic and heterotrophic growth, have contribution in the final biomass and metabolite production; Eqs. (1)–(12) may be used to describe mathematically the secondary metabolite formation. An important stoichiometric relationship exists on carbon metabolism, with the pH changes driven by consumption of carbon source. Bicarbonate consumption increases pH, whereas glucose consumption decreases the pH due to CO_2 production, being the reason that pH is kept almost constant during mixotrophic growth, CO_2 consumption by photosynthesis, this balance may reflect the type of predominant metabolism.

3. Photobioreactor systems: open and closed systems

Despite certain variability in the shape of open and closed systems, technical designs for open systems are the type race track, moved by paddles, usually operating at depths of 15–20 cm. At this depth, the growth rate of microalgae can be 15 g m^{-2} d^{-1}, with a lipid content of 25%. Similar designs in terms of operation are the circular ponds, which are commonly found in Asia and Ukraine [3]. The major disadvantages of open systems are the significant loss of water by evaporation, the loss of CO_2 into the atmosphere, the pollution, and the need for considerable surface for cultivation. Since the 1990s, in certain parameters such as the selection of species with efficient incident light utilization, the path of the incident light through the photobioreactor (PBR), the thickness of the wall, the mixing regime, and release of O_2 via degassing, CO_2 supply, have been focus on several developments [12]. Closed or semi-closed PBRs, based on different design concepts, have been implemented and tested at a pilot level. The latest developments seem to be directed toward tubular or plate-type compact configurations as well as combinations of these major designs in the form of distributing light over an expanded surface [13].

4. Main problems in closed photobioreactors: light supply, temperature, and oxygen accumulation

Microalgae need enough quality and quantity of light supply, and it should be taken into account as a primary critic factor to design proper PBR. Cell density can increase from 10^3 cells ml^{-1} to densities above 10^8 cells ml^{-1}; it produces a reduction of the distance among cells over 250 times, and the cell size can reduce its size 10 times as well. By improving mix capabilities of the PBR, hydrodynamic shearing stress over the cells can be increased; also, it can reduce growth or even cell death at high stress conditions [14]. The temperature has a greater influence on respiration and photorespiration than photosynthesis; when CO_2 or light is limiting for photosynthesis, the influence of temperature is negligible. In contrast, an increase in the temperature will increase significantly the respiration, but flow of carbon through the Calvin cycle increases marginally. In other words, the net efficiency of photosynthesis declines at high temperatures. This effect can worsen in culture suspension by the difference in the solubility of CO_2 and O_2 at high temperatures. Normal temperatures for the growth of microalgae ranged between 25 and 30°C; an increment in the temperature affects the lipid production; at higher temperatures saturated free fatty acids are produced, while low temperatures favor unsaturated free fatty acid formation [15]. High concentration of O_2 can build up in closed PBR; if this happens photosynthesis can be damaged by decreasing microalgae growth, and an improvement in the PBR should be implemented as an effective gas exchange [16].

5. Photobioreactor design and scale-up

The first generation of closed PBR finds limitations over 50–100 L of culture volume; this was not effective for light supply to produce higher biomass density. Several designs of light distribution over the PBR, mainly underwater lamps, optical fiber, and column-shaped photobioreactors, have been used to provide an efficient production system; however, not much success has been obtained [12]. This is the main challenge in the future to find the appropriate scaling criteria for a larger irradiate surface, mass transfer, and coupled steps upstream and downstream processes [17]. The difficulty to scale up PBRs is to establish the inherent relationship among physical parameters involved in the design and the physiology of the microalgae to be cultured. An important design rule is to define quantitatively parameters to describe the interactions between incident light, the light distribution in the PBR, cell growth, and secondary metabolite production.

To encourage the use of microalgae, it is necessary to implement a step-by-step system at different levels. The first step is the bioprospecting for selecting the most promising strain to produce a specific secondary metabolite and is the interaction of various disciplines, such as the analytical chemistry, biochemistry, molecular biology, and microbiology. The second step is the development of the culture medium, applicable to the largest volume. The third step is the strategy to analyze the scaling-up; biochemical or bioprocess engineers play an important role at this point. Strain and medium selection is carried out at flask level; the type of metabolism for the desired metabolite production, namely, mixotrophic, heterotrophic, or autotrophic growth, is also defined in this step. Operation parameters are fixed at small PBR scale; once the critic factors are overcome, PBR is ready to apply a scale-up procedure, from pilot to industrial production [18]. At the same time, recovery and purification steps should be performed. The last step of scale-up process should be a feasibility economic and technological analysis, in which production costs are obtained [19]. Quinn et al. constructed and validated a scalable growth model with species-specific variables, such as light and temperature; it can be used with PBR dimensions to accurate growth modeling for life cycle analysis.

6. Energy efficiency received by microalgae in photobioreactors

Many aspects should be considered to obtain high concentration of biomass and secondary metabolites. Microalgae need energy from light to drive photosynthesis and growth. However, many of these organisms are able to use organic compounds as a source of chemical energy from respiratory mechanism. Although the terms of mixotrophy, autotrophy, photoheterotrophy, and heterotrophy are not well-defined, the influence of organic carbon energy and incident light energy can be quantitatively described in terms of biomass and secondary metabolite production.

Assuming that autotrophy growth occurred in cells absorbing incident light on the irradiate surface of the reactor, growth depends on the specific energy yield (Y_{kJA}), in other words, the amount of energy required to produce an amount of biomass; it can be defined by the following equation for continuous cultures:

$$Y_{kJA} = \frac{DX_A V}{I_O A} \tag{13}$$

Moreover, microalgae growing in mixotrophic mode and energy from carbon source can be included in Eq. (13), as follows:

$$Y_{kJM} = \frac{DX_M V}{I_o A + \Delta H_S DV(S_o - S)} \tag{14}$$

Yield equations can be achieved in continuous cultivation, where D is the dilution rate (h^{-1}). Energy efficiency in batch and continuous cultures for *Spirulina* was calculated in values of 5.0×10^{-3} g biomass kJ^{-1} [11] and 2.4–4.8 $\times 10^{-3}$ g biomass kJ^{-1} [20]. There are two different points of view concerning energy efficiency which should be mentioned: one recently, a photocolor spirometer has been used for direct measurements of photosynthesis (calorimetry) and oxygen evolution at different light intensities [21], and the other one uses a photobioreactor to measure the overall light and carbon energy necessary to produce biomass and secondary metabolites [5]. The first can be useful to provide a potential energetic yield measurement because it considers the metabolic energy flows in the cells; the second provides data necessary for bioengineering purposes, specifically for the photobioreactor design and scaling-up procedures to follow.

Mixotrophic growth can be described according to **Figure 1**. On the left-hand side, in the photosynthetic growth by the consumption of CO_2 from culture medium, in the presence of light, biomass is produced, and oxygen is produced as well. On the other side, biomass is also produced but from organic carbon source consumption. The photosynthesis and respiration rates depend on several factors, such as microalgae species, O_2 and CO_2 availability, light supply, organic carbon source availability, temperature, pH, etc., but the main factor is the ability of microalgae to use O_2 and CO_2 at the same time [5].

The balance for energetic yield ($Y_{X/kJ}$) can be described as follows: (i) autotrophic growth; total energy comes from incident light, and the biomass is formed from inorganic carbon:

$$\Delta ATP_T = \Delta ATP_{hv} \tag{15}$$

$$\frac{\Delta X_A}{kJ_{hv}} \tag{16}$$

Heterotrophic growth, energy, and biomass are provided by an organic carbon source:

$$\Delta ATP_T = \Delta ATP_{Glu} \tag{17}$$

$$\frac{\Delta X_H}{kJ_{Glu}} \tag{18}$$

Then, in mixotrophic growth, energy is supplied by both incident light and chemical energy from the organic carbon source and biomass from both inorganic and organic carbon sources.

$$\Delta ATP_T = \Delta ATP_{hv} + \Delta ATP_{Glu} \tag{19}$$

$$Y_{M_{kJ}} = \frac{\Delta X_M}{kJ_{hv} + kJ_{Glu}} \tag{20}$$

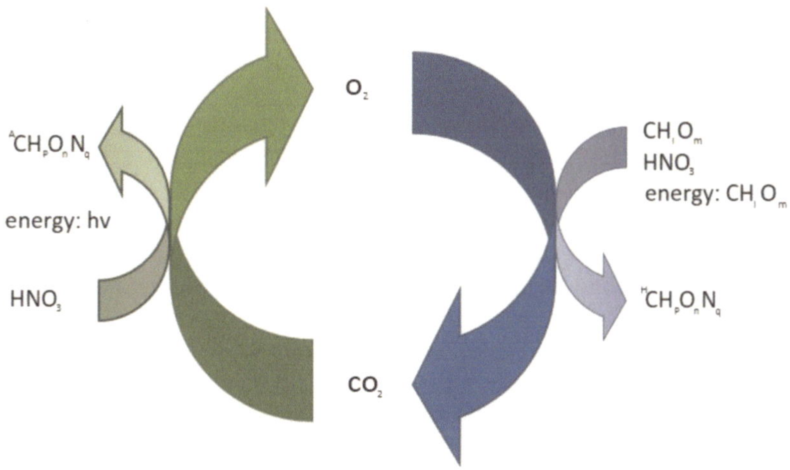

Figure 1.
Drawing describing the interaction of heterotrophy and autotrophy during mixotrophic growth. A, biomass from autotrophy; H, biomass from heterotrophy, modified from [5].

7. Secondary metabolite production

In the mixotrophic growth mode, certain molecules are accumulated, and there is a need to elucidate which metabolite is able to accumulate under specific growth conditions, but in general mixotrophic growth, it seems to be an efficient way for secondary metabolite accumulation [15, 22].

7.1 Fine chemicals

Several high valuable products have been described to be produced by photosynthetic microorganisms: antitumor agent from *Amphidinium* sp., food supplements from *Dunaliella* and *Isochrysis galbana*, antioxidants from *Phaeodactylum tricornutum*, and elastase inhibitor from *Oscillatoria agardhii* [17]. Algae biomass can accumulate or produce (i) bioenergy-based products, such as ethanol, methanol, biodiesel, biohydrogen, biogas, and long-chain hydrocarbons; (ii) staple food and vitamins such as yellow-white proteins, β-carotene, and phycobiliproteins, such as phycocyanin; (iii) polyunsaturated fatty acids, such as linolenic acid and arachidonic acid, that is, omega-3 fatty acids [23, 24]; (iv) base compounds for cosmetic industry and plant growth regulators; and (v) compounds with anticancer, antimicrobial, and antiviral activities.

Spirulina platensis showed higher antioxidant activity than other microalgae tested; *Nostoc muscorum* and *Oscillatoria* sp., moreover, have an important increment of phycobiliproteins by increasing nitrogen to the culture medium. It produces an important increment of the antioxidant activity in aqueous extracts of these microalgae. These extracts exhibit anticancer activity as well; in the extracts phenolic compounds, terpenoids, and alkaloids have been detected which can be responsible for several biomedical activities [24]. Water extracts of *S. platensis* have shown vulvovaginal antifungal activity on *Candida* and antifungal activity on several strains of *Candida* sp.; this can be the basis for therapeutic treatments, where secondary effects seem to be absent [25].

7.2 Pigments

Dietary supplements have been produced from biomass of microalgae; they include pigments and colorants from *Haematococcus pluvialis*, *Chlorella* sp., *Dunaliella*, red algae, cyanobacteria, and *S. platensis* [23]. A profile of natural pigments in dietary supplements of *Spirulina* including 51 pigments has been found in commercial products [26]. Pre-column reaction with DPPH radical followed by fast UHPLC-PDA separation revealed different classes of pigments grouped among carotenes, xanthophylls, and chlorophylls. Diadinoxanthin, alloxanthin, canthaxanthin, diatoxanthin, zeaxanthin, and echinenone were found in powder and tablets as minor components, in addition to β-carotene as a major component of dietary supplements [26]. Astaxanthin from *H. pluvialis*, c-phycocyanin from *Limnothrix* sp., and phycoerythrin from *Phormidium* have been produced [27–29], respectively.

Production of pigments is affected by the amount of light supplied, and in combination with mixotrophic growth mode, phycocyanin, chlorophyll-a, and carotenoid concentrations, increased as light intensity increased, the concentration increased at least 30% in *S. platensis* [4, 11]. Production of chlorophylls and carotenoids increases 1.5 fold in *Chlorella vulgaris* in stirred tank photobioreactor [4]. Carotenoid accumulation and composition seem to be induced by light intensity, nitrogen starving, and salt stress. Higher light and salt stresses active synergistically carotenogenesis, mechanism such as esterification of astaxanthin and adonixanthin in *Scenedesmus* sp. [30].

Red marine microalga has proven their ability to produce pigments and hydrocolloids, due to their diversity, and perhaps produce a diversity of high valuable compounds. Fine chemicals are used as cosmetics, nutraceutics, and therapeutic agents; some are used in the food industry, diagnostic, biomedical research, and biosensor. Carbon source such as glucose, sucrose, glycerol, or acetate in the culture medium can help for accumulation of β-carotene and zeaxanthin by red microalga [31].

7.3 Antioxidants

Microalgae produce a wide range of antioxidants, some of them involved in the scavenging machinery of photosynthesis, respiration, and oxidative protection mechanisms. Pigments (carotenoids, chlorophylls, phycobiliproteins) play an important role in the photosynthetic mechanism in tocopherols including α, δ, and γ tocopherols, and pigments are accumulated as secondary metabolites at different amounts. Tocopherols have been found in *Nannochloropsis oculata*, *Tetraselmis suecica*, *Spirulina maxima*, *Chaetoceros* sp., *Synechococcus*, and *Porphyridium cruentum*. These compounds are formed depending several factors, such as microalga specie, growth phase, nutrient availability, light supply, and oxygen concentration, but also their production is affected by the processes of extraction and purification [32–34].

7.4 Biofuels

Lipid metabolism can be induced by a nitrogen-limiting condition; nitrogen obtained from amino acid catabolism is assimilated via the glutamate-glutamine pathway; then, it is stored as an amino acid. The excess of carbon obtained from photosynthesis or glycolysis is redistributed into carbon-containing compounds. Carbon enters lipid metabolism via gamma-aminobutyrate pathway, glycolysis, and the tricarboxylic acid cycle [35]; malonyl-CoA is formed via acetyl-CoA from respiration; then, lipogenesis proceeds [15]. Supplementing microalgae cultures with an organic carbon source increases the productivity of biomass, lipid, and carbohydrates, enhancing the production of biodiesel, ethanol, starch, and polyunsaturated fatty acids. However, organic carbon source addition has limitations, for example, the cost and the bacterial contamination during cultivation. Progress on biorefineries has been focused on mixotrophic cultivation to enhance either secondary metabolite accumulation or fine chemicals [36]. Triacylglycerol content in *Neochloris oleoabundans*, *Dunaliella* sp., and *Botryococcus braunii* is more abundant when glycerol was used as organic carbon source than with autotrophic cultures. Profile of free fatty acids is also different. Saturated free fatty acids increase significantly in the presence of glycerol, but unsaturated free fatty acids decrease in general [37]. Biomass productivity and also the lipid productivity increased with the addition of acetate, glucose, and glycerol; although lipid content is smaller than other cultures, light supply also affected the content of lipids [38]. In contrast, lipid concentration in *Chlorella protothecoides* was as high as 55%, four times those obtained in autotrophic growing cells. Microalgae metabolic pathways for lipid accumulation are influenced by nitrogen-limiting conditions and carbon metabolism, where distribution pathways contribute to lipid biosynthesis [39, 40]. Biomass is considered as a renewable fuel source and does not affect the overall balance of CO_2 in the atmosphere. Algal biofuel production coupled to a biomass power plant waste can serve as a cost-effective process to enhance microalgae biomass and biofuel productivity by sequestration of the CO_2 produced in the power plant [35]. Productivity of biodiesel from oily plant crops, in terms of produced oil by surface production, varies from 27.57 to 972 L per ha, whereas that from microalgae cultivation is 7688–23,067 L per ha [23].

Biofuels derived from algal biomass depend on algal species: for biodiesel, *Cladophora fracta*, *C. protothecoides*, and *B. braunii*; for biohydrogen, *C. protothecoides*, *S. platensis*, and *Chlamydomonas reinhardtii*; for bioethanol *Palmaria*, *Porphyra*, *Ascophyllum*, *Ulva lactuca*, *Tetraselmis* sp., and *Chlorococum* sp.; and for biogas *C. reinhardtii*, *Chlorella kessleri*, and *Spirogyra neglecta* [23, 41–44]. To produce biodiesel from microalgae, it is very important to select strains with oil content over 50% to improve biodiesel yield. With respect to oil content, microalgae can be divided into low, medium, and high oil content strains [45].

7.5 Proteins

Soluble proteins have been used as nutritional supplements and personal care products or insoluble proteins for animal feeds [36]. Protein production has been reported in *S. platensis* using beet vinasse-supplemented culture media, in tubular photobioreactor biomass, which reached to 6.5 g L^{-1} and 168 mg L^{-1} d^{-1} of protein productivity. Continuous cultivation was also suitable for protein production from *S. platensis* using a medium supplemented with beet vinasse [46].

Incorporation of carbon from an organic carbon source, the type of carbon source, the amount supplemented to the culture, and the specie of microalgae are important for lipid accumulation in the cells of microalgae. The content of protein mostly increases by the addition of an organic carbon source, but lipid content decreases, although productivity of biomass, protein, and lipids increases substantially in the presence of organic carbon source [38].

Figure 2 represents the secondary metabolite production along with biomass that should be included in balance equations. The main components are carbon, hydrogen, oxygen, and nitrogen; in other words, a secondary metabolite can be a fraction of the total biomass, and it can be defined as $\Theta\Delta X$, where is the fraction corresponding to the secondary metabolite produced, for chlorophyll accumulation, and it depends on the availability of carbon and nitrogen sources [40].

Figure 2.
Drawing describing the production of secondary metabolites under mixotrophy. P, metabolite from autotrophy and heterotrophy; A, biomass from autotrophy; H, biomass from heterotrophy. Modified from Ref. [5].

Mixotrophy is coupled with three metabolic mechanisms, glycolysis, Calvin-Benson-Bassham, and the tricarboxylic acid cycle, where ATP is formed in the tricarboxylic acid cycle helping to drive electron flux on the light reactions of the photosynthesis to generate NADH, which is needed in the tricarboxylic acid cycle. These mechanisms are focal point to perform metabolic engineering, which open new routes to enhance the synthesis of fine chemicals by microalgae [47].

8. Concluding remarks

In the past, microalgae cultures were used as components of aquaculture feeds and human food supplements. Recently, new alternatives have been opened for the production of fine chemicals and biofuels. However, production costs have been a concern; several efforts have been made to reduce processing costs to construct a profitable process. In this context, Allen et al. propose an integration of biology, ecology, and engineering topics for a sustainable biofuel and bioproduct production from microalgae [48].

The potential markets of value-added products from microalgae are nutraceuticals for human applications and nutraceutical with applications for animal and fish feed, bulk chemicals, and biofuels, with commercial costs of 100 €/kg biomass, 5–20 €/kg biomass, <5 €/kg biomass, and <0.4 €/kg biomass, with a volume market of 60 million, 3–4 billion, >50 billion, and >1 trillion €, respectively [49].

High value-added products such as antiviral, anticancer, and antioxidants are target products to be obtained from microalgae, since it is an alternative process that can be continuously cultivated of axenic cultures in a closed photobioreactor adapted with a special light source of irradiation, such as fiber-optic or halogen lamps. In this case, biomass increases as long as microalgae receive light and the broth hydrodynamic allows enough movement to reach the illuminated surface (see **Table 1**), in continuous cultivation. Once the light limitation occurred and due to the effect of washing out, biomass starts to decrease to a new dilution rate. When an organic carbon source has a positive effect on the growth, continuous cultivation can be used as well, to produce an increment in biomass density (**Table 1**) and secondary metabolite formation as well, producing an increment of biomass and in the metabolites. Productivity also has a substantial increment at same light intensity and same dilution rate (D, h^{-1}). Productivity and biomass concentration have been obtained in semicontinuous cultivation with a biomass of 5.31 g L^{-1} and productivity of 1.32 g L^{-1} d^{-1} [50]. Therefore, semicontinuous cultivation seems to be a good strategy as well.

Secondary metabolite production can be effectively improved, by three advantages,(i) using a continuous process (up- and downstream processes),

I_o (J cm^{-2} h^{-1})	ΔX_A (g L^{-1})	$\Delta X_A D$ (g L^{-1} h^{-1} × 10^{-3})	ΔX_M (g L^{-1})	$\Delta X_M D$ (g L^{-1} h^{-1} × 10^{-3})
3.22	0.050	1.74	0.079	2.76
5.85	0.092	3.21	0.136	4.75
11.11	0.175	6.11	0.241	8.41
18.98	0.301	10.81	0.405	14.13

Modified from Ref. [51].

Table 1.
Biomass concentration and productivity in continuous culture, in autotrophic and mixotrophic conditions (D = 0.03 h^{-1}).

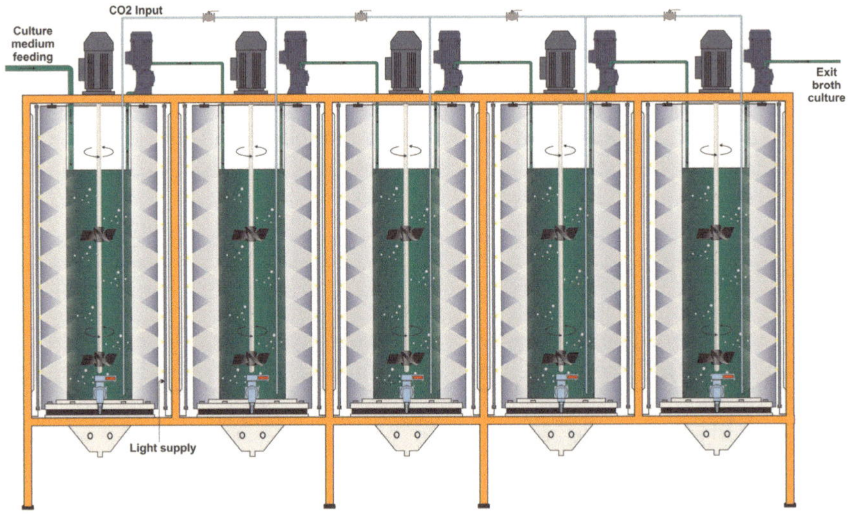

Figure 3.
Schematic representation of a series of photobioreactors to operate in continuous cultivation to produce fine chemicals.

(ii) implementing mixotrophic cultivation, and (iii) recycling broth medium at least three times (**Figure 3**).

Acknowledgements

The authors thank Mr. Mauricio Ramos for the drawing of figures and the Instituto Politécnico Nacional for funding this work.

Conflict of interest

The authors declare that there is no known conflict of interest associated with this publication.

Notes/thanks/other declarations

Thanks to Lada Bozic for helping in an efficient communication with IntechOpen.

Author details

Facundo J. Márquez-Rocha[1]*, Diana Palma-Ramírez[2], Pedro García-Alamilla[3], Jenny F. López-Hernández[4], Ivonne S. Santiago-Morales[5] and Abelardo I. Flores-Vela[2]

1 Regional Center for Cleaner Production, National Polytechnic Institute, Cunduacan, Tabasco, Mexico

2 Mexican Center for Cleaner Production, National Polytechnic Institute, Ciudad de México, Mexico

3 Academic Division of Agriculture and Livestock Science, Juarez Autonomous University of Tabasco, Tabasco, Mexico

4 Academic Division of Biological Sciences, Juarez Autonomous University of Tabasco, Mexico

5 Institute of Industries, University of the Sea, Puerto Angel, Oaxaca, Mexico

*Address all correspondence to: fjmrocha97@gmail.com

IntechOpen

© 2019 The Author(s). Licensee IntechOpen. This chapter is distributed under the terms of the Creative Commons Attribution License (http://creativecommons.org/licenses/by/3.0), which permits unrestricted use, distribution, and reproduction in any medium, provided the original work is properly cited.

References

[1] Bumbak F, Cook S, Zachleder V, Hauser S, Kovar K. Best practices in heterotrophic high-cell-density microalgal processes: Achievements, potential and possible limitations. Applied Microbiology and Biotechnology. 2011;**91**:31-46. DOI: 10.1007/s00253-011-3311-6

[2] Madkour FF, Kamil AEW, Nasar HS. Production and nutritive value of *Spirulina platensis* in reduced cost media. Egyptian Journal of Aquatic Research. 2012;**38**:51-57. DOI: 10.1016/j.ejar.2012.09.003

[3] Rogers JN, Rosemberg JN, Guzman BJ, Oh VH, Mimbela LE, Ghassen A, et al. A critical analysis of paddlewheel-driven raceway ponds for algal biofuel production at commercial scales. Algal Research. 2014;**4**:76-88. DOI: 10.1016/j.algal.2013.11.007

[4] Benavente-Valdés JR, Méndez-Zavala A, Morales-Oyervides L, Chisti Y, Montañez Y. Effects of shear rate, photoautotrophy and photoheterotrophy on production of biomass and pigments by *Chlorella vulgaris*. Journal of Chemical Technology and Biotechnology. 2017;**92**: 2453-2459. DOI: 10.1002/jctb.5256

[5] Chojnacka K, Marquez-Rocha FJ. Kinetic and stoichiometric relationships of energy and carbon metabolism in the culture of microalgae. Biotechnology. 2004;**3**(1):21-34. DOI: 10.3923/biotech.2004.21.34

[6] Chojnacka K, Noworyta A. Evaluation of *Spirulina* sp. growth in photoautotrophic, heterotrophic and mixotrophic cultures. Enzyme and Microbial Technology. 2004;**34**: 461-465. DOI: 10.1016/j.enzmictec.2003.12.002

[7] Melis A. Photosynthetic H_2 metabolism in *Chlamydomonas reinhardtii* (unicellular green algae). Planta. 2007;**226**(5):1075-1086. DOI: 10.1007/s00425-007-0609-9

[8] Ogbonna JC, McHenry MP. Cultures systems incorporating heterotrophic metabolism for biodiesel oil production by microalgae. In: Moheimani NR, McHenry MP, de Boer K, Bahri PP, editors. Biomass and Biofuels from Microalgae, Biofuel and Biorefinery Technologies 2. Switzerland: Springer International Publishing; 2015. p. 63-74. DOI: 10.1007/978-3-319-16640-7_4

[9] Sforza E, Pastore M, Barbera E, Alberto Bertucco A. Respirometry as a tool to quantify kinetic parameters of microalgal mixotrophic growth. Bioprocess and Biosystems Engineering. 2019;**42**:839-851. DOI: 10.1007/s00449-019-02087-9

[10] Márquez-Rocha F, Sasaki K, Kakizono T, Nishio N, Nagai S. Growth characteristics of *Spirulina platensis* in mixotrophic and heterotrophic conditions. Journal of Fermentation and Bioengineering. 1993;**76**(5):408-410. DOI: 10.1016/0922-338X(93)90034-6

[11] Marquez FJ, Nishio N, Nagai S, Sasaki K. Enhancement of biomass and pigment production during growth of *Spirulina platensis* in mixotrophic culture. Journal of Chemical Technology and Biotechnology. 1995;**62**:159-164. DOI: 10.1002/jctb.280620208

[12] Pruvost J, Cornet JF, Goetz V, Legrand J. Theoretical investigation of biomass productivities achievable in solar rectangular photobioreactors for the cyanobacterium *Arthrospira platensis*. Biotechnology Progress. 2012; **28**(3):699-714. DOI: 10.1002/btpr.1540

[13] Markou G. Fed-batch cultivation of *Arthrospira* and *Chlorella* in ammonia-rich wastewater: Optimization of nutrient removal and biomass

production. Bioresource Technology. 2015;**193**:35-41. DOI: 10.1016/j.biortech.2015.06.071

[14] Michels MHA, van der Goot AJ, Norsker NH, Wijffels RH. Effects of shear stress on the microalgae *Chaetoceros muelleri*. Bioprocess and Biosystems Engineering. 2010;**33**:921-927. DOI: 10.1007/s00449-010-0415-9

[15] Mohan SV, Rohit MV, Chiranjeevi P, Chandra R, Navaneeth B. Heterotrophic microalgae cultivation to synergize biodiesel production with waste remediation: Progress and perspectives. Bioresource Technology. 2015;**184**:169-178. DOI: 10.1016/j.biortech.2014.10.056

[16] Kazbar A, Cogne G, Urbain B, Marec H, Le-Gouic B, Tallec J, et al. Effect of dissolved oxygen concentration on microalgal culture in photobioreactors. Algal Research. 2019;**39**:101432. DOI: 10.1016/j.algal.2019.101432

[17] Molina GE, Acién FFG, García CF, Chisti Y. Photobioreactors: Light regime, mass transfer, and scaleup. Journal of Biotechnology. 1999;**70**(1-3):231-247. DOI: 10.1016/S0168-1656(99)00078-4

[18] Posten C. Design principles of photo-bioreactors for cultivation of microalgae. Engineering in Life Sciences. 2009;**9**(3):165-177. DOI: 10.1002/elsc.200900003

[19] Quinn J, de Winter L, Bradley T. Microalgae bulk growth model with application to industrial scale system. Bioresource Technology. 2011;**102**:5083-5092. DOI: 10.1016/j.biortech.2011.01.019

[20] Ogawa T, Aiba S. Bioenergetic analysis of mixotrophic growth in *Chlorella vulgaris* and *Scenedesmus acutus*. Biotechnology and Bioengineering. 1981;**23**:1121-1132. DOI: 10.1002/bit.260230519

[21] Maskow T, Rothe A, Jakob T, Paufler S, Wilhelm C. Photocalorespirometry (photo-CR): A novel method for access to photosynthetic energy conversion efficiency. Scientific Reports. 2019;**9**:9298. DOI: 10.1038/s41598-019-45296-8

[22] Perez-García O, Escalante F, de Bashan LE, Bashan Y. Heterotrophic culture of microalgae: Metabolism and potential products. Water Research. 2011;**45**:11-36. DOI: 10.1016/j.watres.2010.08.037

[23] Kothari R, Pandey A, Ahmad S, Kumar A, Pathak VV, Tyagi VV. Microalgal cultivation for value-added products: A critical enviro-economical assessment. 3 Biotech. 2017;7:243-257. DOI: 10.1007/s13205-017-0812-8

[24] Shanab SMM, Mostafa SSM, Shalaby EA, Mahmoud GI. Aqueous extracs of microalgae exhibit antioxidant and anticancer activities. Asian Pacific Journal of Tropical Biomedicine. 2012;**2**(8):608-615. DOI: 10.1016/s2221-1691(12)60106-3

[25] Marangoni A, Foschi C, Micucci M, Palomino RAN, Toschi TG, Vitali B, et al. In vitro activity of *Spirulina platensis* water extract against different *Candida* species isolated from vulvo-vaginal candidiasis cases. PLoS One. 2017;**12**(11):e0188567. DOI: 10.1371/journal.pone.0188567

[26] Sommella E, Conte GM, Salviati E, Pepe G, Bertamino A, Ostacolo C, et al. Fast profiling of natural pigments in different *Spirulina* (*Arthrospira platensis*) dietary supplements by DI-FT-ICT and evaluation of their antioxidant potential by pre-column DPPH-UHPLC assay. Molecules. 2018;**23**:1152-1166. DOI: 10.3390/molecules23051152

[27] Reyes FA, Mendiola JA, Ibañez E, Del Valle JM. Asthaxantin extraction from *Heamatococcus pluvialis* using CO_2-expanded ethanol. Journal of Supercritical Fluids. 2014;**92**:75-83. DOI: 10.1016/j.supflu.2014.05.013

[28] Gantar M, Simovic D, Dijilas S, Gonzalez WW, Miksovska J. Isolation, characterization and antioxidant activity of c-phycocyanin from *Limnothrix* sp. strain 37-2-1. Journal of Biotechnology. 2012;**159**:21-26. DOI: 10.1016/j.biotec.2012.02.004

[29] Parmar A, Singh NK, Kaushal A, Madamwar D. Characterization of an intact phycoerythrin and its cleaved 14 kDa functional subunit from marine cyanobacterium *Phormidium* sp. A27DM. Process Biochemistry. 2011;**46**: 1793-1799. DOI: 10.1016/j.procbio. 2011.06.006

[30] Aburai N, Sumida D, Abe K. Effect of light level and salinity on the composition and accumulation of free and ester-type carotenoids in the aerial microalgal *Scenedesmus* sp (*Chlorophyceae*). Algal Research. 2015;**8**: 30-36. DOI: 10.1016/j.algal.2015.01.005

[31] Gaignard C, Gargouch N, Dubessay P, Delattre C, Pierre G, Laroche C, et al. New horizons in culture and valorization of red microalgae. Biotechnology Advances. 2019;**37**:193-222. DOI: 10.1016/j. biotechadv.2018.11.014

[32] Santiago-Morales IS, Trujillo-Valle L, Márquez-Rocha FJ, López Hernández JF. Tocopherols, phycocyanin and superoxide dismutase from microalgae: As potential food antioxidants. Applied Food Biotechnology. 2018;**5**(1):19-27. DOI: 10.22037/afb.v%vi%i.17884

[33] Kelman D, Kromkowski PE, McDermid KJ, Tabandera NK, Patrick R, Wright PR, et al. Antioxidant activity of Hawaiian marine algae. Marine Drugs. 2012;**10**:403-416. DOI: 10.3390/md10020403

[34] Bong SC, Loh SP. A study of fatty acid composition and tocopherol content of lipid extracted from marine microalgae, *Nannochloropsis oculata* and *Tetraselmis suecica*, using solvent extraction and supercritical fluid extraction. International Food Research Journal. 2013;**20**(2):721-729

[35] Chen H, Wang J, Zheng Y, Zhan J, He C, Wang Q. Algal biofuel production coupled bioremediation of biomass power plant wastes based on *Chlorella* sp. C2 cultivation. Applied Energy. 2018;**211**:296-305. DOI: 10.1016/j. apenergy.2017.11.058

[36] Perez-Garcia O, Bashan Y. Microalgal heterotriphic and mixotrophic culturing for bio-refining: From metabolic routes to techno-economics. In: Prokop A, Bajpai RK, Zappi ME, editors. Algal Biorefineries. Vol. 2. Products and Refinery Design. Switzerland: Springer International Publishing; 2015. pp. 61-131. DOI: 10.1007/978-3-319-20200-6_3

[37] Choi HJ, Lee SM. Biomass and oil content of microalgae under mixotrophic conditions. Environmental Engineering Research. 2015;**20**(1):25-32. DOI: 10.4491/eer.2014.043

[38] Liang Y, Sarkany N, Cui Y. Biomass and lipid productivities of *Chlorella vulgaris* under autotrophic, heterotrophic and mixotrophic growth conditions. Biotechnology Letters. 2009; **31**:1043-1049. DOI: 10.1007/ s10529-009-99757

[39] Chen H, Zheng Y, Zhan J, Wang Q. Comparative metabolic profiling of the lipid-producing green microalga *Chlorella* reveals that nitrogen and carbon metabolic pathways contribute to lipid metabolism. Biotechnology for Biofuels. 2017;**10**:153-172. DOI: 10.1186/ s13068-017-0839-4

[40] Adesanya VO, Davey MP, Scott SA, Smith AG. Kinetic modelling of growth and storage molecule production in microalgae under mixotrophic and autotrophic conditions. Bioresource Technology. 2014;**157**: 293-304. DOI: 10.1016/j.biortech.2014.01.032

[41] Sydney EB, da Silva TE, Tokarski A, Novak AC, de Carvalho JC, Woiciecohwski AL. Screening of microalgae with potential for biodiesel production and nutrient removal from treated domestic sewage. Applied Energy. 2011;**88**:3291-3294. DOI: 10.1016/j.apenergy.2010.11.024

[42] Laurinavichene TV, Fedorov AS, Ghirandi ML, Seibert M, Tsygankov AA. Desmostration of sustained hydrogen photoproduction by immobilized, sulfur-deprived *Chlamydomonas reinhardtii* cells. International Journal of Hydrogen Energy. 2006;**31**:659-667. DOI: 10.1016/j.ijhydene.2005.05.002

[43] Ross AB, Jones JM, Kubacki ML, Bridgeman T. Classification of microalgae as fuel and its thermos chemical behavior. Bioresourse Technology. 2008;**99**:6494-6504. DOI: 10.1016/j.biortech.2007.11.036

[44] Baltrenas P, Misevicius A. Biogas production experimental research using algae. Journal of Environmental Health Science and Engineering. 2015;**13**:18-24. DOI: 10.1186/s40201-015-0169-z

[45] Amalina-Kadir WN, Lam MK, Uemura Y, Lim JW. Harvesting and pre-treatment of microalgae cultivated in wastewater for biodiesel production: A review. Energy Conversion and Management. 2018;**171**:1416-1429. DOI: 10.1016/j.enconman.2018.06.074

[46] Coca M, Barrocal VM, Lucas S, González-Benito G, García-Cubero MT. Protein production in *Spirulina platensis* biomass using beet vinasse-supplemented culture media. Food and Bioproducts Processing. 2015;**94**: 306-312. DOI: 10.1016/j.fbp.2014.03.012

[47] Sun H, Zhao W, Mao X, Li Y, Wu T, Chen F. High-value biomass from microalgae production platforms: Strategies and progress based on carbon metabolism and energy conversión. Biotechnology for Biofuels. 2018;**11**: 227-249. DOI: 10.1186/s13068-018-1225-6

[48] Allen J, Unlu S, Demirel Y, Black P, Riekhof W. Integration of biology, ecology, and engineering for sustainable algal-based biofuel and bioproduct biorefinery. Bioresource and Bioprocess. 2018;**5**:47-75. DOI: 10.1186/s40643-018-0233-5

[49] Wijffels RH. Potential of sponges and microalgae for marine biotechnology. Trends in Biotechnology. 2008;**26**(1):26-31. DOI: 10.1016/j.tibtech.2007.10.002

[50] Pedrosa BR, Erika Ortiz MY, Sato S, Perego P, Monteiro de Carvalho JC, Converti A. Effects of light intensity and dilution rate on the semicontinuous cultivationof *Arthrospira* (*Spirulina*) *platensis*. A kinetic Monod-type approach. Bioresource Technology. 2011;**102**:3215-3219. DOI: 10.1016/j.biortech.2010.11.009

[51] Márquez-Rocha FJ. Reassessment of the bioenergetic yield of Arthrospira platensis using continuous culture. World Journal of Microbiology and Biotechnology. 1999;**15**(2):235-238. DOI: 10.1023/A:1008841605798

Chapter 12

Physiological Limitations and Solutions to Various Applications of Microalgae

Manoj Kamalanathan and Antonietta Quigg

Abstract

Despite over a century of research, the various applications of microalgae have only been realized only since the 1940s. With a repertoire of valuable products like biodiesel, astaxanthin, canthaxanthin, lutein, β-carotene, phycocyanin, chlorophyll a, polyunsaturated fatty acids, exopolymeric substance (EPS), and biohydrogen, the commercial importance of microalgae and demand for its product is gaining increasing attention. However, successful transition of the synthesis of microalgal products from laboratory to industries has yet to be realized, even after over 70 years of extensive research. This failure of commercial success of microalgal products can be attributed to the lack of understanding of the physiological role of the products and biological constraint placed by the bioenergetics and physiology, which has been largely ignored. This chapter focuses on the physiological limitations behind synthesis of microalgal products, highlights the crucial unknowns behind the role and synthesis of these products, and hints strategies to overcome the limitations to realize the commercial dream of microalgal products.

Keywords: microalgae, biofuels, PUFA, EPS, hydrogen

1. Introduction

The biological reaction of photosynthesis has played a significant role in shaping the course of evolution through the great oxygenation event converting atmospheric CO_2 into organic form that is being currently extracted as modern day fossil fuels and finally balancing the elemental cycles (C, N, P, and Si) on Earth. The magnitude of the impact photosynthesis reaction has on humans has encouraged scientists, engineers, and others to wonder about exploiting the potential of this reaction in different ways. Factors such as simplicity, metabolic elasticity, diversity, shorter life cycles, and the range of commercially important products synthesized make microalgae the most attractive and practical option of all the other photosynthesizing life forms. Some of the commercially important products produced by microalgae include biofuels (i.e., biodiesel, biohydrogen), high value products (i.e., astaxanthin, canthaxanthin, β-carotene), polyunsaturated fatty acids (i.e., omega-3-fatty acids), farming products (i.e., animal and fish feeds), fertilizer, single-cell proteins, and miscellaneous products such as exopolymeric substances (EPS) that have various applications. Despite the potential of microalgae in producing such a variety of commercially important products, the idea of microalgae mega factories has yet to become a reality. Several factors impede the commercial success of microalgal products. This includes existing sources (fossil fuels), technological development, and physiological limitation. In this chapter, we will describe the various known physiological limitations associated with productions of various microalgal products in a commercial scale and list possible solutions.

2. Biodiesel from microalgae

Spoehr and Milner in 1949 demonstrated that the lipid content increased in the green microalga *Chlorella* under nitrogen starvation [1]. However, it was not until the early 1970s, with the oil embargo and the increase in energy prices, that the concept of lipids from microalgae for biofuel production gained attention [2]. Despite decades of research, biodiesel from microalgae has yet to be introduced to the gas station near you. It has been reported that microalgae can produce around 20–50% of their dry weight as lipids [3, 4], with only one research article suggesting as high as 80% [5]. The amount of lipid produced by a variety of species under optimum growth conditions is summarized in **Table 1**. The ability to synthesize a higher percentage of their dry weight as lipids makes microalgae the highest producer of

Microalga	Taxa	%Lipid	%Protein	%CHO	Growth rate d^{-1}	Source
Amphora sp.	Bacillariophyceae	13.6	17.3	74.9	5.1	[6]
Amphiprora hyaline (ENTOM3)	Bacillariophyceae	22–37			2.3	[6]
Anabaena cylindrical	Cyanophyceae	4–7	43–56	25–30		[7]
Ankistrodesmus falcatus	Chlorophyceae	40.3	14.3	18.3	2.89	[6]
Boekelovia sp.	Chrysophyceae	20.7			3.43	[6]
Botryococcus braunii	Chlorophyceae	54.2	20.6	14.3	1.8	[6]
Chaetoceros sp.	Bacillariophyceae	22.2	31.9	43	4.3	[6]
Chaetoceros calcitrans	Bacillariophyceae	16	34	6		[8]
Chaetoceros gracilis	Bacillariophyceae	7	12	5		[8]
Chaetoceros muelleri	Bacillariophyceae	31	59	10		[9]
Chlamydomonas reinhardtii	Chlorophyceae	21	48	17		[7]
Chlorella sp.	Chlorophyceae	34–48	19–31		1.33	[6]
Chlorella ellipsoidea	Chlorophyceae	8.9	26.1	26.3	5.3	[6]
Chlorella protothecoides	Chlorophyceae	15–56				[6]
Chlorella pyrenoidosa	Chlorophyceae	2	57	26		[7]
Chlorella vulgaris	Chlorophyceae	14–22	51–58	12–17		[7]
Cyclotella sp.	Bacillariophyceae	42.1	16.4	10.2	5.1	[6]
Dunaliella bioculata	Chlorophyceae	8	49	4		[7]
Dunaliella salina	Chlorophyceae	6	57	32		[7]
Dunaliella tertiolecta	Chlorophyceae	15–43			2.58	[6]
Euglena gracilis	Chlorophyceae	14–20	39–61	14–18		[7]
Isochrysis aff. Galbana	Prymnesiophyceae	26	23.3	20.5	2.83	[6]
Monoraphidium sp.	Chlorophyceae	17.9–24.4		25.5	3.1	[6]
Nannochloris	Chlorophyceae	31–63			3.19	[6]

Microalga	Taxa	%Lipid	%Protein	%CHO	Growth rate d^{-1}	Source
Nannochloropsis	Chlorophyceae	46–68			1.05	[6]
Nannochloropsis granulata	Chlorophyceae	24–28	18–34	27–36		[10]
Nannochloropsis oculata	Chlorophyceae	17.8–39.9				[6]
Nannochloropsis salina	Chlorophyceae	59.8	24.3	15.9	1.05	[6]
Nannochloropsis sp.	Chlorophyceae	64			1.04	[6]
Navicula acceptala	Bacillariophyceae	32–48			3.8	[6]
Nitzschia sp.	Bacillariophyceae	27	36	16		[6]
Nitzschia dissipata	Bacillariophyceae	66	12.6	9.3	1.32	[6]
Nitzschia closterium	Bacillariophyceae	13	26	10		[8]
Oocystis pusilla	Chlorophyceae	10.5	39	37		[6]
Phaeodactylum tricornutum	Bacillariophyceae	16–50	31–35	11–17	1.96	[6, 8, 10]
Porphyridium cruentum	Rhodophyceae	9–14	28–39	40–57		[7]
Prymnesium parvum	Prymnesiophyceae	22–38	28–45	25–33		[7]
Scenedesmus obliquus	Chlorophyceae	12–14	50–56	10–17		[7]
Scenedesmus quadricauda	Chlorophyceae	1.9	47			[7]
Scenedesmus dimorphus	Chlorophyceae	16–40	8–18	21–52		[7]
Skeletonema costatum	Bacillariophyceae	10	25	5		[8]
Spirogyra sp.	Chlorophyceae	11–21	6–20	33–64		[7]
Spirulina maxima	Cyanophyceae	6–7	60–71	13–16		[7]
Spirulina platensis	Cyanophyceae	4–9	46–63	8–14		[7]
Synechococcus sp.	Cyanophyceae	11	63	15		[7]
Tetraselmis sp.	Chlorophyceae	18	46	36	2.1	[6]
Tetraselmis chuii	Chlorophyceae	20			2.1	[6]
Tetraselmis maculate	Chlorophyceae	3	52	15		[7]
Tetraselmis succia	Chlorophyceae	15–33			2.1	[6]
Thalassiosira pseudonana	Bacillariophyceae	19	34	9		[8]

Table 1.
Summary of macromolecular composition and growth rate of microalgae belonging to various taxa.

lipid per unit mass and arguably the most practical choice for the production of biodiesel. Theoretical estimates of maximum biomass production vary from 24 to 65 g $m^{-2} \cdot day^{-1}$ [7]. Benedetti et al. have reported a theoretical maximum up to 280 tons ha^{-1} $year^{-1}$ of biomass but also noted that the actual cultivation record never exceeded 100 tons ha^{-1} $year^{-1}$ [11]. Even assuming the theoretical maximum estimates of biomass and 80% of the biomass composition as lipid, the maximum amount of 224 tons ha^{-1} $year^{-1}$ of lipids is what can theoretically be produced. However, the lipid produced has to be transesterified to fatty acid methyl esters, which then can be used as biodiesel, which can result in some loss. However, earlier studies have reported up to 96% recovery of lipids through direct transesterification [12]. With 4% loss by transesterification, a maximum of 215 tons ha^{-1} $year^{-1}$ of biodiesel can be produced. It is important to note that a recent study on *Scenedesmus accuminatus* using open and polyhouse raceway ponds estimated around 2.1 tons ha^{-1} year^{-1}, which is 100 times lower than the theoretical maximum [13]. Another study with *Chlorella* sp. L1 and *Monoraphidium dybowskii* Y2 using batch and semicontinuous mode in a raceway pond resulted in lipid productivities of 13.91 and 14.45 ton ha^{-1} year^{-1} in semicontinuous mode [14], tenfold lower than the theoretical estimates. This discrepancy between theoretical estimates and laboratory and field tests are primarily due to physiological limitations, especially carbon fixation and light absorption [15, 16].

The current demand of fossil fuel is approximately 100 million barrels per day (~11,563 million liters per day), which is only expected to increase with time (105 million barrels per day by 2021) [17]. To meet this magnitude of global demand, assuming 15 ton ha^{-1} year^{-1} lipid productivity from microalgae, around 276 million hectares of land would be required for microalgal cultivation every day. This huge demand can be reduced by increasing the lipid productivity of microalgae by minimizing energy and carbon wasting physiological processes in the cell and redirecting it towards lipid synthesis. RUBISCO, the enzyme catalyzing the dark reaction of photosynthesis, wherein the atmospheric or dissolved CO_2 is converted into organic triose phosphate, has a lower affinity to CO_2. In addition, oxygen is a competitive substrate for RUBISCO, catalyzing the reaction of photorespiration wherein energy and NADPH are utilized and fixed CO_2 is released. Therefore, during CO_2 limitation, some species of microalgae utilize a carbon concentration mechanism (CCM) that increases the concentration of CO_2 at the site of RUBISCO, leading to the catalysis of the dark reaction of photosynthesis. Although the CCM is an ingenious mechanism evolved to address the lower CO_2 affinity of RUBISCO and the competition by oxygen, the process consumes one to two molecules of ATP [18]. This energy can instead be redirected towards lipid synthesis by simply bubbling CO_2 into the cultures [19] or by selection CCM lacking microalgae species.

Finkel et al. found that the median macromolecular composition of nutrient-sufficient exponentially growing microalgae is 17.3% lipid, 32.2% protein, and 15% carbohydrates [20]. For biofuels, lipids are the most important cellular fraction. *Botryococcus braunii* has the highest lipid as percent dry weight, 43%, while *Tetraselmis suecica* and *Dunaliella tertiolecta* have the least (<10%). Also in species-specific differences (**Table 1**), Finkel et al. showed that there are some phylogenetic differences [20]. Cyanobacteria have the lowest lipid content (11.7%; dry wt), while the Bacillariophyta (diatoms) had the highest (18.6–21.3%; dry wt). In fact, in most cases, these median values are all significantly lower than those values reported in **Table 1**, which are up to 60% (w/w) for lipid, protein, and carbohydrate. The values at the higher end of the spectrum are typically induced after some kind of stress (e.g., light, nutrient) is placed on the microalgae. Nutrient stress is most often used to induce lipid production, particularly nitrogen [5, 21, 22]. These large lipid stores are thought to provide a growth advantage under variable resource supplies [23].

Light harvesting systems of microalgae are usually capable of absorbing between 350 and 700 nm; however much of the radiation extending out to 1100 nm remains unused. In addition, Raven et al. calculated a 22% loss of energy even within the spectrum of 350–700 nm [24]. However, anoxygenic photosynthetic bacteria possess bacteriochlorophylls capable of absorbing light up to 1050 nm and perform photochemistry [19]. These bacteriochlorophyll-based photosystem can be engineered into microalgae. By replacing photosystem I, it would extend the range of absorbed photons that can be used for photochemistry [25], therefore the energy available for lipid synthesis. It has been proposed that smaller cells can photosynthetically perform better larger cells [26, 27], primarily attributing to the "package effect." Package effect refers to the inverse relationship between light harvesting ability of the cells and the cell volume. However, studies have shown that larger cells tend to decrease the antennae size, thereby counteracting the negative impacts of package effect [28, 29]. Moreover, several studies have shown that decreasing the antennae size significantly increased the photosynthetic efficiencies [30, 31]. This could be due to decreased photoinhibition and decreased allocation of resources towards light harvesting systems [19]. However, the hypothesis that larger cells would be less photosynthetically efficient than smaller cells due to package effect was recently disproved by a recent study by Malerba et al. [32], which showed that larger cells developed compensatory mechanisms by reducing the antenna size, increasing the connectivity between the photosynthetic units, and decreasing the levels of photo-protective pigment β-carotene. This in turn minimized the negative significances of larger cell volume-induced package effects. Exposure to extreme conditions such as photoinhibitory light levels, UVA and UVB radiation, and nutrient and temperature stress could result in oxidative damage to cellular components especially proteins. These kinds of damage result in breakdown of the damage component and re-synthesis of these components or adaptive changes to the cellular components. These changes however have a cost in terms of energy (i.e., ATP, NADPH) and resources which could have been directed towards biomass and lipid synthesis. Protein turnover during scotophase has been shown to consume about one-third of the total respiratory ATP production [33]. Exposure to nitrogen-limiting conditions could result in upregulation of nitrogenase in diazotrophic cyanobacteria, which in turn could result in 10% decrease in growth rates [34, 35]. These are avoidable energy and resource expenditure, which can be easily prevented by providing optimal light, nutrient, and temperature during mass cultivation.

Lipids are not secreted/excreted actively by all microalgae, with exception of *Botryococcus* sp. [36]. This constrains the maximum amount of lipid that is synthesized and stored in a microalgal cells, as the largest reported cell volume for a microalgae is for species *Noctiluca scintillans* belonging to the class Dinophyceae (83,700,000 μm^3) [37]. The most commonly used microalgal species in biotechnology belong to the class Chlorophyceae such as *Chlorella* sp., *Haematococcus* sp., and *Scenedesmus* sp., which have cell volumes of 34.5, 48, and 26,200 μm^3, respectively [38], and the species belonging to class Bacillariophyceae such as *Phaeodactylum tricornutum* and *Thalassiosira pseudonana* which have cell volumes of ~60–100 μm^3 and 28–200 μm^3 [37, 39]. On the other hand, the maximum cellular density has an inverse relationship with the volume of a cell, by a factor of 3/4 [38]. This further constrains the maximum amount of lipids that can be synthesized in a microalgal mass culture. However, if the lipids were secreted in a manner similar to EPS, physiological limitations such as cell volume or maximum density would not play a constraining role on the maximum amount of lipid that can be synthesized in an algal mass culture. Microalgal species like *Botryococcus* sp. has already been shown to actively secrete lipids that can be converted into fuel-grade biodiesel. However, their slower growth rates limit the maximum lipids that can be synthesized. Microalgal species like *Scenedesmus* has been shown to grow as fast as 1.53 per day [40]. Therefore, genetic engineering techniques to integrate the lipid-secreting trait into the fast-growing *Scenedesmus* sp. could be a possible solution.

Phototrophic modes of cultivation depend on two critical factors, namely, light and carbon dioxide [16]. Therefore, microalgae grown phototrophically can either be limited in light or carbon dioxide or both, limiting the maximum cellular density one can achieve using this mode of cultivation. In addition, various inefficiencies described above with light-harvesting abilities of microalgae, lower affinity for CO_2 of RUBISCO, cost associated with protein turnover, and photoinhibition can be simply avoided by growing microalgae heterotrophically and mixotrophically. Extensive amount of research suggests enhanced biomass production under heterotrophic and mixotrophic modes of cultivation, increasing the cellular density to as much as 4 to 5 times [41, 42]. Further modification in these modes of cultivation such as using fed-batch mode has resulted in further increase in biomass as much as two- to five-fold [43, 44]. Therefore, by growing a hybrid strain of lipid-secreting *Botryococcus* sp. and fast-growing *Scenedesmus* sp., in a fed-batch heterotrophic or mixotrophic cultivation system, one can possibly overcome the physiological limitations of the maximum amount of lipids that can be synthesized in a microalgal system.

3. High valued products

Apart from lipids for biodiesel, microalgae synthesize other products of value such as astaxanthin, canthaxanthin, lutein, β-carotene, phycocyanin, chlorophyll *a*, and polyunsaturated fatty acids (PUFAs) such as γ-linolenic acid, docosahexaenoic acid, and eicosapentaenoic acid. A subset of these, astaxanthin, canthaxanthin, lutein, and β-carotene, have antioxidant properties and are commercially high valuable pharmaceutical products. Phycocyanin is a high value natural food dye, and products such as γ-linolenic acid and eicosapentaenoic acid are considered animal-free based sources of essential fatty acids. Some of these products are also considered to boost the immune system. **Table 2** summarizes the common high value products, their use, the species of microalgae most commonly grown to harvest the products, and the global market values.

Product	Use	Species (dominant)	Global market value*
Carotenoids			Total US$1.2 billion in 2010
β-carotene	*Antioxidant, anti-inflammatory*	*Dunaliella salina, Dunaliella bardawil*	US$ 300–1500 kg^{-1}
Astaxanthin	Feed additive for farmed fish; pigmenter of the fish flesh	*Haematococcus pluvialis*	US$ 2500–7000 kg^{-1}
Phycobilins	Food coloring, cosmetics coloring		US$ 60 million
C-Phycocyanin	*Antioxidant, anti-inflammatory*	*Spirulina*, *Porphyridium*, *Rhodella, Galdieria*	US$ 500 to 100,000 kg^{-1}
Fatty acids *DHA* *EPA* *PUFA*	*Health food supplements*	*Nannochloropsis, Tetraselmis, Isochrysis, Thalassiosira, and Chaetoceros*	
Docosahexaenoic acid		*Crypthecodinium cohnii*	US$140 kg^{-1}
Omega-3 oils			US$1.5 billion or US$80–160 kg^{-1}
Sterols	Pharmaceutical applications or in functional foods		US$ 300 million
Polyhydroxyalkanoates	Production of biodegradable plastics	*Spirulina, Synechocystis*	

**(from Borowitzka - 2013, Chew et al. 2017) [45, 46].*

Table 2.
Summary of the main high value products derived from microalgae.

The major fraction of studies on microalgal PUFA have primarily focused on screening for microalgal species and optimization of conditions that lead to enhanced PUFA products. However, we have no clear understanding on the physiological role of PUFAs in microalgae. It is hypothesized that PUFAs might play a role in homeoviscous adaptation [47], i.e., the enhanced fluidity provided by increased unsaturation membrane fatty acids. Additionally, the fatty acid composition and therefore the PUFA concentrations appear to have some level species specificity, suggesting varying roles [48]. However, the physiological roles of PUFAs are not yet confirmed in microalgae, and therefore strategic optimization of their synthesis in microalgae has not been realized. Nevertheless, PUFAs make a significant portion of the neutral lipid content synthesized in microalgae; therefore optimization strategies discussed above to enhance biomass and lipid synthesis should also increase the PUFA content in microalgae.

Astaxanthin is another promising product synthesized by microalgae that has already achieved a profit of $200 million per year [49]. Astaxanthin is a red keto-carotenoid pigment ubiquitous in nature and has antioxidative, anti-inflammatory, and anti-apoptotic properties. It is also proposed as a potential therapeutic agent for cardiovascular and neurological diseases [50]. Apart from its pharmaceutical use, this product is also used as a pigment source in aquaculture of salmon and trout [51]. Despite the relatively vast amount of research into its properties and applications, the mass production of astaxanthin is still unable to meet its huge market demand [49]. This problem can be primarily attributed to the lack of knowledge on the underlying mechanisms of why these algae accumulate astaxanthin similar to PUFAs and hence the lack of better strategies to optimally produce the pigment. The most prominent hypothesis regarding its production includes a multifunctional photoprotective response to stress induced by exposure to unfavorable conditions (excess light, UV-B radiation, nutrient deprivation) leading to ROS formation [52–54]. However, this hypothesis does not explain the synthesis of astaxanthin under heterotrophic conditions [55, 56]. Therefore, more studies are needed to have a better understanding of the physiological roles of astaxanthin.

Nutrient limitation such as nitrogen, phosphorus or sulfur are widely used strategies for inducing astaxanthin accumulation in *Haematococcus pluvialis* [57, 58]. However, nutrient limitation reduces the maximum amount of biomass one can achieve, thereby reducing the total amount of astaxanthin that can be produced. Additional strategies for inducing astaxanthin production include high salt stress and high light exposure, but these external stressors also lower the biomass yield and therefore the associated pigment production. Using stress as a mechanism to induce astaxanthin synthesis has the fundamental problem of stopping cell cycle and therefore reducing biomass and astaxanthin production. Hence, new strategies to boost astaxanthin and biomass production with the use of stressors need to be discovered.

Phycocyanin is a phycobiliprotein exclusively produced by cyanobacteria and commercially important as a high value natural blue coloring agent for food. Being a part of light-harvesting pigment-protein complex in cyanobacteria, phycocyanin is very critical in light capture and therefore is an indispensable element in the growth and survival of cyanobacteria. Phycocyanin has been shown to constitute up to 60% of the total cellular protein content [59]. As it is involved in light capture, its synthesis is tightly regulated by the wavelength of light the cells are exposed to. Green light has been shown to stimulate the synthesis of phycocyanin, whereas red light has the opposite effect [60]. In addition, low light levels have been also shown to induce accumulation of phycocyanin [61, 62]. Other factors such as glucose and salt have also been shown to enhance the synthesis of phycocyanin [61]. Despite knowing their physiological roles, the commercial success of phycocyanin from microalgae is limited as their production is strictly dependent on the maximum amount of biomass that can be generated. A suggested final concentration of higher than 10% of cell dry weight of phycocyanin is required to make a profit over the cost of pigment downstream separation [63]. With factors such as blue light and low light levels required to induce maximum cellular synthesis of phycocyanin, the growth is significantly attenuated under these low energy light conditions, thereby limiting the maximum biomass and hence the maximum quantity of phycocyanin that can be synthesized. Therefore, a dual phase production approach to maximize the biomass production under mixotrophic conditions with cheaper organic carbon source such as molasses [41] followed by the second phase of low levels of blue light to stimulate the synthesis phycocyanin can significantly maximize the total amount of phycocyanin that be produced.

4. Exopolymeric substances from microalgae

Studies show that microalgae actively release from 3 to 40% of the fixed carbon into the surrounding environment as exopolymeric substances, mostly polysaccharides and proteins but also nucleic acids, DNA, RNA, and other macromolecules [64]. Although initially presumed as experimental artifact or a product of dead and decaying phytoplankton [65], EPS is now universally accepted as a product that is actively secreted by microalgae. The relatively higher percentage of fixed carbon released extracellularly has led the physiologist to question the reasons behind this phenomena. Several hypotheses have been put forward, including carbon overflow, photoprotection of the over-reduced photosystems, motility, self-defense mechanisms, active selection of phycosphere residents, and passive excretion due to osmosis and permeability. The hypothesis of carbon overflow and photoprotection has been discredited due to the presence of proteins, amino acids, and vitamins in the released substances and due to the secretion of EPS during the night [66]. The hypothesis of EPS secretion as a self-defense mechanism, motility, active selection of phycosphere residents, and passive excretion due to osmosis and permeability needs to be experimentally tested.

Experimental studies have shown contrasting results in the secretion of EPS in response to environmental factors such as temperature, nutrient (N, P, and S) limitation, salinity, and heterotrophy/mixotrophy [67]. The results vary depending on the species of microalgae being tested. In addition, EPS secretion during various phases of growth was species dependent, with some showing an increased secretion during stationary phase and others in exponential phase. Overall, with no universal explanation behind the mechanisms of EPS secretion by phytoplankton, and multiple hypotheses explaining the phenomena, strategic means to regulate the production and composition of EPS release by microalgae is clearly lacking. Although more research is needed, EPS are usually composed of carbohydrates, nitrogenous compounds, lipids, and organic acids [68]. Polysaccharides usually could account for 80–90% of the EPS composition even under healthy conditions [69]. Nitrogenous compounds, such as amino acids and proteins, on the other hand, only make up to 4–7% of the total EPS secreted [70, 71]. These protein fractions can include exoenzymes like phosphatase, β-glucosidase [72], and siderophores such as ISIP2a [73]. Characterization of EPS involves quantification of organic matter released as polysaccharides, proteins, lipids, neutral sugars, and/or uronic acids. EPS characterization of these macromolecules is often performed under the assumption these are the dominant molecules, however, possibilities of the same molecules possessing both sugar chain and a protein moiety are quite certain. Moreover, rarely are the monomers that make up these polymers investigated (e.g., sugars that make up these polysaccharides), and therefore the polymeric composition and the structure and physical properties remain unknown. Therefore, more in-depth characterization of EPS from microalgal species of potential and under different growth conditions and growth phase are needed to integrate EPS from microalgae into the algal biotechnology market. Needless to say, these characterization studies of EPS have to be performed with appropriate controls within the context of its application, whether as a surfactant, lubricant, antioxidant, anti-inflammatory, antivirus, antibacterial, antifungal, or anticoagulant.

Xiao et al. [74] in their review compared the emulsifying activity of EPS from *Dunaliella salina* (88% retention) reported by Mishra et al. [75] to commercially available surfactants Tween 20 (65% retention), Tween 80 (60% retention), and Triton X-100 (65% retention). Furthermore, being extracellularly secreted, EPS is not constrained by physiological feedback inhibition mechanisms, or retention capacity dictated by the maximum cell volume, or the need of expensive procedures to lyse the cell to release the products from the cell, unlike other microalgal products (such as lipid for biodiesel, PUFAs, and high valued pigments). Therefore, with better-focused research, EPS has a relatively greater potential of becoming a commercial reality as a microalgal product.

5. Hydrogen from microalgae

With the raising awareness of cleaner and sustainable fuel, development of hydrogen fuel cell-powered cars, and the high cost and greenhouse gas emission associated with thermochemical hydrogen production, microalgae is increasingly becoming an attractive source for the fuel. First observed by Hans Gaffron in 1939 [76], this phenomenon of hydrogen production has been extensively studied since then. The production of hydrogen by microalgae only occurs during anaerobic conditions, due to the sensitivity of hydrogenase (the enzyme catalyzing the reversible reaction of hydrogen production) to oxygen. Three major enzymes that lead to the production of hydrogen in microalgae include (1) reversible/classical hydrogenases, (2) membrane-bound uptake hydrogenases, and (3) nitrogenase enzyme [77, 78]. Of all three, the reversible/classical hydrogenase is the most studied enzyme. Located in the chloroplast, the primary electron donor for this enzyme is photosystem I (PS I). However, the generation of molecular oxygen through photolysis by photosystem II (PS II) inhibits the activity of hydrogenase. Therefore, this enzyme only functions when the rate of photosynthesis is below the compensation point (rate of photosynthesis = rate of respiration). Past studies have achieved this condition by either flushing the system with argon or nitrogen [79] or using PS II lacking mutants [80], or selective excitation of PS I through far red light (>710 nm) [81], or more commonly through sulfur deprivation [82]. During a combination of anaerobiosis and below compensation point conditions, supply of electron from PSI to hydrogenase has been shown to either come from the excitation of PS I, and/or through the photolysis of water, and/or through non-photochemical reduction of the plastoquinone pool through type II NAD(P)H dehydrogenase that mediates the transfer of electron derived from anaerobic catabolism of cellular carbon reserve, primarily thought to be proteins at this point [83, 84].

It is proposed that under ideal conditions, one should expect a generation of 2:1 H_2/O_2 per 8 photons [85]. However, only around 20% of this efficiency is practically achieved [86]. The discrepancy between theory and practical estimations could be due to several physiological reasons unaccounted for in the theoretical estimation. (1) The physiological role of hydrogenase was although a mystery for the most part [86], discuss it's to act as an electron sink and hence oxidation of reducing equivalents derived through anaerobic catabolism of cellular carbon reserve. Therefore, all the physiological and thermodynamic limitations that apply to anaerobic catabolism of cellular would in turn limit the supply of electrons and protons to hydrogenase. (2) Hydrogenase catalyzes a reversible reaction, and Kessler [84] has shown a "photoreduction" process of uptake of hydrogen gas to reduce molecular oxygen or CO_2 similar to the photosynthate derived under aerobic conditions. Whether the reduction of CO_2 through uptake of hydrogen gas by hydrogenase under anaerobic conditions involves the Calvin cycle remains to be demonstrated. Regardless, the uptake of hydrogen by hydrogenase should contribute to the reduced efficiency of the hydrogenase. (3) Direct oxidation of photosysnthetically derived reducing equivalents would mean a net production of 3.1:0 ATP/NADPH per 8 photons (4 molecules of H_2O) and generation of 2 H_2. This would provide no anabolic advantage to the cells producing hydrogen, as this would lead to a big imbalance in the ATP/NADPH levels in the cells. (4) Often the yields of hydrogen derived under sulfur deprivation are compared to the theoretical yield of 2:1 H_2/O_2 per 8 photons; however, during sulfur deprivation, the anaerobiosis is created by the failure to regenerate the high sulfur photo-damaged D1 protein; therefore the hydrogen gas derived is not directly through photolysis of water but instead through catabolism of carbon reserves [84, 87]. These carbon reserve can be starch or proteins [82], and every molecule of glucose derived from the breakdown of starch, is then catabolized to pyruvate through glycolysis yielding 2 NADH and ATP, based on the observation by Melis et al. [82] of increase in acetate levels beyond 120 hour incubation in sulfur deprived anaerobic condition in the presence of light and no ethanol of formate secretion, hints towards catabolism of pyruvate to acetate possibly via phosphotransacetylase and acetate kinase, which would generate an additional molecule of NADH and ATP. Atteia et al. [88] have detected phosphotransacetylase and acetate kinase in the species *Chlamydomonas reinhardtii* that was used in Melis et al. (2000) study [82]. Overall, this process of sulfur deprivation-induced anaerobiosis leads to a total of 3 NADH and 3 ATP, and 2 molecules of acetate from 1 molecule of glucose; therefore a net loss of 8 protons and hence $4H_2$ in the form of 2 acetate molecules and formation of $6H_2$ occurs in this biochemical pathway. This, when expressed per photon cost, would result in $6H_2$ per 32 photons (the ideal required amount to make a molecule of glucose), which would result in 75% efficiency compared to the ideal generation of $2H_2$ per 8 photons directly through photolysis of water via photosynthesis. The discrepancy of the observed 20% efficiency vs. calculated 75% via fermentation can be due to the existence of other pathways competing for the same NADH. Oxidative pentose phosphate pathway has been shown to be upregulated under sulfur deprivation-induced anaerobiosis [89, 90]; however, whether reductive pentose phosphate pathway is upregulated remains unknown. This is probably due to the cellular demand for ribulose 5-phosphate to synthesize nucleic acids and NADP and NAD to maintain the integrity/protect the DNA and cellular metabolism. Nevertheless, if the glucose was catabolized only via oxidative pentose phosphate pathway, only 2 NADPH would be synthesized, which would yield just two molecules of H_2, which would match the observed 20% efficiency of H_2 generation. However, despite nucleic acid synthesis, further catabolism of ribulose 5-phosphate is inevitable and should lead to generation of more NADH. Furthermore, catabolism via oxidative pentose phosphate pathway solely would not explain the increase in acetate levels observed by Melis et al. [82]. On the other hand, if proteins served as carbon source during sulfur-deprived anaerobiosis, we hypothesize only amino acid with three or more carbon chains would be utilized, based on the absence of acetate consumption during sulfur-deprived anaerobiosis observed by Melis et al. [82] and the absence of H_2 production and methyl viologen reduction when supplied with α-ketoglutarate during anaerobiosis as observed by Noth et al. [91]. Both these studies indicate that TCA cycle was not active during anaerobiosis, which explains the absence of acetate and α-ketoglutarate uptake under such conditions, therefore making catabolism of amino acid smaller than the three-carbon chain (glycine) and serving as a source for reducing equivalents for impractical H_2 production. However, Noth et al. [91] did observe H_2 production and methyl viologen reduction when *Chlamydomonas reinhardtii* was grown on oxaloacetate, which suggest glyoxylate pathway might be still active under anaerobiosis. It is important to note the absence of acetate uptake in sulfur deprivation-induced anaerobiosis as seen in Melis et al. [82] study and the opposite phenomena observed by Gibbs et al. [87] where acetate was readily uptaken when anaerobiosis was established via flushing with N_2. This suggests that glyoxylate pathway is inhibited under sulfur deprivation-induced anaerobiosis, probably due to the iron–sulfur (Fe—S) cluster of aconitase [92], thereby limiting the ability of cells to use external acetate as a carbon source. Therefore, sulfur deprivation-induced anaerobiosis would lead to relatively lower H_2 production than N_2 flushing-induced anaerobiosis, which is due to the extra one NADH and FADH produced via glyoxylate pathway that could potentially yield an additional two molecules of H_2 per molecule of glucose and the two molecules of H_2 per molecule of external acetate metabolized. Nevertheless, calculation of H_2 per photon derived through protein catabolism is complicated by the presence of three to five carbon substrates and the various pathways through which they can be broken down to acetate. Isotope labelling studies will definitely help shed light into which amino acids are preferentially degraded during anaerobiosis and allow for a more accurate determination of H_2 per photon. Studies have suggested that using genetic engineering to develop mutants lacking the ability to carry out state transitioning, cyclic electron transport, and mutants that have a smaller antennae size produced more hydrogen [93, 94].

Hydrogen is also produced in the dark and therefore in the absence of excitation and transfer of electrons by PS I. Noth et al. [91] under anaerobic conditions generated by N_2 flushing observed around 0.422 H_2 μg Chl^{-1} in the dark compared to 3.128 H_2 μg Chl^{-1} in *Chlamydomonas reinhardtii*. Along with the H_2, fermentative metabolism in the dark also leads to the production of formate, acetate, and ethanol [87]. Atteia et al. [88] revealed the presence of pyruvate-formate lyase in *Chlamydomonas reinhardtii*, suggesting mixed fermentative metabolism via pyruvate-formate lyase, aldehyde-alcohol dehydrogenase, phosphotransacetylase, and acetate kinase, leading to the formation of formate, ethanol, and acetate [91]. Furthermore, Noth et al. [91] suggested pyruvate ferrodoxin oxidoreductase and not hydrogenase being responsible for H_2 production in the dark. The combination of mixed acid fermentation and pyruvate ferrodoxin oxidoreductase suggests the operation of a different set of pathways in dark anaerobic conditions compared to light. More studies are needed to study to confirm this hypothesis and also to test whether these pathways are mutually exclusive. Even though the hydrogen produced were around 87% lower than that observed in the light, the absence of photosystem involvement eliminates the complication of oxygen-induced inhibition of hydrogenases and competition for NADPH by photorespiration. In addition, the complete heterotrophic nature of this production and the valuable co-products such as ethanol and acetate allows for further optimization and scaling up the fermentative reactions, leading up to hydrogen production. Overall, the process of hydrogen production from microalgae via anaerobiosis whether in the light and/or dark clearly has a strong potential, especially with the development of hydrogen fuel cell-powered cars. Given the advantage of being a cleaner source of hydrogen than thermochemical production, hydrogen production from microalgae can definitely benefit more from more research.

6. Conclusions

Overall, the successful transition of microalgal products from laboratory to industry largely depends on addressing various physiological limitations of microalgae. Biodiesel production from microalgae, although requires further research, can achieve commercial success by simple modifications such as heterotrophic and mixotrophic cultivation of microalgae in combination with genetic engineering to gain properties such as fast-growing and high lipid-secreting ability in microalgae. High valued products from microalgae can also benefit from similar modification such as heterotrophic and mixotrophic cultivation; however, the lack of knowledge on the physiological role and the biochemical pathway regulating the synthesis of these products demands further research to strategically optimize the production to its maximal potential. EPS production by microalgae, on the other hand, has a benefit of the product being naturally secreted by microalgae; however, its commercial success is hindered by the lack of knowledge of its physiological role and the nebulous nature of its applications. Hydrogen production from microalgae is a promising candidate for being a cleaner source of energy over other alternatives; however, its failure to gain commercial attention is primarily due to the limited research invested.

Acknowledgements

The Open Access publishing fees for this article have been covered by the Texas A&M University Open Access to Knowledge Fund (OAKFund), supported by the University Libraries and the Office of the Vice President for Research.

Conflict of interest

The authors declare no conflict of interest.

Acronyms and abbreviations

CO_2	Carbon dioxide
H_2	hydrogen
O_2	oxygen
N_2	nitrogen
H_2O	water
Fe-S	iron–sulfur
C	carbon
N	nitrogen
P	phosphorus
S	sulfur
Si	silica
EPS	exopolymeric substances
RUBISCO	ribulose bisphosphate carboxylase oxidase
NAD	nicotinamide adenine dinucleotide
NADH	(reduced) nicotinamide adenine dinucleotide
NADP	nicotinamide adenine dinucleotide phosphate
NADPH	(reduced) nicotinamide adenine dinucleotide phosphate
FAD	flavin adenine dinucleotide

FADH	(reduced) flavin adenine dinucleotide
CCM	carbon concentration mechanism
ATP	adenosine triphosphate
ADP	adenosine diphosphate
UVA	ultraviolet rays A
UVB	ultraviolet rays B
PUFAs	polyunsaturated fatty acids
DNA	deoxyribonucleic acid
RNA	ribonucleic acid
PS-I	photosystem I
PS-II	photosystem II

Author details

Manoj Kamalanathan and Antonietta Quigg*
Texas A&M University at Galveston, Galveston, Texas, United States

*Address all correspondence to: quigga@tamu.edu

IntechOpen

© 2019 The Author(s). Licensee IntechOpen. This chapter is distributed under the terms of the Creative Commons Attribution License (http://creativecommons.org/licenses/by/3.0), which permits unrestricted use, distribution, and reproduction in any medium, provided the original work is properly cited.

References

[1] Spoehr HA, Milner HW. The chemical composition of chlorella; effect of environmental conditions. Plant Physiology. 1949;**24**(1):120

[2] Energy Efficiency and Renewable Energy. National Algal Biofuels Technology Roadmap. 2010. Available from: https://www1.eere.energy.gov/bioenergy/pdfs/algal_biofuels_roadmap.pdf

[3] Araujo GS, Matos LJ, Gonçalves LR, Fernandes FA, Farias WR. Bioprospecting for oil producing microalgal strains: Evaluation of oil and biomass production for ten microalgal strains. Bioresource Technology. 2011;**102**(8):5248-5250

[4] Sun XM, Ren LJ, Zhao QY, Ji XJ, Huang H. Microalgae for the production of lipid and carotenoids: A review with focus on stress regulation and adaptation. Biotechnology for Biofuels. 2018;**11**(1):272

[5] Chisti Y. Biodiesel from microalgae. Biotechnology Advances. 2007;**25**(3): 294-306

[6] Sheehan J, Dunahay T, Benemann J, Roessler P. A look back at the US Department of Energy's aquatic species program: Biodiesel from algae. National Renewable Energy Laboratory. 1998;**328**:1-294

[7] Becker EW. Microalgae: Biotechnology and Microbiology. Vol. 10. Cambridge University Press; 1994

[8] Brown MR. The amino-acid and sugar composition of 16 species of microalgae used in mariculture. Journal of Experimental Marine Biology and Ecology. 1991;**145**(1):79-99

[9] Velasco LA, Carrera S, Barros J. Isolation, culture and evaluation of Chaetoceros muelleri from the Caribbean as food for the native scallops, Argopecten nucleus and Nodipecten nodosus. Latin American Journal of Aquatic Research. 2016;**44**(3):557-568

[10] Tibbetts SM, Milley JE, Lall SP. Chemical composition and nutritional properties of freshwater and marine microalgal biomass cultured in photobioreactors. Journal of Applied Phycology. 2015;**27**(3):1109-1119

[11] Benedetti M, Vecchi V, Barera S, Dall'Osto L. Biomass from microalgae: The potential of domestication towards sustainable biofactories. Microbial Cell Factories. 2018;**17**(1):173

[12] Lepage G, Roy CC. Improved recovery of fatty acid through direct transesterification without prior extraction or purification. Journal of Lipid Research. 1984;**25**(12):1391-1396

[13] Koley S, Mathimani T, Bagchi SK, Sonkar S, Mallick N. Microalgal biodiesel production at outdoor open and polyhouse raceway pond cultivations: A case study with Scenedesmus accuminatus using low-cost farm fertilizer medium. Biomass and Bioenergy. 2019;**120**:156-165

[14] He Q, Yang H, Hu C. Culture modes and financial evaluation of two oleaginous microalgae for biodiesel production in desert area with open raceway pond. Bioresource Technology. 2016;**218**:571-579

[15] Kenny P, Flynn KJ. Physiology limits commercially viable photoautotrophic production of microalgal biofuels. Journal of Applied Phycology. 2017;**29**(6):2713-2727

[16] Beardall J, Raven JA. Limits to phototrophic growth in dense culture: CO2 supply and light. In: Algae for

Biofuels and Energy. Dordrecht: Springer; 2013. pp. 91-97

[17] Available from: https://www.eia.gov/outlooks/steo/report/global_oil.ph

[18] Raven JA, Beardall J, Giordano M. Energy costs of carbon dioxide concentrating mechanisms in aquatic organisms. Photosynthesis Research. 2014;**121**(2-3):111-124

[19] Raven JA, Ralph PJ. Enhanced biofuel production using optimality, pathway modification and waste minimization. Journal of Applied Phycology. 2015;**27**(1):1-31

[20] Finkel ZV, Follows MJ, Liefer JD, Brown CM, Benner I, Irwin AJ. Phylogenetic diversity in the macromolecular composition of microalgae. PLoS One. 2016;**11**(5):e0155977. DOI: 10.1371/journal. pone.0155977

[21] Myklestad S. Production of carbohydrates by marine planktonic diatoms. 1. Comparison of nine different species in culture. Journal of Experimental Marine Biology and Ecology. 1974;**15**:261-274

[22] Hu Q, Sommerfeld M, Jarvis E, Ghirardi M, Posewitz M, Seibert M, et al. Microalgal triacylglycerols as feedstocks for biofuel production: Perspectives and advances. The Plant Journal. 2008;**54**:621-639. DOI: 10.1111/j.1365-313X.2008.03492.x

[23] Talmy D, Blackford J, Hardman-Mountford N, Polimene L, Follows M, Geider RJ, et al. N ratio enhances metabolism of large phytoplankton when resource supply is intermittent. Biogeosciences. 2014;**11**:4881-4895

[24] Raven JA, Donnelly S. Brown dwarfs and black smokers: The potential for photosynthesis using radiation from low-temperature black bodies. In: Habitability of Other Planets and Satellites. Dordrecht: Springer; 2013. pp. 267-284

[25] Blankenship RE, Tiede DM, Barber J, Brudvig GW, Fleming G, Ghirardi M, et al. Comparing photosynthetic and photovoltaic efficiencies and recognizing the potential for improvement. Science. 2011;**332**(6031):805-809

[26] Morel A, Bricaud A. Theoretical results concerning light absorption in a discrete medium, and application to specific absorption of phytoplankton. Deep Sea research part a. Oceanographic Research Papers. 1981;**28**(11):1375-1393

[27] Geider RJ, Osborne BA. Light absorption by a marine diatom: Experimental observations and theoretical calculations of the package effect in a small Thalassiosira species. Marine Biology. 1987;**96**(2):299-308

[28] Agusti S. Allometric scaling of light absorption and scattering by phytoplankton cells. Canadian Journal of Fisheries and Aquatic Sciences. 1991;**48**(5):763-767

[29] Berner T, Dubinsky Z, Wyman K, Falkowski PG. Photoadaptation and the "package" effect in Dunaliella tertiolecta (Chlorophyceae) 1. Journal of Phycology. 1989;**25**(1):70-78

[30] Melis A, Neidhardt J, Benemann JR. Dunaliella salina (Chlorophyta) with small chlorophyll antenna sizes exhibit higher photosynthetic productivities and photon use efficiencies than normally pigmented cells. Journal of Applied Phycology. 1998;**10**(6):515-525

[31] Neidhardt J, Benemann JR, Zhang L, Melis A. Photosystem-II repair and chloroplast recovery from irradiance stress: Relationship between chronic photoinhibition, light-harvesting chlorophyll antenna size and photosynthetic productivity in Dunaliella salina

(green algae). Photosynthesis Research. 1998;**56**(2):175-184

[32] Malerba ME, Palacios MM, Palacios Delgado YM, Beardall J, Marshall DJ. Cell size, photosynthesis and the package effect: An artificial selection approach. New Phytologist. 2018;**219**(1):449-461

[33] Quigg A, Beardall J. Protein turnover in relation to maintenance metabolism at low photon flux in two marine microalgae. Plant, Cell & Environment. 2003;**26**(5):693-703

[34] Raven JA. Protein turnover and plant RNA and phosphorus requirements in relation to nitrogen fixation. Plant Science. 2012;**188**:25-35

[35] Raven JA. The cost of photoinhibition. Physiologia Plantarum. 2011;**142**(1):87-104

[36] Jin J, Dupré C, Yoneda K, Watanabe MM, Legrand J, GrizeauD. Characteristicsof extracellular hydrocarbon-rich microalga Botryococcus braunii for biofuels production: Recent advances and opportunities. Process Biochemistry. 2016;**51**(11):1866-1875

[37] Harrison PJ, Zingone A, Mickelson MJ, Lehtinen S, Ramaiah N, Kraberg AC, et al. Cell volumes of marine phytoplankton from globally distributed coastal data sets. Estuarine, Coastal and Shelf Science. 2015;**162**:130-142

[38] Agusti S, Kalff J. The influence of growth conditions on the size dependence of maximal algal density and biomass. Limnology and Oceanography. 1989;**34**(6):1104-1108

[39] Allen AE, LaRoche J, Maheswari U, Lommer M, Schauer N, Lopez PJ, et al. Whole-cell response of the pennate diatom Phaeodactylum tricornutum to iron starvation. Proceedings of the National Academy of Sciences. 2008;**105**(30):10438-10443

[40] Gardner RD, Cooksey KE, Mus F, Macur R, Moll K, Eustance E, et al. Use of sodium bicarbonate to stimulate triacylglycerol accumulation in the chlorophyte Scenedesmus sp. and the diatom Phaeodactylum tricornutum. Journal of Applied Phycology. 2012;**24**(5):1311-1320

[41] Kamalanathan M, Chaisutyakorn P, Gleadow R, Beardall J. A comparison of photoautotrophic, heterotrophic, and mixotrophic growth for biomass production by the green alga Scenedesmus sp.(Chlorophyceae). Phycologia. 2018;**57**(3):309-317

[42] Heredia-Arroyo T, Wei W, Hu B. Oil accumulation via heterotrophic/ mixotrophic Chlorella protothecoides. Applied Biochemistry and Biotechnology. 2010;**162**(7):1978-1995

[43] Shi XM, Chen F. High-yield production of lutein by the green microalga chlorella protothecoidesin heterotrophic fed-batch culture. Biotechnology Progress. 2002;**18**(4):723-727

[44] Chen F, Zhang Y. High cell density mixotrophic culture of Spirulina platensis on glucose for phycocyanin production using a fed-batch system. Enzyme and Microbial Technology. 1997;**20**(3):221-224

[45] Borowitzka MA. High-value products from microalgae—Their development and commercialisation. Journal of Applied Phycology. 2013;**25**(3):743-756

[46] Chew KW, Yap JY, Show PL, Suan NH, Juan JC, Ling TC, et al. Microalgae biorefinery: High value products perspectives. Bioresource Technology. 2017;**229**:53-62

[47] Sinensky M. Homeoviscous adaptation—A homeostatic process that regulates the viscosity of membrane lipids in Escherichia coli. Proceedings of the National Academy of Sciences. 1974;**71**(2):522-525

[48] Jónasdóttir SH. Fatty acid profiles and production in marine phytoplankton. Marine Drugs. 2019;**17**(3):151

[49] Zhang C, Zhang L, Liu J. The role of photorespiration during astaxanthin accumulation in Haematococcus pluvialis (Chlorophyceae). Plant Physiology and Biochemistry. 2016;**107**:75-81

[50] Chen YT, Kao CJ, Huang HY, Huang SY, Chen CY, Lin YS, et al. Astaxanthin reduces MMP expressions, suppresses cancer cell migrations, and triggers apoptotic caspases of in vitro and in vivo models in melanoma. Journal of Functional Foods. 2017;**31**:20-31

[51] Lorenz RT, Cysewski GR. Commercial potential for Haematococcus microalgae as a natural source of astaxanthin. Trends in Biotechnology. 2000;**18**(4):160-167

[52] Lemoine Y, Schoefs B. Secondary ketocarotenoid astaxanthin biosynthesis in algae: A multifunctional response to stress. Photosynthesis Research. 2010;**106**(1-2):155-177

[53] Fan L, Vonshak A, Zarka A, Boussiba S. Does astaxanthin protect Haematococcus against light damage? Zeitschrift für Naturforschung. Section C. 1998;**53**(1-2):93-100

[54] Li Y, Sommerfeld M, Chen F, Hu Q. Consumption of oxygen by astaxanthin biosynthesis: A protective mechanism against oxidative stress in Haematococcus pluvialis (Chlorophyceae). Journal of Plant Physiology. 2008;**165**(17):1783-1797

[55] Kobayashi M, Kakizono T, Yamaguchi K, Nishio N, Nagai S. Growth and astaxanthin formation of Haematococcus pluvialis in heterotrophic and mixotrophic conditions. Journal of Fermentation and Bioengineering. 1992;**74**(1):17-20

[56] Sun N, Wang Y, Li YT, Huang JC, Chen F. Sugar-based growth, astaxanthin accumulation and carotenogenic transcription of heterotrophic chlorella zofingiensis (Chlorophyta). Process Biochemistry. 2008;**43**(11):1288-1292

[57] Grewe C, Griehl C. Time- and media-dependent secondary carotenoid accumulation in Haematococcus pluvialis. Biotechnology Journal: Healthcare Nutrition Technology. 2008;**9-10**:1232-1244

[58] García-Malea MD, Brindley C, Del Río E, Acién FG, Fernández JM, Molina E. Modelling of growth and accumulation of carotenoids in Haematococcus pluvialis as a function of irradiance and nutrients supply. Biochemical Engineering Journal 2005:**26**(2-3):107-114

[59] Bogorad L. Phycobiliproteins and complementary chromatic adaptation. Annual Review of Plant Physiology. 1975;**26**(1):369-401

[60] Luzardo AD, Niell FX, López-Figueroa F. Photocontrol of the synthesis of chlorophyll a and phycocyanin in the cyanobacterium Calothrix crustacea Schousboe. Revista Española de Fisiología. 1991;**47**(3):109-114

[61] Hsieh-Lo M, Castillo G, Ochoa-Becerra MA, Mojica L. Phycocyanin and phycoerythrin: Strategies to improve production yield and chemical stability. Algal Research. 2019;**42**:101600

[62] Kuddus M, Singh P, Thomas G, Al-Hazimi A. Recent developments in production and biotechnological applications of C-phycocyanin. BioMed Research International. 2013:1-9

[63] del Rio-Chanona EA, Zhang D. A Bilevel programming approach to optimize C-phycocyanin bio-production under uncertainty. IFAC-PapersOnLine. 2018;**51**(18):209-214

[64] Quigg A, Passow U, Chin WC, Xu C, Doyle S, Bretherton L, et al. The role of microbial exopolymers in determining the fate of oil and chemical dispersants in the ocean. Limnology and Oceanography Letters. 2016;**1**(1):3-26

[65] Sharp JH. Excretion of organic matter by marine phytoplankton: Do healthy cells do it? 1. Limnology and Oceanography. 1977;**22**(3):381-399

[66] Bjørrisen PK. Phytoplankton exudation of organic matter: Why do healthy cells do it? 1. Limnology and Oceanography. 1988;**33**(1):151-154

[67] Delattre C, Pierre G, Laroche C, Michaud P. Production, extraction and characterization of microalgal and cyanobacterial exopolysaccharides. Biotechnology Advances. 2016;**34**(7):1159-1179

[68] Biddanda B, Benner R. Carbon, nitrogen, and carbohydrate fluxes during the production of particulate and dissolved organic matter by marine phytoplankton. Limnology and Oceanography. 1997;**42**(3):506-518

[69] Myklestad SM. Release of extracellular products by phytoplankton with special emphasis on polysaccharides. Science of the Total Environment. 1995;**165**(1-3):155-164

[70] Mague TH, Friberg E, Hughes DJ, Morris I. Extracellular release of carbon by marine phytoplankton; a physiological approach 1. Limnology and Oceanography. 1980;**25**(2):262-279

[71] Myklestad S, Holm-Hansen O, Vårum KM, Volcani BE. Rate of release of extracellular amino acids and carbohydrates from the marine diatom Chaetoceros affinis. Journal of Plankton Research. 1989;**11**(4):763-773

[72] Aaronson S. Excretion of organic matter by phytoplankton in vitro. Limnology and Oceanography. 1978;**23**(4):838

[73] Morrissey J, Sutak R, Paz-Yepes J, Tanaka A, Moustafa A, Veluchamy A, et al. A novel protein, ubiquitous in marine phytoplankton, concentrates iron at the cell surface and facilitates uptake. Current Biology. 2015;**25**(3):364-371

[74] Xiao R, Zheng Y. Overview of microalgal extracellular polymeric substances (EPS) and their applications. Biotechnology Advances. 2016;**34**(7):1225-1244

[75] Mishra A, Jha B. Isolation and characterization of extracellular polymeric substances from micro-algae Dunaliella salina under salt stress. Bioresource Technology. 2009;**100**(13):3382-3386

[76] Gaffron H. Reduction of carbon dioxide with molecular hydrogen in green algae. Nature. 1939;**143**(3614):204

[77] Tamagnini P, Axelsson R, Lindberg P, Oxelfelt F, Wünschiers R, Lindblad P. Hydrogenases and hydrogen metabolism of cyanobacteria. Microbiology and Molecular Biology Reviews. 2002;**66**(1):1-20

[78] Benemann JR. Feasibility analysis of photobiological hydrogen production. International Journal of Hydrogen Energy. 1997;**22**(10-11):979-987

[79] Gaffron H, Rubin J. Fermentative and photochemical production of

hydrogen in algae. The Journal of General Physiology. 1942;**26**(2):219-240

[80] Makarova VV, Kosourov S, Krendeleva TE, Semin BK, Kukarskikh GP, Rubin AB, et al. Photoproduction of hydrogen by sulfur-deprived C. reinhardtii mutants with impaired photosystem II photochemical activity. Photosynthesis Research. 2007;**94**(1):79-89

[81] Stuart TS, Kaltwasser H. Photoproduction of hydrogen by photosystem I of Scenedesmus. Planta. 1970;**91**(4):302-313

[82] Melis A, Zhang L, Forestier M, Ghirardi ML, Seibert M. Sustained photobiological hydrogen gas production upon reversible inactivation of oxygen evolution in the green alga Chlamydomonas reinhardtii. Plant Physiology. 2000;**122**(1):127-136

[83] Kessler E. Hydrogenase, photoreduction and anaerobic growth. Algal Physiology and Biochemistry. 1974;**10**:456-473

[84] Kessler E. Effect of anaerobiosis on photosynthetic reactions and nitrogen metabolism of algae with and without hydrogenase. Archiv für Mikrobiologie. 1973;**93**(2):91-100

[85] Benemann JR, Berenson JA, Kaplan NO, Kamen MD. Hydrogen evolution by a chloroplast-ferredoxin-hydrogenase system. Proceedings of the National Academy of Sciences. 1973;**70**(8):2317-2320

[86] Melis A, Happe T. Hydrogen production. Green algae as a source of energy. Plant Physiology. 2001;**127**(3):740-748

[87] Gibbs M, Gfeller RP, Chen C. Fermentative metabolism of Chlamydomonas reinhardii: III. Photoassimilation of acetate. Plant Physiology. 1986;**82**(1):160-166

[88] Atteia A, Van Lis R, Gelius-Dietrich G, Adrait A, Garin J, Joyard J, et al. Pyruvate formate-lyase and a novel route of eukaryotic ATP synthesis in Chlamydomonas mitochondria. Journal of Biological Chemistry. 2006;**281**(15):9909-9918

[89] Nguyen AV, Thomas-Hall SR, Malnoë A, Timmins M, Mussgnug JH, Rupprecht J, et al. Transcriptome for photobiological hydrogen production induced by sulfur deprivation in the green alga Chlamydomonas reinhardtii. Eukaryotic Cell. 2008;**7**(11):1965-1979

[90] Zhang Z, Shrager J, Jain M, Chang CW, Vallon O, Grossman AR. Insights into the survival of Chlamydomonas reinhardtii during sulfur starvation based on microarray analysis of gene expression. Eukaryotic Cell. 2004;**3**(5):1331-1348

[91] Noth J, Krawietz D, Hemschemeier A, Happe T. Pyruvate: Ferredoxin oxidoreductase is coupled to light-independent hydrogen production in Chlamydomonas reinhardtii. Journal of Biological Chemistry. 2013;**288**(6):4368-4377

[92] Ruzicka FJ, Beinert HE. The soluble "high potential" type iron-sulfur protein from mitochondria is aconitase. Journal of Biological Chemistry. 1978;**253**(8):2514-2517

[93] Bernát G, Waschewski N, Rögner M. Towards efficient hydrogen production: The impact of antenna size and external factors on electron transport dynamics in Synechocystis PCC 6803. Photosynthesis Research. 2009;**99**(3):205-216

[94] Kruse O, Rupprecht J, Bader KP, Thomas-Hall S, Schenk PM, Finazzi G, et al. Improved photobiological H2 production in engineered green algal cells. Journal of Biological Chemistry. 2005;**280**(40):34170-34177